www.brookscole.com

www.brookscole.com is the World Wide Web site for
Thomson Brooks/Cole and is your direct source to dozens of
online resources.

At *www.brookscole.com* you can find out about supplements
demonstration software, and student resources. You can
also send e-mail to many of our authors and preview new
publications and exciting new technologies.

www.brookscole.com
Changing the way the world learns®

2ND EDITION

HUMAN SERVICE AGENCIES

An Orientation to Fieldwork

Lupe Alle-Corliss
California State University, Fullerton

Randy Alle-Corliss
California State University, Fullerton

THOMSON

BROOKS/COLE

AUSTRALIA • BRAZIL • CANADA • MEXICO • SINGAPORE • SPAIN • UNITED KINGDOM • UNITED STATES

Human Service Agencies: An Orientation to Fieldwork, 2nd Edition

Lupe Alle-Corliss and Randy Alle-Corliss

Editor: *Lisa Gebo*
Assistant Editor: *Monica Arvin*
Editorial Assistant: *Sheila Walsh*
Technology Project Manager: *Barry Connolly*
Marketing Manager: *Caroline Concilla*
Marketing Communications Manager: *Tami Strang*
Project Manager, Editorial Production: *Emily Smith*
Creative Director: *Rob Hugel*
Art Director: *Vernon Boes*

Print Buyer: *Doreen Suruki*
Permissions Editor: *Stephanie Lee*
Production Service: *Sara Dovre Wudali, Buuji, Inc.*
Copy Editor: *Patricia Tompkins*
Illustrator: *Jill Wolf, Buuji, Inc.*
Cover Designer: *Andy Norris*
Cover Image: © 2005 gettyimages®
Compositor: *Cadmus Professional Communications*
Text and Cover Printer: *Edwards Brothers, Incorporated*

Printed in the United States of America
1 2 3 4 5 6 7 09 08 07 06 05

For more information about our products, contact us at:
Thomson Learning Academic Resource Center
1-800-423-0563

For permission to use material from this text or product, submit a request online at
http://www.thomsonrights.com.

Any additional questions about permissions can be submitted by e-mail to
thomsonrights@thomson.com.

Thomson Higher Education
10 Davis Drive
Belmont, CA 94002-3098
USA

Library of Congress Control Number: 2005926849

ISBN 0-534-51610-6

To our children, Jessica and Justin

You are growing up to be such special and beautiful young adults.

*We admire your courage to speak
your mind and your inquisitiveness and creativity.*

*We are proud of your giving natures
and welcome your continued emotional growth.*

Our love for you is unconditional.

*We wish you all the success
and blessings you most certainly deserve.*

*We hope you will continue to live your lives fully and encourage
you to take healthy risks that will keep you open to change and new ideas.*

*We honor you as special human beings who in your own way
will surely make a unique contribution to helping others through your lives.*

Contents

Preface

This text is subtitled *An Orientation to Fieldwork* because it aims to help beginning professionals deal with agency life. The scope of this guidebook is comprehensive; a wide array of issues relevant to working in human services have been addressed. The text introduces the beginning helper to the most important concerns human service providers face. Certain issues have been treated in greater depth to orient the new practitioner to difficult or complex aspects of agency work. Concrete examples and practical tips complement the theoretical and factual material presented.

OUR MOTIVATION FOR WRITING THIS TEXT

Our intent has been twofold: to provide a sound knowledge base with regard to salient issues in the field, and to contribute to building professional skills, which are equally necessary to the effective helper. A person is most likely to become a qualified, competent, and skilled professional helper when both knowledge and practical skills have been mastered.

Working in human services can be a rewarding and exciting career choice. However, one must be aware of common hazards and dilemmas encountered in helping others. Helpers who are aware are better equipped to face ongoing challenges, to take preventive steps to avoid encapsulation and burnout, and to benefit from the rewards inherent in the helping professions.

This book is the culmination of over 20 years of teaching students and beginning helpers about the issues facing human service workers. Our commitment to helping students has long motivated us to be active in course development and textbook reviews. We have always endeavored to contribute to human services education directly or indirectly. Having the opportunity to write this text was a great learning adventure. Similarly, revising and adding new material to this 2nd edition has been equally rewarding and challenging.

The concept of this text originated at a conference in human services in 1994. As we reviewed a text with a colleague, we realized that many years of agency and teaching experience had provided us with a wealth of information about the field. It suddenly seemed feasible to incorporate our experiences and knowledge into a text for beginning helpers. Our initial excitement about his endeavor was fostered by colleagues and editors at Brooks/Cole. This encouragement was the catalyst for our first steps in researching and writing. Now, ten years later, we are pleased to present the 2nd edition of this text that represents years of persistence and active involvement in the human service field. It has been exciting to experience the growth and advancements of the 21st century as they relate to the helping professions.

Many issues in professional helping seem to cut across disciplines. By clearly articulating these common issues in this text, we hope to aid beginning helpers in developing a positive sense of their professional self, as well as to promote a strong identification with the larger community of helpers. Although this book is written primarily for beginning helpers, we believe that it also has relevance for professionals working in agencies. Although some of the material is introductory and of interest primarily to the beginning helper, many lifelong professional issues are illuminated here. In our own professional development, we continue to ponder many of the issues written about in this text.

Another reason for initially writing this text has to do with our years of teaching undergraduate and graduate fieldwork courses. We have always resorted to a combination of reading materials and texts, because we never encountered a single text that offered a practical approach to working in agencies. We decided to develop our own. We believe strong in learning through practical experience obtained from fieldwork and training. By developing a text that addressed both the technical aspects of agency work and the basic concepts, we felt that we could pave an innovative path for learning.

This text remains most appropriate for use in practicum, fieldwork, internship, and practice courses in which direct fieldwork experiences are examined. For the beginning helpers, entering fieldwork can sometimes be confusing and anxiety-producing. Given a knowledge base and opportunities for skill building, helpers can more easily engage in experiential learning. Perplexing issues may come clear; helpers may become self-aware, able to manage their anxieties, and confident in their role as helping professionals. Whether used as a reference tool, a workbook, or a guidebook, this text gives readers a framework for understanding themselves and their experiences in fieldwork settings.

Human Service Agencies: An Orientation to Fieldwork has multiple applications. It can be used by helping professionals embarking on careers in social work, counseling, psychology, human resources, public administration, and management. This text is recommended primarily for use in upper-division undergraduate courses and graduate programs, but it can also serve as a resource for established human service workers who want to further their professional development.

WHAT'S NEW IN THIS REVISION?

In addition to a complete update of information presented in the original, this revised edition includes:

- New information on diversity such as cultural competence, use of cultural stories in assessment, and practical suggestions for working with diverse populations.
- An expanded discussion of primary characteristics of human service workers, including the impact of early development on the individual helper.
- Updates on key topics such as ethical decision-making, informed consent, confidentiality, suicidal risk factors for children and teens, and technology issues of privacy that provide students with current information important to them during their practicum/internship.
- A discussion of current trends in client populations.
- New content on active listening and nonverbal communication, as well as multicultural and gender issues which enhances the importance of these core skills and competencies.
- A dialogue about the impact of the Internet on human services; compassion fatigue and vicarious traumatization.
- More in-depth coverage of intimate partner violence that once was referred to as spousal abuse.

ORGANIZATION OF THE BOOK

A unique feature of this text remains the manner in which theory and knowledge are blended with practical aspects of helping. Each chapter begins by introducing the main topic areas in "Aim of Chapter". There follow in-depth discussions of the relevant subtopics. At the close of each chapter are "Concluding Exercises", with an emphasis on self-awareness. Fieldwork examples, "Scenarios", and "Thinking Things Through" exercises are integrated throughout the text. Appropriate case studies and quotations help to personalize the issues discussed. Helpful interventions are sometimes suggested. Many of the core issues of human services are presented. The reader is consistently encouraged to engage in honest self-assessment.

Material for the numerous Scenarios comes from various sources. Some are derived from our own experiences as clinicians and supervisors in public and private agencies; others are based on our work with students, interns engaged in fieldwork, and colleagues. In all cases, care has been taken to maintain confidentiality. The many Thinking Things Through exercises aim to engage the reader by asking thought-provoking questions.

We have sought to provide information on a range of issues in a succinct and accessible manner. Each chapter contains references to relevant literature. (See the references and index at the back of the book.) We encourage readers to pursue additional learning in areas that are of special interest to them.

In sum, we have attempted to present important material in a clear, readable, and straightforward manner. In conjunction with providing valuable information, we challenge readers with real-life scenarios and probing inquiries. We hope that both beginning and seasoned helpers will consider this text a helpful and instructional supplement to their agency experience

ACKNOWLEDGMENTS

A first thanks goes to all those who participated in our personal and professional development through the years. Without those special individuals and mentors to help us along, our journey would not have been as rich and fulfilling as it has been. A special thanks goes to one such mentor and colleague, Gerald Corey, who has been instrumental in our professional growth as instructors and authors. We thank him for his ongoing support and challenging and caring spirit.

We are deeply indebted to Harry Lewis, not only for his photo shoot, but also for his inspiration in helping us learn to relax and have fun; without those jazz concerts, trips to Amoeba, shopping sprees, and for just being there to listen, we might have succumbed to burnout. There are many other colleagues and friends that have been supportive and encouraging through the years; these include: Martha Sibbens, Leo Juarez, Chris Rubel, Christine Williams, Susan Larsen, and David Rodriguez, to name a few.

Much gratitude must also go out to our many students who through the years have taught us so much. Their curiosity, tenacity, creativity, zest for learning, and dedication to helping others have continually energized us to maintain our commitment to human service education. It's clear that without their help, we would not have gained the knowledge and experience that has allowed us to dedicate ourselves to teaching others who are also devoted to helping.

This revision would have not been possible without the help of our reviewers who so generously gave of their time to read and critique our manuscript and to provide us with constructive criticism and encouragement. In specific, our gratitude goes out to Karen Barrett, University of Maine at Farmington; Molly Duggan, Old Dominion University; Sheri Goik-Kurn, Baker College Muskegon; Nancy Janus, Eckerd College; Ara Lewellen, Texas A & M University-Commerce, and Kathylene Siska, University of Missouri. Their feedback has been instrumental in helping us realize our goal of writing a text that is up to date and relevant to real life practice within human services.

Our appreciation goes to the entire staff at Brooks/Cole, who have made this revision a reality. We greatly appreciate Monica Arvin for all her assistance and Sara Dovre Wudali at Buuji, Inc., for being a great production manager. Most of all, we are forever indebted to Lisa Gebo, our editor and greatest supporter. Her continual support and belief in our work has been no less than inspirational.

Lastly, we would like to offer our deepest gratitude to our children, Jessica and Justin, who are growing up to be such caring, insightful, creative individuals of fine character and integrity. We are so proud of their development and only hope we can continue to have a positive impact in their lives in the years to come. We thank them for their patience and for being a motivating force in both our personal and professional lives.

About the Authors

Lupe Alle-Corliss, formerly Lupe Alley is a bilingual/bicultural licensed clinical social worker who has worked in many areas of human services. She attended the University of Southern California from 1974 to 1980, obtaining a bachelor's in social gerontology (1978) and a master's in social work (1980). Lupe's professional positions have included medical social worker, minority outreach coordinator lead clinician, graduate fieldwork supervisor, private practitioner/EAP, manage care provider, outreach/crisis intervention counselor, and lecturer. This work has taken place in a variety of settings: a rehabilitation hospital, a family services agency, a county-contracted mental health clinic, a country department of mental health, and the psychiatry department at Kaiser Permanente. For the past 20 years, Lupe has also served as an instructor in the human services program at California State University, Fullerton. She has also been a faculty liaison and professor in the graduate school of social work at California State University, San Bernardino. Lupe has also taught at the University of Southern California in the social work program. Currently, she teaches at both California State University, Fullerton and San Bernardino, remains active in clinical work and frequently provides in-services on a variety of topics to diverse audiences and agencies.

Randy Alle-Corliss, formerly Randy Corliss, is a licensed clinical social worker with a broad range of professional experience. He has worked in a variety of community agencies, with many client populations. At California State University, Fullerton, he earned a bachelor's degree in psychology. After he had graduated with his psychology degree, he began taking courses in the human services department at California State University, Fullerton, where he received considerable training and experience in facilitating groups. After working several years as a community worker in a county and state funded mental health center, he entered the graduate school of social work at the University of Southern California in 1978, concentrating in mental health. The professional positions Randy has since held include clinical therapist/psychiatric social worker, community outreach coordinator, outreach/crisis intervention counselor, group counselor, fieldwork coordinator, and field instructor. He has worked at both county-contracted and state mental health centers and is currently a therapist for adult clients in the psychiatry department at Kaiser Permanente. At a National Organization of Human Service Providers annual conference, he was a featured speaker on men's issues and men's groups. In addition to helping clients therapeutically, Randy is committed to human service education. For the last 21 years, he has taught undergraduate courses in the human services department at California State University, Fullerton. For two of those years he taught full time and more recently, he has divided his time between direct clinical practice and part time teaching. For several years Randy also served the human services department and the community as the fieldwork coordinator. Randy currently also has a small private practice, helping children, adults, and families with a range of therapies.

In addition to the revision of this text, both Lupe and Randy Alle-Corliss have continued to write, as noted with the publication of their second text, *Advanced Practice in Human Service Agencies: Issues, Trends and Treatment Perspectives* (1999).

Our Background and Experiences

We commence our journey by orienting you to our background in the human services field. Quite symbolic, these very experiences have been the major impetus for the evolution of this text. We have been fortunate to have interned and worked in a variety of community agencies that have allowed us to gain invaluable hands-on experience and observe how different human service agencies function. We have worked in public, government-contracted, and private settings that have broadened our awareness of the intricacies of these divergent organizational systems.

PUBLIC AGENCIES

Between the two of us, we have worked in both county and state organizations, including: county medical/rehabilitative hospitals, county department of mental health, and the state university system. By working in these settings, we have experienced the challenges of successfully working within the parameters of governmental policies and regulations. Furthermore, we have encountered the breadth of human needs through our direct social work with those individuals, most commonly of lower socioeconomic and/or minority status.

NONPROFIT AGENCIES

Our training and experience in working in family service agencies and mental health centers that contract with state, county, and fereral agencies for additional funding has educated us regarding the need for negotiation, flexibility, and compromise. We have also learned to be receptive and able to adapt to the ongoing changes inherent in the political and economically changing environment that directly and indirectly impacts the operation of human services.

FOR PROFIT/PRIVATE AGENCIES

Our most recent experiences in working in large health maintenance organizations and with managed care systems in our own private practice has provided us with new insights and frustrations concerning the changing world of health care in America. Although we, as have many others, aspired to work in the private sector, hoping to be more autonomous and free from agency restrictions, we have discovered this to be an erroneous perception. Because of the economic state of flux of our health-care system, working in the private, for-profit arena can be equally as demanding and restrictive, and requires the human service provider to be just as flexible and open-minded as those working in community agencies.

As noted, our agency experiences have been diverse. In the following text we will address many of the most relevant issues we have found important to fieldwork within human services.

CHAPTER ONE

Introduction to Fieldwork in Human Services

▧ AIM OF THE CHAPTER

Fieldwork is an essential component of professional development for anyone pursuing a career in human services. The focus of this text will be on providing the helper with a comprehensive perspective on the many aspects of fieldwork within the human services profession. We will discuss a broad range of issues that are relevant to working in agency settings and give an overview of many topics. We will devote more in-depth coverage to certain areas that we have found to be most vital to your overall learning. The text has a dual purpose: providing a sound *knowledge base* followed by *building important skills*. We believe both of these facets of learning are crucial to the development of an effective and successful helper. This book also serves as a fieldwork guide to help you in your professional development.

To understand fieldwork issues in particular, you must first have a full orientation in the human services profession in general. To this end, this chapter will give some historical background and an overview of the various career paths in human services. Also, we will discuss the importance of values and self-awareness in human services work. Learning as a lifelong process will be emphasized. In this introductory chapter, we hope to provide you with a foundation to build on as you begin your professional career.

We have found it helpful to view a career as a journey filled with many adventures—some more exciting and invigorating than others. In our careers, each experience has given us insights about ourselves and others and added to our knowledge base and skill development. We encourage you to consider your fieldwork as the beginning of a journey. We hope you will be willing to take risks to learn as much as you can from each experience, regardless of how negative or insignificant it may seem. In retrospect, these early adventures are often the most memorable. Beginning with this chapter, try to learn from your fieldwork and to consider your experiences adventures that may one day prove to have been instrumental in your growth as a professional helper.

▧ HUMAN SERVICES DEFINED

The human services field was born out of an increased sense of social responsibility toward the poor, minorities, women, and people with mental illness (Neukrug, 2004). In the literature, there are various definitions of *human services*. A commitment to protecting and enhancing the personal well-being of individuals is a basic premise of human services as proposed by Hansenfeld (1983). Others view human services as organized activities for helping people in health care, mental health, developmentally and physically challenged, social welfare, criminal justice, housing, recreation, education, and child care (Hepworth, Rooney, & Larsen, 2002; Neukrug, 2002, 2004; Roberts, 2000; and Schmolling, Youkeles, & Burger, 1993).

Brill and Levine (2002) point out that human services do not exist in a vacuum; they are shaped by our changing social system. Increased poverty and decay of our cities and a rise in violence and its impact on personal safety are cited as variables directly or indirectly affecting human services. The AIDS crisis and the need for an increased focus on domestic violence and child welfare are other targets for human services.

A more extensive list of groups served by human service professions follows:

Persons who are homeless
Families that have problems of child neglect or sexual, physical, or spousal abuse
Couples that have serious marital conflicts
Families, including single-parent families, that have serious conflicts manifested by runaways, delinquency, or violence
People afflicted with AIDS
Individuals and families whose lives have been disrupted by punishment for violations of the law
Unwed, pregnant teenagers
Persons who are gay, lesbian, or bisexual who face personal or family difficulties
Individuals or families whose lives are disrupted by physical or mental illness or disability
Persons who abuse drugs or alcohol
Foster parents and children whose parents are deceased or otherwise unable to safely care for them
Immigrants and persons from oppressed groups who lack essential resources
Persons with several physical or developmental disabilities and their families
Aging persons no longer able to care for themselves adequately
Persons who are migrants and transients lacking essential resources
Children (and their families) who have school-related difficulties
Persons who experience high stresses related to traumatic events or to major life transitions such as retirement, death of loved ones, children who leave home, or coming out for gay, lesbian, and bisexual persons (Hepworth, Rooney, & Larsen, 2002; Neukrug, 2004).

In considering current trends in the new millennium, there is a recognition that human services is ever changing. According to various authors (Gladding, 2004; Bass & Yep, 2002; Hackney & Wren, 1990; Lee & Waltz, 1998), this evolution is related to changes in the roles of men and women, breakthroughs in media and technology, and an increase in poverty, homelessness, trauma, loneliness, and advanced aging.

Our view of human services favors a comprehensive perspective. We see the helping field as encompassing professional service provided to those in need as listed earlier. Victims of natural (earthquakes, hurricanes, tsunamis, fires) and man-made disasters (terrorism such as 9/11) and those directly or indirectly affected by war (for example, the war in Iraq in 2004–2005) are additional client populations to consider.

In conjunction with our views, consider Abraham Maslow's hierarchy of needs (Box 1.1). This paradigm shows the broad range of needs that we believe can be met by human service providers.

Developed in 1968, this hierarchy places *physical needs* (housing, food, and other needs for general survival) on the bottom because Maslow believed them to be the most basic needs, demanding immediate attention. *Safety needs* (security, stability, freedom from anxiety and chaos) follow, and the *need for belongingness and love is* next. *Esteem needs* (self-esteem, the respect of others, achievement, recognition, dignity) are higher on the hierarchy, and on top is the *need for self-actualization* (the ability to direct one's own life, a sense of belonging and fulfillment in life). As noted, Maslow stresses a hierarchy of needs: Lower-order needs must typically be first satisfied before pursuing

Box 1.1 Maslow's Hierarchy of Needs

Need for Self-Actualization
Justice, spontaneity, truth, meaningfulness, aliveness, self-sufficiency, perfection, goodness, flexibility
The ability to direct one's own life

Self-Esteem Needs
Self-confidence, independence, achievement, competence, knowledge, status, personal recognition, respect
Dignity in oneself and esteem of others

Social Needs: Belongingness and Love
Love and affection, of oneself and of others
Need for a sense of belonging; need for family, friends, and affiliations

Safety Needs
Physical and psychological safety
Security, stability, freedom from anxiety and chaos
Need for structure, order, and protection from threats to psychological or economic well-being

Physiological Needs
Homeostasis
Specific requirements for food, water, air, shelter, and general survival

Source: Adapted from Maslow (1968); Neukrug (1994); and Woodside and McClam (1994).

more higher-order needs (Neukrug, 2004). We agree that Maslow's hierarchy "has great implications for working with individuals" (Neukrug, 2004, p. 74).

HUMAN SERVICE WORKERS

Just as the meaning of human services varies, so do descriptions of the human service worker. Neukrug (2004, p. 3) insists that "the human service worker is a person who has an associate's or bachelor's degree in human services or a closely related field." By contrast, Corey and Corey (2003, p. 3) use the terms *helper* and *human services professional* interchangeably to refer to a wide range of practitioners, including social workers, clinical and counseling psychologists, marriage and family therapists, pastoral counselors, community mental health workers, and rehabilitation counselors.

Gladding (2004) describes three levels of helping relationships as nonprofessional, generalist human service workers, and professionals. *Nonprofessional helpers* possess varying degrees of wisdom and skill. Such helpers include friends, colleagues, untrained volunteers, or supervisors. *Generalist human service workers* often possess some formal educational training in human relationships and usually engage in teamwork rather than practicing independently. Included in this group are mental health technicians, child-care workers, probation personnel, and youth counselors. *Professionals* are those who are educated to provide assistance on both a preventative and treatment level. Counselors, psychologists, psychiatrists, social workers, psychiatric nurses, and marriage and family therapists all belong in this group. "Workers on this level have a specialized advanced degree and have had supervised internships to help them prepare to deal with a plethora of situations" (Gladding, 2004, p. 39). We will expand on the roles of these various professionals in the following pages.

Besides level of education, other factors that distinguish professional helpers from nonprofessional helpers are "their identification with a professional organization, their use of an ethical code and standards of practice, and an accrediting body that governs training, credentialing, and licensing of their practice" (Cormier & Hackey, 2005, p. 1).

Certainly, there is great variance among human service workers. Although we agree that everyone employed in the human services is in fact a human service worker, our focus in this text will be on professionals who are somehow involved in providing services to people in need. This includes people who have undergraduate or graduate degrees and work in human services. Here are three major categories of helpers, according to professional degree.

AA Degree

Paraprofessionals who perform a variety of duties involving case management, discharge planning, child development, and community work

BA or BS Degree

Outreach worker
Child development specialist
Social worker (BSW level)
Teacher/educator (BA level)
Case manager administrator (BA level)
Human service manager
Community organizer

Graduate Degree

Psychiatrist
Counseling/clinical psychologist
Marriage, family, and child counselor
Social worker (MSW, PhD)
Teacher/educator (MA)
Nurse (MS)
Psychiatric nursing administrator (MA or higher)
Human service manager
Community organizer

This list is by no means exhaustive. Many other human service professionals might fit in these categories; overlap is also possible. Next we present an example of the intricacies of professional degree status among social workers. Similar complexities occur in other helping professions such as counseling and psychology.

SCENARIO: Baccalaureate-Level Social Worker (BSW) and Masters-Level Social Worker (MSW)

Both the BSW and MSW degrees are recognized as professional in status. Both are governed by the National Association of Social Workers (NASW), which determines criteria for the social work profession. The NASW stipulates that the minimum requirement for social work professional status is the satisfactory completion of a baccalaureate program in social work and acceptance for membership in the NASW. Further, the Council on Social Work Education (CSWE) must approve the college program for accreditation. In 1981, the NASW developed a classification system to help clarify the various entry points to social work. They defined the educational and practice requirements for each level as follows:

Basic Professional: BSW from a program accredited by CSWE. Prepares individuals for entry-level positions in social work and requires completion of 300 hours of supervised fieldwork.
Specialized Professional: MSW degree from a program accredited by CSWE. Prepares individuals for further advancement in the field and requires two years of supervised fieldwork.
Independent Professional: Requires an accredited MSW and at least two years of postmasters experience under approved supervision.
Advanced Professional: Typically requires a doctoral degree in social work or a closely related discipline. Requires special proficiency in theory, practice, administration, or policy, the ability to conduct research, or the study of social welfare. Prepares individuals for highly specialized positions and eligibility into many tenure-track university teaching positions [National Association of Social Workers, 1981, p. 9].

Despite the apparent precision of the preceding categories, many BSW social workers perform some of the same roles as MSWs (often for less pay) in various settings. This seems to be especially true in rural areas; salary distinctions are often based on geographical location. All in all, regardless of whether a helper is at the bachelor's or graduate level, paraprofessional, or nonprofessional, he or she is equally worthy of respect: Our common mission is to serve people's needs. We agree with the authors and human service practitioners Schmolling, Youkeles, and Burger (1993), who believe that no matter which particular field the human service worker practices in, the emphasis should be on the common ground—the desire to help others.

In our agency experience, we have learned the value of respecting all the professionals on the multidisciplinary team whose goal is to help those in need. The collaboration involved in teamwork can be beneficial to helpers and clients alike. Helpers can share their fears and frustrations, gain validation, support, and feedback, and learn from other team members. "Working in a supportive environment can also create cohesion and a sense of unity. For clients, teamwork, in most instances, leads to the delivery of more complete and higher-quality services" (Alle-Corliss & Alle-Corliss, 1999, p. 85). The following scenario illustrates the value of teamwork.

SCENARIO

A suicidal client enters a mental health center in crisis and is greeted by a receptionist. Her first task is to locate the on-call social worker, who is to assess the client's situation. Once the social worker completes the assessment, the client gets a referral to a psychiatrist to evaluate the need for medication. The doctor then sends the client to a nurse for medication samples, and the secretary schedules a follow-up appointment.

In all, various workers, some professional (social worker, psychiatrist, nurse) and some not (receptionist, secretary) have been involved in treatment of this client. Without the concentrated effort of all these individuals, the client might not receive appropriate care in a timely manner. This scenario emphasizes the value of teamwork and a multidisciplinary approach.

THINKING THINGS THROUGH

Can you think of a situation within an organizational setting in which different levels of human service workers engage in teamwork? Describe how you see the members of the team cooperating to achieve a unified outcome. Can you imagine a situation where such teamwork does not occur? Describe how you see the workers being uncooperative. How do you believe that would affect the outcome of the intervention with the client? Describe your recommendations to improve that situation.

Roles

Human service workers play various roles, according to three factors:

> The educational level of the helper: nonprofessional, paraprofessional (AA or more), or professional (BA/BSW, MA/MSW, or EdD/PhD)
> The particular job at hand
> The specific interests of the individual helper

In general, roles of human service workers all have the goal of serving people's needs. For some roles, helping the client in the change process is also a common theme. In our view, the term *human service professional* also covers counselors and psychotherapists, although it is not limited to such clinically licensed specialists. We agree with the following list of 13 roles and functions of human service workers (Mandel & Shram, 1985).

> *Outreach Worker:* works on community level with clients
> *Broker:* helps client locate and use existing services (e.g., case manager)
> *Advocate:* one who champions client's cause
> *Mobilizer:* organizes client and community support to facilitate provision of service
> *Behavior Changer:* facilitates behavior change in the client, using intervention strategies and counseling skills

Teacher/Educator: acts as mentor and role model to students

Evaluator: assesses programs in the interest of accountability for desired results

Administrator: supervises community service programs

Caregiver: offers encouragement, hope, and direct support to clients

Community Planner: responsible for designing, implementing, and organizing new programs to meet clients' needs

Data Manager: develops systems to gather facts, as a means of evaluating programs

Consultant: offers knowledge and support to other professionals and also meets with clients and community groups

Assistant to Specialist: acts as an aide to a highly trained professional in serving clients

Obviously, many helpers perform several of these roles as part of their overall job descriptions. For instance, an undergraduate intern at a group home or a battered women's shelter might serve as advocate, caregiver, and mobilizer and could assist a specialist as well. Similarly, someone with a BA might fill the roles of case manager, outreach worker, and community planner.

CAREER-RELATED SCENARIOS

The following profiles of human service professionals illustrate differences in roles, competencies, and professional degrees.

Community Worker

With an AA degree, Randy became a community worker for a local mental health center. His primary duties are to be an advocate and liaison for the agency and local community. On a typical day, Randy meets with a particular community agency, such as the Red Cross, Boys and Girls Club, YMCA, or local school, and informs them of the available mental health services and the referral process. He is also involved in conducting community studies designed to identify existing mental health needs within the community that were not being met. One specific study involved researching the mental health issues within the preschool setting in hopes of developing prevention strategies. Additionally, Randy often is engaged in public speaking and acts as a gatekeeper between the mental health center and the community. To perform his job effectively, he needs sound communication skills along with flexibility and risk taking to work with a variety of individuals and settings.

Case Manager

Susan works as a case manager at a large rehabilitation facility. Her job description is multidimensional given the many roles she often performs daily. Typically, when patients are first admitted to the facility, Susan conducts a psychosocial evaluation that will assist her in determining the most important needs of each patient and the patient's immediate family. Because of the critical situations of many of her patients, she often must engage in crisis intervention services with both patients and their families. Susan needs sound communication skills that enhance her ability to form strong relationships with her co-workers, patients, and the community. Frequently she is involved in information and referral services that require her to act as a broker, an advocate, and at times, a community organizer. She also provides education to patients and their families on various aspects of rehabilitation. Similarly, Susan offers in-service training to staff on the psychosocial considerations of patient care. In addition, her pivotal role in successful discharge planning for patients requires sensitivity and skill. Susan has a bachelor's degree in human services.

Child Protective Service Worker

Rick specialized in child welfare studies in graduate school. He is now employed by the county department of social services where his primary function is to investigate child-abuse allegations and assess service needs. He must assist individuals and families by recognizing dysfunctional behaviors and help clients take corrective actions to ameliorate these. He must be skilled at conducting ongoing assessments of his clients' needs and be able to provide continuous case management. He is knowledgeable of community resources so that he can provide appropriate client and family referrals to other programs or agencies. He is often called on to monitor minors in out-of-home placements as well as to prepare periodic reports for the courts. Although his role is not that of a clinician, he uses clinical skills in child-abuse risk assessment and family interventions. Because he is the primary professional managing the case, he may be the first one on the scene when problems erupt and immediate action is necessary. Skill in emergency response is critical as is the ability to work with families who have been in the system for an extended time and may resist following the stipulated case plan. Overall, Rick has found that his prior skills in interviewing, referring, counseling, maintaining records, practicing crisis intervention, and providing other social services have served him well in his current position. Rick has a master's degree in social work (MSW).

Clinical Administrator

Mark works as a clinic administrator for a small psychiatric facility that is part of a larger health maintenance organization. Although each day is an adventure in ensuring that the clinic runs smoothly, Mark's most consistent roles are supervisor, mediator, and manager. In each of these functions Mark must use sound clinical skills to perform his job effectively. He is often called on to resolve conflicts between staff and clients or among co-workers, so his assessment skills are essential. His knowledge of human nature, problem solving and conflict resolution is crucial in his daily efforts to help the facility run smoothly. Additionally, Mark acts as a buffer between upper management and his staff, a role that requires keen insight, sensitivity, and effective communication skills. Other administrative functions he performs, such as budget preparation, personnel evaluation and hiring, and program design, are equally important in his position. Mark has a master's degree in social work (MSW) and a clinical license (LCSW).

Clinical Social Worker

Jane is a clinical social worker whose primary role is to provide therapeutic services to clients seeking counseling for mental health concerns. Because of the broad range of clients she treats, Jane's job requires a great deal of flexibility and creativity; she may engage in traditional psychotherapy with some clients, brief counseling with others, or crisis intervention with those requiring immediate services. Because of the high demand for rapid access to services and the cost-effectiveness of group therapy, her skills as a group therapist are also valuable in Jane's daily professional activities. Her knowledge of assessment and treatment planning is essential to her success in working with her clients. Jane must also be adept at working with staff in her agency as well as with other professionals and agencies in the community. In addition to counseling or therapy, case management services are part of her job description. Sound writing skills—to record a detailed psychosocial evaluation, document symptoms in a behavioral manner, or correspond with numerous community agencies—are also important to her advocacy role in helping her clients. Other activities Jane engages in on a weekly basis include peer review, consultation, and in-service training for interns and newly hired clinicians. Jane has a master's degree in social work (MSW) and a clinical license in social work (LCSW).

Career Counselor

Joe is a career counselor at a local community college. His primary role is to provide students with information about educational programs and career opportunities, including available academic and financial resources. Joe is often involved in crisis intervention with students who are struggling with personal, academic, or social issues. He has found that his assessment skills and ability to relate to his students in a compassionate way are vital to helping them. He also realizes the importance of maintaining positive community contacts and knowledge of existing community referrals. Because he does not provide ongoing counseling, he often helps students find additional counseling at a local agency or in private practice. Joe also may refer students to other agencies that provide health services, chemical dependency treatment, legal aid, and other services. Joe has a bachelor's degree in human services.

Chemical Dependency Counselor

Sally is a counselor at a chemical dependency unit at a major community hospital. She provides support, education, and other services through individual, group, and family therapy to clients in her caseload and their families. Her knowledge about addiction and recovery is central to her ability to assess and treat her clients appropriately. She must also be aware of differential and dual diagnosis. Sally often leads psychoeducational groups with clients and their families. In her work with clients, she emphasizes relapse prevention by educating them about the various aspects of their addictions, including the triggers that commonly lead to relapse. In her work with families, she teaches about the family dynamics often in place and the nature of co-dependency. Sally has found that crisis intervention skills are crucial in managing volatile and suicidal clients. Sally has a master's degree in counseling and a special certificate in drug and alcohol counseling.

Marriage, Child, and Family Counselor

Tom is a marriage, family, and child counselor who works at a family service agency where he treats children, adolescents, and adults for various behavioral and emotional problems. Depending on the particular case, Tom may use one of several modalities, including individual, marital, or family therapy. Regardless of the type of treatment, Tom's initial and ongoing assessment skills are an essential aspect of his role. Knowledge of child development, child abuse, systems theory, and family therapy techniques is important in his work with his clients. Additionally, he provides counseling and consultation in local schools and often works with students, teachers, and school administrators directly on the school site where he can observe problem behaviors firsthand. Tom has a master's degree in counseling (MS).

Crisis Worker

Michelle is a crisis worker in an outpatient psychiatric facility. In addition to knowledge of crisis theory, Michelle needs excellent assessment skills, especially in regard to child abuse and risk of suicide or homicide. Given the nature of emergency services, Michelle has learned that she must not only be quick but also assertive with the people she often comes into contact with—including family, police, child protective service workers, chemical dependency unit staff, and inpatient psychiatric facility personnel. Crisis intervention skills and the ability to defuse potentially dangerous situations are critical to her job performance. Equally important is her ability to practice triage with others, both within her facility (that is, with psychiatrists, mental health specialists, clinical social workers, and psychiatric nurses) and in the community (with police, inpatient facility staff, outpatient counseling or mental health center staff). Knowledge of community services and available resources is also necessary for Michelle to ensure that clients follow through with appropriate referrals and treatment planning. Michelle has a master's degree in psychology.

Psychiatric Nurse

Lisa is a psychiatric nurse who works closely with others on the psychiatric unit of a large county facility. She is responsible for monitoring clients and ensuring that they are following their medication management and ongoing clinical treatment protocols. She is often involved in crisis work, some ongoing counseling, and much patient education. Besides conducting weekly medication groups with the psychiatrist, she is available to provide case management services to individual clients as well. Lisa is an integral part of the therapeutic treatment team whose members work together to treat clients in a holistic manner. Lisa has a master's degree in nursing.

Psychologist

Brian is a clinical psychologist in a community mental health center. His work activities include testing, counseling, psychotherapy, research, supervising, teaching, and consulting. He is especially valuable to the treatment team because he administers and interprets the various psychological tests that are often key to better understanding client dynamics and personality traits. Brian collaborates with other treatment team members and, at times, with other professionals within the community. Brian has a doctorate in psychology.

Psychiatrist

Kathryn is a psychiatrist who heads up the clinical treatment team at a major behavioral health corporation. Of the many functions she performs, a key one is investigating, diagnosing, and treating mental, emotional, and behavioral disorders. As a medical doctor, Kathryn is the only person on the treatment team who is legally authorized to prescribe psychotropic medication; consequently, she spends a major part of her day evaluating and managing medications. She also consults with other workers on cases for which a psychiatric opinion is considered necessary. Kathryn is a medical doctor whose specialty is psychiatry.

Because of the variety of roles these professionals practice in the human services field, we refer to them in various ways; such terms as *helper, human service worker/provider, clinician,* and *therapist* are all used interchangeably and in an appropriate manner throughout the text.

Mental Health Professionals

The area of mental health is a major division of human services that deserves further elaboration here. Since the term *clinical* is so often connected to mental health, it is important to first clarify a helper's role in a "clinical" capacity. The term *clinical* can be viewed on a continuum encompassing a broad range of services. Conservatively, clinical refers to providing psychotherapeutic interventions to clients in mental health, psychiatric, and counseling agencies by licensed clinicians (that is, licensed professional counselors, clinical social workers, marriage and family counselors, psychologists, and psychiatrists). A more liberal stance contends that such services as case management can also be clinical and can be provided by both licensed and unlicensed human service professionals who practice in such diverse settings as public child welfare agencies, medical and chemical dependency units, rehabilitation and mental health centers, and senior centers. Furthermore, "the term *clinical* and *therapist* are commonly used to describe professionals in the fields of counseling, psychology, and clinical social work who most typically engage in clinical practice" (Alle-Corliss & Alle-Corliss, 1999, p. 7). Clinical skills may also be used by case managers, child development specialists, community organizers, career and occupational counselors, human service administrators, managers, outreach workers, psychiatric nurses, psychiatrists, and residential treatment workers. Indeed, there are many different levels and types of helpers, yet despite our differences, "we are all part of the larger family of human service professionals. We all deserve respect and validation for our individual contributions in the helping profession" (Alle-Corliss and Alle-Corliss, 1999, p. 7).

Characteristics of Competent Practitioners

There is a wealth of literature about the characteristics of competent practitioners. Carl Rogers (1951), founder of client-centered therapy, emphasized the importance of empathy, genuineness, and unconditional positive regard. In the following list, we elaborate on these and other traits we believe are equally as important. Although some of these may seem simplistic, they are nonetheless vital to becoming an effective practitioner. Too many times, beginning helpers overlook these characteristics in their pursuit of more advanced skills; they may fail to cultivate those abilities that form the foundation of helping.

Some Primary Characteristics of Human Service Workers

We begin by elaborating on the concept of empathy because it is one of the core conditions of helping relationships (Capuzzi & Gross, 2003; Corey, 2005; Cormier & Hackney, 2005; Gladding, 2004; Neukrug, 2004; Young, 2001).

Empathy The word *empathy* is derived from Greek words that mean "feeling within" or "feeling oneself into" another person's experience. Rogers (1959, pp. 210–211) emphasizes being able to "perceive the internal frame of reference of another with accuracy and with the emotional components and meanings which pertain thereto as if one were the person, but without ever losing the 'as if' condition." Empathic understanding involves two primary steps:

1. Accurately sensing the client's world—being able to see things the way he or she does
2. Verbally sharing your understanding with the client

Empathic understanding can be seen as a simple phenomenon, but has also been described as a complex process. Simply stated, it is "the ability to feel *with* clients as opposed to feeling *for* clients" (Capuzzi & Gross, 2003, p. 9). Capuzzi and Gross (2003, p. 9) describe it as "the ability to understand feelings, thoughts, ideas, and experiences by viewing them from the client's frame of reference. The counselor [helper] must be able to enter the client's world, understand the myriad of aspects that make up that world, and communicate this understanding so that the client perceives that he or she has been heard accurately."

Egan (1998, p. 9) has identified primary and advanced levels of empathic understanding. At a primary level, being empathic involves being able to understand, identify, and communicate feelings and meaning that are at the surface level of clients' disclosures—in other words, being aware and responsive to overt client expressions. A more advanced form involves recognizing more covert feelings and meanings that are buried, hidden, or beyond the immediate reach of a client.

Besides understanding the client's verbal message and any underlying meanings, taking into consideration the impact of gender and cultural backgrounds is just as important. Cormier and Hackney (2005, p. 19) state, "There is increasing evidence that men and women talk in different ways and that cultural/historical backgrounds affect the meaning the client conveys as well as the meaning the therapist [helper] interprets." Lastly, a helper's own personal history and experiences of empathy in his or her own family will likely have an impact on how empathic he or she may be. What is often modeled to us in our childhood affects us in our adult life.

Respect and ***positive regard*** Rogers considered **these** characteristics necessary antecedents to empathy. Basically, respect and positive regard mean "accepting other people's rights to their own unique individualities and perspectives" (Long, 1996, p. 71. The term *unconditional positive regard* has been replaced by *nonpossessive warmth,* which seems more realistic and appropriate to today's helper-client relationships. Basically, positive regard involves believing in "each client's innate worth and potential and the ability to communicate this belief in the helping relationship. This belief, once communicated, provides clients with positive reinforcement relative to their innate ability to take responsibility for their own growth, change, goal determination, decision making, and eventual problem solving" (Capuzzi & Gross, 2003, p. 8). In the empowering process to facilitate change, respect and positive regard are key ingredients for setting the stage for the use of various helping strategies and interventions (Cormier & Hackney, 2005).

Genuineness Rogers was also instrumental in introducing this concept: being honest in the relationship with clients, being real. He believed that genuineness was a crucial element in all life's relationships; this is certainly true of the helping relationship. Peck (1998) advocates the importance of modeling realness to clients and emphasizes that a helper who is genuine also encourages clients to be real and genuine. Being genuine may require a certain amount of self-disclosure on the part of the helper. Use of appropriate self-disclosure is essential in ensuring that the needs of the client remain primary. In Chapter 4, we will elaborate further on the issue of self-disclosure.

Congruence, a term Rogers used to describe genuineness, is a condition that reflects honesty, transparency, and openness to the client. "Congruence implies that therapists [helpers] are real in their interactions with clients" (Cormier & Hackney, 2005, p. 24). Being congruent in our thoughts, feelings, and actions is therefore critical in demonstrating congruent behavior.

Trustworthiness goes hand in hand with being genuine and congruent. Trustworthiness includes being reliable, responsible, ethical, and predictable. "Trustworthiness is essential, not only in establishing a base of influence with clients but also in encouraging clients to self-disclose and reveal often very private parts of their lives. The counselor [helper] cannot *act* trustworthy; he or she must *be* trustworthy" (Hackney & Cormier, 2005, p. 18).

Acceptance is part of being open and genuine and having positive regard. "People who have a high regard for others can accept people in their differences regardless of dissimilar cultural heritage, values, or belief systems" (Neukrug, 2004, p. 11). There is an important distinction between acceptance of others versus agreement with others. Accepting clients and being empathic does not necessarily mean condoning their behaviors or agreeing with their beliefs or views.

Warmth communicated by a helper is closely tied to demonstrating genuineness and caring that enhances the helper-client relationship. Often such nonverbal behaviors as a smile, a touch, or tone of voice, convey genuine caring and warmth that fosters acceptance.

Relationship building, or being an "alliance builder," lies at the foundation of the helping relationship. Safran and Muran (2000), Sexton and Whiston (1991), and Whiston and Coker (2000) emphasize the value in establishing a solid helper-client relationship that allows for appropriate goal setting and positive client change. Neukrug (2004, p. 14) concurs that "such a relationship is closely related to the ability of the client and helper to build an emotional bond and to work on setting attainable goals. The impact of the helper's ability to build an emotional bond is felt throughout the helping relationship, regardless of whether it is acknowledged by the counselor [helper] or client."

Being open or *open-mindedness* allows helpers to encourage others to express their points of view. Individuals who are open minded "do not feel as if they need to change others to their way of viewing the world, they are open to criticism, open to change, and open to hearing and understanding the views of others in a way that will allow them to reflect on their own values and beliefs" (Neukrug, 2002, p. 5). Helpers who are open are more apt to foster a positive working relationship with their clients; in turn, clients are encouraged to self-disclose more readily. Open-mindedness allows helpers to accept client feelings, attitudes, and behaviors that may be different from their own, allows interaction with a wide range of clients, and is a prerequisite for honest communication (Hackney & Cormier, 2005).

Immediacy, considered as "the ability to deal with the here-and-now factors that operate within the helping relationship" (Capuzzi & Gross, 2003, p. 11), is a prime example of being open. Helpers who are willing to directly address issues with their clients as they arise demonstrate both courage and competence. Often, clients can benefit by gaining insight into how their personal behavioral patterns can hinder growth.

Objectivity on behalf of the helper is equally as important. Objectivity has been described as a component of empathy whereby helpers are able to see a client's problem as if it were their own without losing the "as if" condition (Rogers, 1957, p. 699). According to Hackney and Cormier (2005, p. 16), "[o]bjectivity refers to the ability to be involved with a client and, at the same time, stand back and see accurately what is happening with the client and in the relationship." They believe that helpers who attempt to remain objective provide their clients with "an additional set of eyes and ears that are needed in order to develop a greater understanding or a new perception (or reframe) of the issues" (p. 17).

Self-awareness and *self-understanding* are critical for anyone embarking in a career in helping. Self-awareness is described as "a knowledge of self, including attitudes, values, and feelings and the ability to recognize how and what factors affect oneself" (Gladding, 2004, p. 36). Hackney and Cormier (2005, p. 14) agree. "On the road to becoming an effective counselor [helper], a good starting place for most counselors [helpers] is a healthy degree of introspection and self-exploration." They cite the importance of helpers being aware of their own needs, their motivations for helping, and their personal strengths, limitations, and coping skills. Self-awareness allows helpers to be more objective and avoid blind spots in their relationship with clients. Unhealthy projection and defensiveness on the part of the helper can result when self-awareness is lacking.

Competence, also of critical importance, has been linked directly to client success (Winston & Coker, 2000). Competent helpers have the necessary information, knowledge, and skills to be effective in their work with clients, are well aware of current ethical and legal issues, and strive to have multicultural skills. Competent helpers build confidence and hope in clients, are able to work with a greater variety of clients and wide range of problems, and are ultimately more effective. In essence, competent helpers possess an ongoing thirst for knowledge and a desire to learn skills and specific techniques necessary for positive client outcome.

Cognitive complexity is directly tied to competence, acceptance, openness, and empathy. According to Neukrug (2004, p. 12), helpers who have cognitive complexity "view learning as a mutual and reciprocal process whereby experts can share knowledge with others and learn from others," and they "have a strong sense of self, can hear feedback from others, and are willing to learn, teach, share, and be open with others." Along with self-awareness, a qualified practitioner continually works toward attaining knowledge and cultivating and maintaining essential practice skills.

Being an experiencer of life, feeling alive, and making choices that are life oriented are advantageous in helping others. According to Neukrug (2002), helpers who allow themselves to experience life tend to be more accepting of their feelings and thoughts and are thereby more patient in allowing their client relationships to unfold naturally; they serve as positive role models.

Concreteness allows helpers to see beyond the incomplete picture clients often paint with their words. Helpers who can effectively communicate their perceptions with their clients often help clients view their "situation in a more realistic fashion, clarify vague issues, focus on specific topics, reduce degrees of ambiguity, and channel their energies into more productive avenues of problem solution" (Capuzzi & Gross, 2003, p. 10). Helpers who are objective, adept at abstract thinking, and patient in helping clients reach their own truths are typically more successful.

Psychological health of the helper is a given. Although, helpers endure struggles and need not be perfect, sound emotional health is helpful. Those who are psychologically adjusted are introspective and therefore more open to looking deep within themselves; they have the courage to receive feedback from others and are more willing to be self-critical (Neukrug, 2004). Kinner (1997) cites several criteria for psychological health that include self-acceptance, self-knowledge, self-confidence, and self-control, a clear (though slightly optimistic) perception of reality, courage and resilience, balance and moderation, love of others, love of life, and purpose of life. The emotional health of the helper has been connected to positive client outcomes (Lambert & Bergin, 1983). Furthermore, of even more significance is the correlation between positive client outcomes and the helper's experience with his or her own personal counseling (Deutsch, 1984; Neukrug, 2002; Neukrug, Milliken, & Shoemaker, 2001; Neukrug & Williams, 1993; Norcross, Strausser, & Faltus, 1988; Pope & Tabachnick, 1994; Prochaska & Norcross, 1983). Helpers who are aware of their issues and willing to seek help when necessary model authenticity and are likely to be respected by their clients all the more.

Personhood of the helper is ultimately just as important, and perhaps even more crucial, than the mastery of knowledge, skills or techniques (Cavanagh, 1990; Rogers, 1961). Additional "personal qualities" of an effective counselor as outlined by Cormier and Cormier (1998), Foster (1996), Gladding (2004), Guy (1987), and Patterson and Welfel, (2000) include curiosity and inquisitiveness, ability to listen and converse, emotional insightfulness, capacity for self-denial, tolerance of intimacy, comfortableness with power, ability to laugh, stability, harmony, constancy, purposefulness, energy, flexibility, support, and goodwill. Corey, who writes extensively about counseling and psychotherapy, offers his own list of "personal characteristics of effective counselors" that is also helpful to consider. Despite some overlap, Corey's views offer a different vantage point: "Effective counselors have an identity, they respect and appreciate themselves, recognize and accept their own power, are open to change, make choices that shape their lives, feel alive and make choices that are life-oriented, are authentic, sincere, and honest, have a sense of humor, make mistakes and [are] willing to admit them, generally live in [the] present, appreciate [the] influence of culture, are sincere in their interest in the welfare of others, become deeply involved in their work and derive meaning from it, and are able to maintain healthy boundaries" (Corey, 2005, pp. 18–19).

Awareness of feelings, dynamism, sensitivity, commitment, responsibility, integrity, motivation, purposefulness, and use of appropriate authority complete our list. You may justifiably consider other traits to be of equal importance. Also note that many of these skills evolve over the course of a person's professional development. Please take care not to become overly self-critical in comparing yourself to others who have more experience or knowledge. Learning is an individual experience; each person's journey is unique.

Consider the preceding list as a framework from which to begin self-assessment. We now invite you to assess your current level of interpersonal functioning by engaging in an honest self-appraisal of your qualities and skills. Early in our fieldwork courses, we encourage students to assess their strengths and limitations in working with people. We often use the two activities that follow to help students begin this self-assessment process.

THINKING THINGS THROUGH

1. Introduction: Self-Assessment Form

Pairs of students are asked to interview one another using the brief outline below. Then they are asked to share information about one another with the class.

> Name of student being introduced
> Placement agency
> Personal information (whatever the student is willing to share—marital status, living arrangements, family information, and so on)
> Strengths as a human service worker
> Limitations as a human service worker
> Goals for fieldwork placement
> Interview performed and reported by

2. Drawing One's Strengths and Limitations

Students are given drawing paper and crayons or markers and are asked to pair off with a classmate. They then use one side of the paper to draw a picture that reflects their strengths, and on the reverse side they draw their limitations. Next they share these with their partner and discuss them in relation to becoming a helper. Once this drawing assignment is completed, each person shares his or her partner's drawings with the class. Besides serving as an introduction, this activity promotes attentive listening and beginning self-assessment. What is amazing about the results is the commonality among students' drawings. Typical strengths depicted are large ears for listening and big hearts for caring. Limitations or weaknesses often involve rescuing, fear of becoming overly involved with clients, and being too emotional.

■ THE IMPORTANCE OF VALUES

In addition to recognizing his or her strengths and limitations, anyone embarking on a career in human services must be aware of his or her values and the impact they may have on the helping process. Values are all important: Much of professional helping consists of making decisions based on values. What exactly are values?

Values Defined

Unlike knowledge, values express what ought to be, not what is. "A value is an idea or way of being that you believe in strongly, something you hold dear and this is visible in your actions" (Sweitzer & King, 2004, p. 41). Another definition of values is that they are "standards or ethical guidelines that influence an individual or group's behavior, attitudes, and decisions" (Nugent, 1990, p. 262). As guidelines for proper behavior, values are much more than just emotional reactions. In fact, they are considered the fundamental necessities for decision making (Morales & Sheafor, 1995, p. 163). The most perplexing aspect of values is the fact that they are often hidden. Although many of our values are clearly evident, others lie deeply embedded at a subconscious level— we ourselves are unaware that they exist. Such hidden values can be the most harmful, allowing our biases and prejudices to be manifested unintentionally.

SCENARIO: Hidden Values

You consider yourself very open minded and liberal in your thinking. You honestly believe that you are open to others whose sexual orientation is different from yours. However, in a day-care setting, you find yourself uncomfortable with a particular little girl who often wants to play "marriage" with other girls. What does this say about your deeper values? If you were unaware of your own values, how might this affect your treatment of this child? Would you discourage such play—give the message that it is unacceptable? Consider all the ramifications of such hidden values.

THINKING THINGS THROUGH

Take a moment to reflect on some of your values. Were any of these ever hidden? Are you aware of any new values that have emerged as you begin in your fieldwork? If so, explain further.

Your fieldwork experience is often the first place where you become aware of some of your values. In the paragraphs that follow, we discuss several areas in which values can have a significant impact.

The formulation of social programs is influenced by the values held by legislators, board members, and directors of agencies and nonprofit organizations. If you are involved in any work at the community or policy level, your values also come into play.

SCENARIO

Due to current economic difficulties, many legislatures are imposing major budget cuts. Many social services are directly and indirectly affected. Board members and administrators of agencies are considering ways to decrease costs and increase the cost-effectiveness of service delivery. In one case, the administrator of a mental health clinic contemplated making a policy of scheduling two clients at the same time. Such double booking would minimize the problem of "no shows," but when both clients did show up for their appointments, their time would have to be cut in half. The human service workers objected strenuously. Due to the unethical and perhaps illegal nature of this plan, it was not implemented.

SCENARIO

You are a mental health worker employed by a large managed care organization. Because of difficult economic times, membership has declined and administrators are pushing for more clients to be seen quickly and for brief treatment. As a result, many clients on your caseload can only be seen every few months, although you have determined that more frequent visits are necessary. You feel frustrated that quantity takes precedence over quality of care.

These are clear examples of values in conflict. One value system was oriented toward economic considerations, whereas the other stressed humanistic concerns. For organizations to remain viable and responsive to changing times, the value systems of both administrators and practicing human service professionals should be taken into account.

THINKING THINGS THROUGH

Can you think of a scenario where agency policies are affected by values? Describe such a situation and offer suggestions on how the differing values can be negotiated.

The operational aspects of human services are also affected by values (Morales & Sheafor, 1995, p. 162). Decisions about the how, when, and why of service delivery are often motivated by the values of administrators, funding sources, and practitioners.

SCENARIO

Due to massive budget cuts, many county health care facilities face unprecedented closures and drastic cutbacks in services. Many people who need health care are unable to obtain it. For instance, consider the child of a low-income and/or undocumented couple; the child may be unable to receive proper immunizations due to the cutbacks. Also, consider what happens when a large county mental health agency closes down. How do mentally ill clients continue to receive care? What if transportation and money are not available for travel to another agency further away? How does this impact the client, the client's family, and the community?

THINKING THINGS THROUGH

From the preceding scenario, what effects do you think this financially motivated value decision could have on the child, the family, and the community at large? Can you think of an example from your own experience where delivery of a service was affected by value decisions?

Direct practice is significantly affected by values as well. How we as helpers feel about different issues, the beliefs we hold, and our overall attitudes make up our personal value system. Consciously or subconsciously, aspects of our value system are bound to spill over into our work with clients. If we recognize this, it need not be a negative. Later in this chapter, we will discuss the importance of exposing values as opposed to imposing them.

SCENARIO

Some adolescent clients enter a group setting dressed in loose attire (baggy jeans, huge T-shirts, unlaced shoes). Based on your value system, what assumptions might you make about these teenagers? Does their clothing indicate that they are gang members, neglected by parents, or wild and rebellious? Or could this just be an age-appropriate fashion statement? Depending on your own value system, you may have different feelings, thoughts, and reactions with respect to these clients. How might your perceptions affect your work with them?

THINKING THINGS THROUGH

Consider your strongest values and how these may affect you in your current fieldwork setting. List two of these, along with possible consequences. Are you surprised by your answers? Is the potential impact greater than you would have imagined?

The Role of Values in Work with Clients

As human service providers, we have values that directly and indirectly affect our work with those we aspire to help. For instance, Brill (1990, pp.24–25) states, "Values and standards are internalized and emotion laden, and have become so much a part of one's attitudes, feelings, thinking, and behaving that one is often unaware of their existence."

According to Corey and Corey (2003, p. 167), "Values are embedded in therapeutic theory and practice. Certain basic values are important for maintaining mentally healthy lifestyles and for guiding and evaluating the course of treatment." Egan believes that values permeate and drive the helping relationship: "Values, expressed concretely through working-alliance behaviors, play a critical role in the helping process. Since it has become increasingly clear that helper's values influence clients' values over the course of the helping process, it is essential to build a value orientation into the process itself" (Egan, 2002, p. 44). Another issue tied to the helping relationship is broached by Corey (2005), who considers a core issue to be the degree to which helper's values should enter into a therapeutic relationship. He says, "It is neither possible nor desirable for counselors [helpers] to be neutral with respect to values in the counseling [helping] relationship. Although our values do influence the way we practice, it is possible to maintain a sense of objectivity" (Corey, 2005, p. 21).

Neukrug (2004) writes, "Value clarification challenges one to understand and embrace our unique perspectives while offering new lenses from which to view the world" (p. 20). Think of the understanding of values to be a two-stage process: first, value awareness and then values clarification. Only when we are aware of the values we hold can we begin to see how they might influence our work with clients—which is what values clarification is all about. Neukrug (1994) defines values clarification as an approach to help individuals understand and accept their own perspectives and worldview while encouraging openness to view the world in other ways.

THINKING THINGS THROUGH

Values Clarification Exercise

First, find a partner—someone you don't know well. Tell him or her a little about your life. You might want to talk about such things as how you perceive your future career path, what is important to you in your life, how you find meaning in your life, people who have influenced you the most, and events that have had a great impact on you. Try to listen carefully to each other, and when you are finished, do the following activity.

1. Of the eight human service professional characteristics listed below, rate how important you think each characteristic is for the effective functioning of the human service professional. (Although we believe these qualities are all essential, you may have a different opinion.)
2. Rate yourself next on how well you think you embody these characteristics.
3. Now, here comes the hard part. Have your partner rate you on how well he or she thinks you embodied the characteristics during your conversation. (Remember that this is to be an honest assessment by your partner, which can be affected by many factors. You are to attempt to remain open-minded to whatever feedback you receive.)
4. Next, rate yourself based on whether you think you exhibited those qualities as you went through your life today.
5. When you have completed your ratings, examine your scores. Are there areas that you state you value but do not exhibit through your behavior? In other words, have you prized and cherished the things you state you value or are there areas you can improve?
6. Because we can all improve parts of our lives, what ideas do you have to work on those areas? Write them in the last column of the chart. In other words, what action can you take to enhance the values that you state you prize?

When rating, use the scale provided below. A high rating indicates that you embody the characteristic, believe it is important, or exhibited the particular characteristic today.

An extremely high rating = 1
A very high rating = 2
A moderately high rating = 3
A somewhat high rating = 4
Neither high nor low = 5
A somewhat low rating = 6
A moderately low rating = 7
A very low rating = 8
An extremely low rating = 9

Professional Values and Rating of Importance ____
Self-Rating ____
Other's Rating ____
Exhibit Today? ____
What can you do to improve scores? ____

Empathy ____
Open-Mindedness ____
Acceptance ____
Cognitive Complexity ____
Psychological Adjustment ____
Genuineness ____
Relationship Building ____
Competence ____

(Adapted with permission from Neukrug, 2004, p. 21.)

Value Conflicts

The existence of possible value conflicts may also come to light through this process of awareness and clarification. Because human service clients are diverse, helpers are very likely to encounter clients who have value systems different from their own. Some values may differ only slightly—for example, divergent views about child rearing or politics. However, there are likely to be pronounced differences regarding controversial subjects such as homosexuality, abortion, and ethnicity. It will be next to impossible to work only with people who share all your views. Similarly, coworkers, supervisors, or an entire agency may have values different from yours. You will have to decide whether you are comfortable enough with these differences to remain objective and effective in your work. Many believe that even when one's values differ greatly from those of their clients, coworkers, or agency, effective treatment can still be provided. If value conflicts are too great, however, referrals or transfers are highly recommended. Certain controversial issues engender many such value conflicts, as detailed in the subsection that follows.

Family Issues

It is not uncommon to encounter clients who have chosen to have extramarital affairs, or who engage in lifestyles different from yours, or who are undergoing the turmoil of divorce, custody battles, remarriage, or forming blended families.

SCENARIO

You value the sanctity of marriage. A particular female client has several children out of wedlock, all by different fathers. How might this value conflict influence your work with this client?

Gender Roles

Clients may challenge your views of sex-role identity. They may take on traditional, stereotypical roles that you are uncomfortable with. Or, quite the opposite, their extreme deviation from the norm may surprise you.

SCENARIO

You are working with a family where the wife is the primary breadwinner. The husband is the main caretaker and homemaker, so he is home more of the time. If you value traditional gender roles, how might you react to this family's lifestyle? How might you view each parent? Would you consider the mother uncaring or neglectful, or should she be commended for her venturesome qualities? Would you view the father as inadequate, passive, and submissive or consider him sensitive and nurturing?

Religion

Depending on your own religious beliefs and practices, some clients may represent a challenge to your value system. Either extreme—the devoutly religious or atheist—can trigger uncomfortable feelings in a helper.

SCENARIO

You were brought up in a traditional Catholic home environment. As you matured, you began to question many of the values of your parents' Catholicism. In your agency work, you encounter a deeply religious client, who asks you to participate in his Bible readings. You feel quite uncomfortable with this request. You recognize that religion is highly important to him, and you are concerned about hurting his feelings or alienating him by not going along with his requests.

How might your conflicting feelings affect your work with this client? In what ways might you proceed?

Abortion

The hot debate between advocates of the "right to life" and the "freedom of choice" may be one you have strong feelings about. Depending where you stand, you may experience conflict in dealing with a client who chooses abortion or one who refuses even to consider that option.

SCENARIO

You are pro-choice. A client wants to end her pregnancy, but she is being pressured by family and friends to keep the child. How might you react?

SCENARIO

Now assume that you are very adamant about the right to life. A young female client is seriously contemplating an abortion because she is not ready to have children, is fearful that her boyfriend will dump her, and does not want to sacrifice her freedom by having a child.

How might you react to this client? Would you be tempted to try to sway her to consider keeping the child or giving it up for adoption? What might be some constructive ways to deal with these conflicting values?

Sexuality

Some helpers are quite open regarding sexual issues and are comfortable with their own sexuality. Others remain conservative when it comes to sexual issues. In addition, some may say that they are open to discussing sexual issues, but in reality they may harbor unresolved issues about sex. Clients representing either extreme may arouse discomfort in a helper who holds the opposing value.

SCENARIO

You work in a teen pregnancy center; you inform young people about the importance of contraception if they are going to be sexually active. You usually have no difficulty with this part of your job because you believe in what you do. However, one day you are surprised when one particular girl openly talks about her own sexuality in group therapy. Although you view yourself as open-minded, somehow you are shaken by this young woman's boldness. The fact that she is so free about sexual issues creates anxious feelings in you.

What might this mean? Could there be a value conflict here? How so?

Sexual Orientation

Although the gay and lesbian movement has won wide acceptance for alternative sexual orientations, there are still many (helpers included) who are opposed to homosexuality. (We will discuss the issues of sexism and homophobia in Chapter 2.) "Many helpers have blind spots, biases, and misconceptions about gay and lesbian issues" (Corey & Corey, 1993, p. 92). Such helpers may feel conflicted when working with a gay or lesbian client.

SCENARIO

Your new supervisor openly announces her lesbian orientation. You find yourself feeling uncomfortable when alone with this person, and you question her competence. Could this perhaps be related to your own sexual values and preferences? How might you handle this situation?

AIDS

AIDS is a perplexing and complex disease that has become a tragic epidemic. Despite increased education and awareness, many people still remain fearful and critical of those who are infected with this serious and often deadly virus.

SCENARIO

You are working in a community mental health center, mandated to provide services to all members of the community. More and more of your clients are infected with the HIV virus. The agency provides in-service training regarding AIDS. Some staff members, although initially open and receptive to treating this population, become hesitant, resistant, and fearful about having direct contact with clients who have AIDS.

What implications might this have in service delivery? As a member of the staff, how might you deal with such a situation? How might your decisions reflect your values about AIDS and people who become infected?

Cultural and Racial Identity

A multicultural perspective is necessary today. We need to take into account how values, beliefs, and actions are conditioned by race, ethnicity, gender, religion, and history. Further, we must be careful of tunnel vision—seeing the world only from our own cultural perspective. Working with clients of different races, ethnicities, and cultures may stir up stereotypical thinking, which is bound to affect one's work.

SCENARIO

You were raised within the mainstream American cultural value system, where independence and self-reliance are encouraged. You are working with a young Latina woman who still lives at home with her parents. She relates conflicts in the home about the traditional values held by her parents—particularly regarding her desire to make independent decisions.

How might you intervene? Would you encourage her to move out of the home? Might you consider the cultural influences involved? How?

Conflicts Between Worker and Agency

Conflicts may also arise between the values of an agency and those of its workers. Such differences can create undue stress and tension, lead to burnout, and ultimately damage the quality of service provided to clients. Although differences may be unavoidable, we must be open to exploring ways to cope with the differences effectively.

SCENARIO

A human service intern is placed at a group home for abused children. He enters with a solid knowledge base regarding child-abuse issues and is looking forward to using his empathic skills in establishing relationships and developing trust with the children. He is surprised to find that many of the workers at this agency do not advocate a humanistic approach. Rather, they value order and control. This worker finds it difficult to engage with the children without being sensitive, empathic, and genuine. He begins to experience conflicts with coworkers, who not only take a totally different approach but are also critical of his.

How can he cope with this value conflict between coworkers and himself?

SCENARIO

Today, human service agencies have reduced staff and resources to provide the needed services. In mental health, this has led to cuts in actual services and reduction in the length of time spent with clients. Administrators may advocate brief service, alternative methods of treatment, and perhaps even referrals. In contrast, the clinicians may feel that ongoing treatment remains appropriate for the clients. In time, such differences of opinion are likely to create stress, frustration, and direct or indirect conflicts between the clinicians and the administrators.

How would you deal with a situation like this? How can quality of care be maintained?

The Challenge of Value Awareness

Clearly, the importance of recognizing and clarifying values cannot be overstated, but the process is not always an easy one.

Barriers

You need to be aware of the obstacles that you may encounter and work toward overcoming them. This requires a willingness to engage in open and honest assessment of your own views and feelings. Morales & Sheafor (1995, pp. 163–164) cite three reasons for this difficulty:

Values are so ingrained in our thought processes that we often are not consciously aware of them, so we cannot recognize their influence on our decisions.

Addressing values in the abstract may be quite different from dealing with them in a real-life situation.

Values are dynamic. Our priorities and strongly felt values can change over time to adapt to a new reality.

THINKING THINGS THROUGH

Can you identify any other reasons why value awareness might be difficult? Consider situations you have experienced in your agency or life, or ones you have observed, where becoming aware of values was challenging. What helped you to become aware of your values? How did you cope?

Name your strongest values and identify both the productive unproductive aspects of holding such values.

Cultural Experience

A person's cultural experience—family, ethnicity, racial identity, community, society in general—always affects the development of his or her values. Culture guides people's development of values and standards of behavior (Brill, 1990; Woodside & McClam, 1994).

SCENARIO

A client was raised in a small, close-knit community where neighbors looked after each other and were integrally involved in each other's daily life. Recently, she moved to a big city where people commute to work and rarely have time to cultivate relationships with their neighbors; a strong sense of community seems lacking.

What do you think the client would feel in this situation? What might the client think?

THINKING THINGS THROUGH

Based on the preceding scenario, how might you feel if you were that person? Would it be difficult to adjust? Would you be apt to criticize your new environment? Consider that many of your clients may be experiencing just such adjustments, as many are from other cities, states, and countries.

Manifestation of Values

Values are revealed in numerous ways. Sometimes they are clearly and directly communicated, at other times more subtly. Our interests, affiliations, styles of dress, philosophies, self-disclosures, and choice of friends often convey our values. Clients can often detect our values by interpreting our verbal and nonverbal communication.

SCENARIO

Your office is decorated with southwestern decor and pictures of your family. When clients walk in for the first time, what type of impression might they have about your values?

THINKING THINGS THROUGH

Take a moment to consider how your office or work space reflects your values. Does the decoration indicate a specific preference? If you are using someone else's office, how does the decor reflect that person's particular value systems? If you could design an office to reflect your values, what would it look like?

Take this activity further. Consider your current fieldwork setting and compare it to your ideal agency. What would be most conducive to the values you hold? What are the differences between your current agency and your fantasy? Is there any way to make some of your wishes reality? Or is this unrealistic?

To Expose Rather Than Impose

There is often great temptation to influence, push, or indoctrinate clients toward our own values. Clients may have sought help because of a value conflict, so they are prime candidates to be swayed by helpers. The role of the helper may in part be to facilitate "their understanding of how the particular set of values they hold influences their goals and decisions" (Morales & Sheafor, 1995, p. 164), but this does not mean that clients should be told how to proceed based on our values. Allowing the inappropriate intrusion of one's values into practice settings is ethically unacceptable and unprofessional. Human service workers "can help clients search out alternative value systems and openly indicate when their own biases may unduly influence a client" (Nugent, 1990, p. 263).

Corey and Corey convey the complexity of this issue in *Becoming a Helper* (2003, p. 169):

> The clients with whom you work ultimately have the responsibility of choosing what values to adopt, what values to modify or discard, and what direction their lives will take. Through the helping process, clients can learn to examine values before making choices. You can play a significant role as a catalyst for this examination if you do not either conceal your values from your clients or try to impose your values on them.

You can experience values in a nonclinical setting as well. In either case, it is neither possible nor desirable for human service workers to keep their values from clients. To reiterate, our role is to help our clients to clarify their own values, which can guide them toward what is best for them. In the process, we no doubt expose our values in some manner, yet we must be careful to not impose them.

According to Yarhouse and VanOrman (1999) it is inevitable for value conflicts to enter into the therapeutic relationship. However, despite these conflicts, it is often possible to successfully work through them. When this is not possible, referrals might be appropriate. In some instances, however, referrals are either contraindicated or just not feasible. Helpers need to carefully examine their reasons for referring and to ask themselves if the referral is appropriate. Will the referral benefit the client or the helper? Simply because one does not like a client or is uncomfortable with a person is not a sufficient reason to refer that person to another helper. Some clients may take the referral personally and may be offended or hurt. Informing clients about such referrals needs to be handled sensitively.

In rural areas where agencies are very small and at times only staffed by one person, referrals are just not realistic. In these cases, helpers will need to help their clients despite their own issues. Care must be taken to remain as objective as is possible.

SCENARIO: Imposing our Values

You are working with a woman who reveals repeated spousal abuse, both physical and verbal. Without any hesitation, you advise her to seek a separation. You neglect to consider her specific situation or the relevant aspects of the cycle of violence.

How might this be an imposition of your values?

SCENARIO: Exposing our Values

In the same situation just cited, you decide to expose your values regarding domestic violence. You inform your client about your concern for her well-being if she remains in the abusive situation. You indicate that she deserves to be treated with respect, and you respect her decision to stay or leave.

How might you further expose your values to help her explore her options?

SCENARIO: Inappropriate Referral

Tom is a counselor at a family service agency in a suburb. He usually works with children and adolescents who are having difficulties at school. Because a coworker from the adult unit is on maternity leave, Tom recently has been asked to help out with new clients. Several of his new clients are women who are incest survivors; he feels uncomfortable working with this population. He approaches his supervisor in hopes of having these cases referred out. When his supervisor asks him why, he shares his discomfort with being a male therapist in these cases. With his supervisor's help, Tom is able to differentiate which clients are truly inappropriate for him to treat and would best be served by a female therapist and those he could work with effectively despite his discomfort. As Tom challenges himself, he is appreciative that he did not prematurely refer clients out just because of his initial discomfort.

What would you have done in this situation? Have you ever referred a client due to your own issues? How might this impact clients?

SCENARIO: Unable to Refer

Sally lives and works in a small rural town in the South. She is the only social worker at the local Salvation Army where she routinely treats clients with a variety of mental health and substance abuse problems. Normally, Sally is at ease with her clients and has never had to turn a client away. On one particular day, Sally interviews an older male with an alcohol problem. As she is conducting the intake she begins to feel very uncomfortable and somewhat fearful. She manages to complete the intake, and after the client departs, she begins to cry uncontrollably. Once she calms down, she is able to better understand her unusual reaction as a form of countertransference. This client reminds Sally of her abusive, alcoholic father whom she has been estranged from since early adolescence. Ideally, Sally wishes she could just refer this client elsewhere, yet because her agency is the only one in town to provide low-cost counseling, this is not an option. She decides to consult with a former supervisor she still keeps in contact with and works through some of her feelings from her past, which allows her to be more objective with this client. In time Sally is able to work effectively with this client despite her initial countertransference.

If you were in a similar situation, how would you handle this?

THINKING THINGS THROUGH

Choose a matter you feel strongly about. Imagine working with a client whose attitude in this regard is directly opposite from yours. Present examples of how you might improperly impose your values in this situation. Similarly, how might you expose your values in a more constructive manner?

Value Assessment

In working with students and beginning helpers, we often approach the issue of values through experiential exercises that tap into two basic areas: personal values examination and identification with professional values.

Personal Values Examination
We stress the need to begin working in this profession by being honest, open, and self-aware. This entails a thorough and ongoing examination of the nature and basis of your own values. Using the following values questionnaire can be helpful in creating deeper thought and dialogue about personal values.

THINKING THINGS THROUGH
Values Questionnaire

Please answer the following questions as succinctly and honestly as possible. They will help you begin to think about values and their relationship to human services.

How would you define values?
Give an example of a value you feel very strongly about, and one that is less important to you.
Describe a situation in which your values turned out to differ greatly from those of a peer, family member, friend, coworker, or client. How did you first feel upon noticing the

difference? How did you react? How did you handle the situation? Was there any conflict as a result of your divergent views?

At your agency, have you experienced any value conflicts? How might you deal with this in a professional manner?

What values were prevalent in your home as you were raised? Did you incorporate any of your parents' values?

Identification with Professional Values

What are the values of the human service professions? Each profession within human services has a code of ethics that specifies its' professional values. In general, acceptance, tolerance, individuality, self-determination, and confidentiality are the most commonly held values in human services (Austin, Skedling, & Smith, 1977). Jensen and Bergin (1988, pp. 290–297) list basic values that are important to maintaining mental health and for optimizing the helping process:

Assuming responsibility for one's actions
Developing effective strategies for coping with stress
Developing the ability to give and receive affection
Being sensitive to the feelings of others
Practicing self-control
Having a sense of purpose for living
Being open, honest, and genuine
Finding satisfaction in one's work
Having a sense of identity and feelings of self-worth
Being skilled in interpersonal relationships, sensitive, and nurturing
Being committed in marriage, family, and other relationships
Practicing good habits of physical health
Having deepened self-awareness and motivation for growth

THINKING THINGS THROUGH

In reviewing the preceding list, take a moment to reflect on whether any of these values are a part of your value system. Which values do you not possess? Do you always adhere to the values you believe in? Be aware that even though we may believe certain values are dear to us, we may not always behave in a manner that is congruent with those values. Do you see any areas where your behavior clashes with your stated values?

◼ SELF-AWARENESS

The last value listed by Jensen and Bergin—having deepened self-awareness and motivation for growth—leads directly to our next topic: self-awareness. Together with a personal commitment to helping others, self-awareness lies at the core of successful helping. According to Collins, Thomlison, and Grinnell (1992), developing self-awareness is a key part of the process becoming an effective professional. In regard to students entering fieldwork, Sweistzer and King (2004, p. 8) add, "To make sense of service-learning experience, you need to understand more than stage theory. You need to understand yourself."

The Importance of Self-Awareness

Self-awareness is an essential area of knowledge because the disciplined and aware self remains one of the profession's major tools that must be developed into a fine instrument. The beginning of the honing process is the heightening of self-awareness—the ability to look at and recognize ME, not always nice, sometimes judgmental, prejudiced, and uncaring. You need to be able to recognize when the judgmental, uncaring self interferes with the ability to reach out, explore, and help others mobilize their coping capacity (Devore & Schlesinger, 1981, p. 82).

This excerpt speaks clearly and succinctly to the need for helpers to launch into a lifelong journey of self-exploration. We believe that all professional helpers, whether novice or experienced, must continue this process if they are to remain useful to others and true to themselves. Learning is an ongoing, ever-changing process that requires constant self-awareness. Faiver, Eisengart, and Colonna (2004, p. 13) agree that "we are in the business of introspection, reflection, analysis, synthesis, and personal action," thus "it certainly behooves us to look at ourselves as we move through the process of creative self-examination with our clients, not only during this internship, but throughout our careers."

Self-awareness is significant in two ways. First, it allows us to use ourselves in helping others. Morales and Sheafor (1995) state that the primary tool for helping others is the self of the helper; this is one of the beauties of being a professional helper. According to Kottler (1986, p. 25), the art of helping others "permits a unique lifestyle in which one's personal and professional roles complement each other. . . . All of our personal experiences, our travels, learning, conversations, readings, or intimate dealings with life's joys and sorrows provide the foundation for everything we do" in our work with clients. By engaging in our own self-examination, we may become more empathic in our work with clients and enrich our lives both personally and professionally (Faiver, Eisngart, & Colonna, 2004).

Self-awareness is important for a second reason: Working with others may trigger our own issues, which can affect the helping process, either positively or negatively. Often when working with individuals and families, themes in your life may surface that previously had been outside your conscious awareness. "If you are unaware of issues stemming from your family experiences, you are likely to find ways to avoid acknowledging and dealing with potentially painful areas with your clients" (Corey & Corey, 2003, p. 67). Morales and Sheafor (1995, p. 198) contend that to be effective in using the self, the helper "must be sensitive to his or her own strengths and limitations, be aware of areas of knowledge and ignorance, and recognize the potential to be both helpful and harmful when serving clients." Corey, Corey, and Callanan (2003, p. 36) share this view: "Practitioners should be aware of their own needs, areas of 'unfinished business,' personal conflicts, defenses, and vulnerabilities and of how these may intrude on their work with their clients."

Thus, part of your journey in human services must be an introspective look at your past history, including issues related to your family of origin. Experiences you had as a child in your family are bound to influence your perceptions and reactions as an adult. If a helper is unaware of this, he or she is likely to be ineffective and might even hinder clients' growth in areas where the helper still has "unfinished business." Failure to understand your family of origin and the patterns of interpersonal behavior you learned means you will be more likely to repeat these patterns with your clients. "The more you are aware of your patterns of interacting with your own family members, the greater is the benefit to your clients" (Corey & Corey, 2003, p. 73).

SCENARIO: A Worker Who Is Unaware of Unfinished Business

You are working in an agency; you perform intake interviews and, depending on the case, may follow up with treatment or provide an appropriate referral. Personally, you have just left your spouse of 30 years and feel great about taking this step; you wish you had done so sooner. One of your clients is a young woman who describes a difficult and unhappy marital situation yet feels ambivalent about leaving. Without any hesitation, you encourage her to separate from her husband. You stress the importance of not "wasting her time" in an unhealthy marriage.

What might this client feel? Could it be that you are revealing your own unfinished business? How might this influence your work with the client?

SCENARIO: A Worker Who Is Aware of Unfinished Business

Lisa has just completed her undergraduate training and has taken a job at a group home for abused children. Herself a victim of physical abuse, Lisa feels that her past has motivated her to work with this vulnerable population. If she can help empower these children and encourage them to feel good about themselves, her goal of making an impact will be fulfilled. At the outset, she discussed her issues with her supervisor and agreed that she would stay alert to the possibility of counter-transference and be willing to seek supervision and personal counseling to deal with any unfinished business that might surface.

Do you think she is aware? What issues might surface? Do you feel she can be effective with these children? What potential problems might she look out for?

■ SELF-AWARENESS EXERCISES

Your Family of Origin and Its Impact on Your Work

Take some time to reflect on the statements listed below regarding family-of-origin issues. We have drawn them from our own experience and from Corey and Corey (2003). Also consider the questions that follow the statements.

> It is essential in this field to examine your family history and understand how your experiences in your family of origin affect your present relationships. Why?

> Your perceptions and reactions to others are influenced by your childhood experiences. Explain how.

> If you are aware of your sensitive areas, you can avoid unnecessary entanglements with your clients or coworkers. How so?

> People who have worked through their own negative and positive family experiences are better able to assist their clients. Elaborate further by giving an example.

To enhance your own functioning, you must undertake the process of identifying and addressing your own problematic family issues. Identify one area related to your family of origin that could benefit your work as well as one area that might hinder your objectivity and effectiveness.

Unmet needs in early family relationships can be manifested later in intense and conflicting personal relationships. Provide an example of this, either from personal experience or from your observations in working with others.

A helper's childhood role in his or her family is often reflected in patterns of dealing with intimacy and conflict in relationships later in life. Give an illustration of this.

Group Exercise: Family of Origin

We present a set of issues and questions (adapted from Corey & Corey, 2004, p. 74) for you to answer. After each answer, briefly state how each might directly or indirectly affect your work in the helping professions. Then share your answers with the group. This may help you discover the differences between yourself and other helpers. It may also stimulate a rewarding process of interactive learning.

Family Structure: What type of family structure did you grow up in? What is your current family structure?

Parental Figures and Relationships with Parents: What was your relationship with each of your parents? How did each parent deal with feelings? How did he or she behave, react, cope? What were their thinking styles like? How did you express your feelings to each of your parents? How did your family as a whole deal with feelings?

Individuation and Separation Issues: In Western culture, it is felt that a psychologically healthy person must achieve a sense of psychological separateness from his or her family of origin. This separateness allows true intimacy to develop. An increasing sense of differentiation promotes the development of our own thoughts, feelings, and actions. How was this individuation process handled in your family? Was it encouraged, discouraged, sabotaged, undermined, or forced? Did any aspects of your cultural background tend to discourage individuation? If so, how?

Existence of Triangular Relationships: To reduce stress between two individuals in a family, the two may pull in a third person. For instance, instead of talking directly to each other, a couple may talk to a child, friend, or relative. The original conflict then remains frozen in place. A child may be put in the role of caretaker and given too much responsibility, which can lead to hurt, anger, confusion, or guilt. Such feelings may not even be recognized. Were you ever involved in a triangular relationship? If so, how did you feel? How did you cope?

Conflict: If you have difficulty with certain feelings in current relationships, you may want to explore how these feelings were dealt with in your family of origin. In specific, how was conflict handled? Is this still a pattern for you today?

Family Rules: Are you aware of any rules that governed the behavior of your family members, such as unspoken injunctions, myths, or secrets? Living by these rules may have helped you survive then but may no longer be functional in your present relationships. What were some of the family rules that you recall as most significant in your family of origin? Do they still exert influence in some form in your current family system? How so?

Significant Life Events: Crisis points—serious illness in the family, divorce, death—are likely to have been influential in your overall development. Describe any events in your family of origin that have been important to your development.

Treat these family-of-origin issues carefully. They may create some emotional turmoil, especially if unresolved feelings are triggered. The realization that there are unresolved issues and patterns may result in resistance or precipitate a crisis. Seek guidance from your supervisor, field instructor, or a counselor if this process becomes too overwhelming.

Unfinished Business and Family of Origin

In this exercise, Satir (1983) demonstrated that the past can be revisited in our current relationships with others, including family, friends, and clients.

1. Imagine someone in your current life (Person A) with whom you're experiencing strong feelings. This person might be a client, colleague, family member, or friend.
2. After visualizing Person A in your mind for a moment, then let a picture of a person from your past (Person B) come to mind.
3. Who comes to mind? How old are you? How old is person B? What relationship did you have with this individual you are remembering?
4. Now reexamine your reactions to Person A. Do you see any connections between what Person A is eliciting and the past feelings elicited by Person B?
5. Consider using this exercise whenever you experience strong feelings in your present relationships.

THINKING THINGS THROUGH

Examine an instance where present interactions with a current relationship have triggered feelings about past relationships. How might being unaware of these feelings affect your work? How can you remain open to exploring past issues as they enter your present work?

Self-Awareness and Self-Acceptance

Also important in relation to self-awareness is the fact that acceptance of oneself is a prerequisite to accepting others. In *The Development of a Professional Self* (1978), Virginia P. Robinson cites several statements from lecture notes of Otto Rank that illuminate the interrelationship between self-awareness and self-acceptance. Rank states, "From self-acceptance, which is the biggest change from self-denial, follows acceptance of others, the world at large, that is life"(p. 27). This notion can have great implications for human service professionals who have not yet accepted themselves. Growth is an ongoing process in which we continuously learn who we are and how we need to change to become more self-actualized. This is true for everyone, but our point here is that human service professionals in particular must be willing to invest in ongoing self-exploration and self-acceptance. We believe that if one does not accept oneself, issues of competence are likely to arise. Specifically, helpers who are not comfortable with who they are may be ineffective in helping others battle their own negativity. We ask our clients to engage in self-exploration and need to model this by being open to our own self-examination. Resistance to this process is a disservice to both our clients and to ourselves and may be a sign of professional burnout.

The road to self-acceptance starts with exploration of one's limitations and sources of low self-esteem. Helpers must continue to work toward self-improvement and resolution of any past issues that still influence their attitudes toward themselves and others they relate to. How can we ask our clients to engage in self-exploration and self-improvement if we are unwilling to do the same? Helpers who are willing to work on themselves are more likely to develop a positive self-concept, to be genuine, and to inspire others to work on themselves, too. Egan (2002) has noted that when helpers feel confident about themselves and their skills, clients perceive them as competent and therefore feel more positive about the helping process.

The Impact of Early Development on the Individual Helper

We have seen how our own family-of-origin issues may impact our work with clients. A look at how early developmental issues affect who we are today is one way to begin to understand our past. In particular, we have found that considering Erik Erikson's work on life-span development is helpful. Sweizer and King (2004, p. 51) concur: "Some of the basic tenets of his work are valuable tools for self understanding. . . . Your early experience with these issues often reverberates into your later life, and we find that some of these issues are especially relevant in internships." Let's take a look at Erikson's theory on stages of psychosocial development.

Erik Erikson's Psychosocial Theory of Human Development

Initially aligned with Freud, Erik Erikson later developed his own model that viewed human development in a very different way. He believed that psychosocial forces are major motivators in an individual's development over the life span. Erikson had faith in human nature because he believed that individuals had the capacity to overcome many of their issues. In essence, he described eight stages of development from infancy to old age that each person passes through. In his theory, Erikson postulated that as we approach each stage, we are confronted with specific age-related developmental tasks or crises to overcome and must establish an equilibrium between ourselves and our social world. When these tasks or developmental milestones are successfully mastered, "a strong ego is formed, a positive identity is created, and the individual is ready to move on to the next level" (Neukrug, 2004, p. 141). In contrast, when an individual is unable to conquer the developmental tasks at hand, a low self-image and bruised ego results, impairing the person's ability to master future developmental stages. These tasks or crises can offer both dangers and opportunities for growth. "At these critical turning points we can achieve successful resolution of our conflicts and move ahead, or we can fail to resolve the conflicts and regress. To a large extent, our lives are the result of the choices we make at each stage" (Corey & Corey, 2002, p. 46).

Erikson's model provides us with a framework to understand the typical developmental tasks of the individual (Erikson, 1968, 1998). In sum, "Erikson's model is holistic, addressing humans inclusively as biological, social, and psychological beings" (Corey & Corey, 2003).

Table 1 describes the major principles of each of the stages of development as presented by Erikson (1963) and elaborated by others (Corey & Corey, 2002; Miller & Stiver 1997; Newman & Newman, 2003). It would be wise for beginning human service workers to review these stages and carefully examine their own development to determine where they may have unresolved issues because these issues may continue to effect their current functioning and their work in agency settings.

The Role of Thoughts, Feelings, and Actions in Self-Awareness

In understanding ourselves, we must be aware of thoughts, feelings, and actions as they relate to human behavior. Hutchins and Cole (1992, p. 4) state, "To be an effective helper, you must first understand major aspects of your own behavior and how you interact with others." They use the model labeled TFA (thinking, feeling, action), which Hutchins (1979) first introduced. In the process of becoming self-aware, such a model is useful to clearly delineate these aspects of human behavior. It is helpful to be able to understand the subtle distinctions between these aspects of your own personality. In our work with students, they are often confused when describing their own thoughts, feelings, and actions in working with others. According to Goldman (1995, p. 43):

> The ability to monitor feelings from moment to moment is crucial to psychological insight and self-understanding. People with greater certainty about their feelings are better pilots of their own lives, having a surer sense of how they really feel about personal decisions.

■ **Table 1. Overview of Psychosocial Developmental Stages**

Age	Development	Psychosocial View	Central Process	Prime Adaptive Ego Quality	Core Pathology	Applied Topic
Prenatal (conception to birth)						Abortion
Infancy (birth to 2 years)	Maturation of sensory, perceptual and motor functions; Attachment; Sensorimotor intelligence and early causal schemes; Creating categories	Trust vs. Mistrust; Basic task is to develop a sense of trust in self, others, and the environment; Infants need to feel a sense of being cared for and loved	Mutuality with caregiver	Hope	Withdrawal; Greediness and acquisitiveness, a view of the world based on mistrust, fear of reaching out, rejection of affection, fear of loving and trusting, inability to be intimate	The role of the parents; Impact of early environment
Toddler-hood (2 and 3)	Elaboration of locomotion; Language development; Fantasy play; Self-control	Autonomy vs. Shame and Doubt; A time for developing autonomy	Imitation	Will	Compulsion; Failure to master self-control tasks may lead to shame and doubt, feelings of hostility, rage, destructiveness and hatred; future difficulties in accepting one's feelings	Childcare: parental attitudes may be expressed verbally or nonverbally; Negative learning experiences tend to lead to feelings of guilt about natural impulses
Early School Age (4 to 6)	Gender identification; Early moral development; Self-theory; Peer play	Initiative vs. Guilt; A time to establish a sense of competence and initiative	Identification	Purpose	Inhibition: If not allowed to make decisions children tend to develop a sense of guilt and sense of inadequacy; Strict parental indoctrination can lead to rigidity, severe conflicts, remorse, and self-condemnation	School readiness

(Continued)

Table 1. (Continued)

Age	Development	Psychosocial View	Central Process	Prime Adaptive Ego Quality	Core Pathology	Applied Topic
Middle Childhood (6 to 12)	Friendship Concrete operations Skill learning Self-evaluation Team play Expand understanding of world Continue to develop appropriate gender-role identity	Industry *vs.* Inferiority Central task is to achieve a sense of industry; failure to do so results in sense of inadequacy	Education	Competence	Inertia Negative self-concept, feelings of inferiority in establishing social relationships, conflicts over values, confused gender-role identity, dependency, fear of new challenges and lack of initiative	Violence in the lives of children
Early Adolescence (12 to 18)	Physical maturation Formal operations Emotional development Membership in peer group Sexual relationship	Group Identity *vs.* Alienation	Peer Pressure	Fidelity to others	Dissociation Confusion and possible loss of sense of self	Adolescent alcohol and drug use
Later Adolescence (18 to 24)	Autonomy from parents Gender identity Internalized morality Career choices	Individual Identity *vs.* Identity Confusion	Role Experimentation	Fidelity to values	Repudiation Absence of a stable set of values can prevent mature development of a philosophy to guide one's life	Challenges of social life
Early Adulthood (24 to 34)	Exploring intimate relationship Child-bearing Career/work Lifestyle	Intimacy *vs.* Isolation Sense of identity tested by challenge of achieving intimacy	Mutuality among peers Challenge is to maintain one's separateness while becoming attached to others	Love	Exclusivity Failure to reach a balance can lead to self-centeredness or to exclusive focus on needs of others Alienation and isolation	Divorce

(Continued)

Life Stage	Tasks	Psychosocial crisis	Central process	Prime adaptive ego quality	Core pathology	Applied topics
Middle Adulthood (34 to 60)	Managing a career Nurturing an intimate relationship Expanding caring relationship Managing the household A time for going outside oneself One challenge is to recognize accomplishments and accept limitations	Generativity vs. Stagnation Individuals become more aware of their eventual death and begin to question whether they are living well The crossroads of life; a time for reevaluation	Person-environment fit and creativity	Care	Rejectivity Failure to achieve a sense of productivity can lead to stagnation Pain can result when individuals recognize the gap between their dreams and what they have actually achieved	Discrimination in the workplace
Later Adulthood (60 to 75)	Accepting one's life Redirecting energy toward new roles Promoting intellectual vigor Developing a point of view about death	Integrity vs. Despair Ego integrity is achieved by those who have few regrets, who see themselves as productive, and who have coped with both successes and failures. Key tasks are to adjust to losses, maintaining outside interests, and adjustment to retirement	Introspection	Wisdom	Disdain Failure to achieve ego integrity often leads to feelings of hopelessness, guilt, resentment, and self-rejection Unfinished business from earlier years can lead to fears of death stemming from sense that life has been wasted	Retirement Caring for older parents
Very Old (75 until death)	Coping with the physical changes of aging Developing a psycho-historical perspective Traveling through uncharted terrain	Immortality vs. Extinction	Social Support	Confidence	Diffidence	Meeting the needs of the frail elderly Senility

(Adapted from Corey & Corey, 2002; Erikson, 1963; Miller & Stiver 1997; Newman & Newman, 2003.)

As you become more accurate in knowing your own feelings, thoughts, and actions, you will be more effective at recognizing these in others. Some examples of these three aspects of human behavior are discussed in the following paragraphs (adapted from Goldman, 1995).

Examples of Feelings

Anger: fury, outrage, resentment, wrath, exasperation, indignation, vexation, acrimony, animosity, annoyance, irritability, hostility

Sadness: grief, sorrow, gloom, melancholy, loneliness, dejection, despair

Fear: anxiety, apprehension, nervousness, concern, consternation, misgiving, wariness, edginess, dread, fright, panic

Enjoyment: happiness, joy, relief, contentment, bliss, delight, amusement, pride, sensual pleasure, thrill, rapture, satisfaction, euphoria, whimsy, ecstasy

Love: acceptance, friendliness, trust, kindness, affinity, devotion

Surprise: shock, astonishment, amazement, wonder

Disgust: contempt, disdain, scorn, abhorrence, aversion, distaste, revulsion

Shame: guilt, embarrassment, chagrin, remorse, humiliation, regret, mortification

Sexual Feelings: attraction, stimulation, excitement, orgasm

THINKING THINGS THROUGH

Using the preceding list of emotions, identify some emotions you have experienced at your fieldwork placement and give specific instances. Did you experience any other emotions not listed here? If so, please explain.

Thinking is a cognitive process involving attitudes, assumptions, and beliefs that guide our behavior. The *American Heritage Dictionary* (1995) defines thinking as "the ability to exercise the power of reason, as by conceiving ideas, drawing inferences, and using judgement." Recognize your thinking as important, because what we think plays a part in what we do. According to Long (1996, p. 36), "Our attitudes affect our words and behaviors, they affect our ability to use helping skills. Our beliefs and attitudes go with us into our helping relationships. It is essential, therefore, that we consider our attitudes if we are to develop helping skills."

SCENARIO: Thoughts

You are a male intern placed in a group home for abused teens. Although you believe that "it is okay for a man to cry," you find yourself very uncomfortable when one of the abused teens starts to cry as he tells you about his father's abuse when he was younger. You notice that your eye contact is poor throughout most of the interaction. Also, when he expresses his sadness, you quickly jump in, saying, "It's okay to cry," which seems to cut off his feelings. Discussing this with your supervisor later, you realized that the thought that it is *not* okay for a man to cry had crept into your interaction with this youth. The automatic, unconscious nature of this thought surprised you.

What do you see as the thinking aspects of this situation? What feelings might you be having?

THINKING THINGS THROUGH

In your fieldwork placement, have you become aware of how your thoughts influence your work? Take a moment to describe a situation where your thoughts, unbeknownst to you at the time, affected your work with a particular client.

Actions, the third aspect of self-awareness, are direct, observable behaviors. Behavior is defined as "the actions or reactions of persons or things in response to external or internal stimuli" (*American Heritage Dictionary*, 1995). Behavior can communicate feelings or thoughts that we have not expressed in other ways. It is important to recognize your actions; they can profoundly affect your interactions with others.

SCENARIO: Behaviors

One morning, you wake up late. In your rush to get to work, you get a speeding ticket. When you finally arrive, your client is angry with you for being late. You find yourself being more confrontational than usual once you finally begin dealing with her issues. You also make no attempt to work on the college forms she wants help with. You are clearly annoyed with her. Later that day, when you have had some time to process your thoughts, you realize that you expressed your angry and hurt feelings through your rigid manner. With the help of your supervisor, you understand that you felt especially upset with this particular client because she is consistently late to appointments, whereas you are normally on time.

In this situation, how might you have felt? Would you have behaved in a different manner? How so?

THINKING THINGS THROUGH

Consider your behaviors with the following individuals at your placement: your supervisor, your coworkers, the clerical staff, and your clients. Are there marked differences in your interactions? How might your behaviors be expressing underlying feelings or thoughts?

Understanding Our Motivations for Becoming Helpers

For your success in the field of helping, it is useful to gain some personal awareness of your motivations for pursuing such a career. This text will help you explore your feelings, thoughts, actions, values, and underlying motivations. By offering you a comprehensive perspective, we hope to give you the opportunity to take a "taste test" of the many aspects of helping and to provide you with some insight into obstacles and dilemmas you are likely to face.

The literature clearly shows the importance of understanding your motivations for wanting to be a helper. Corey and Corey (2003) state, "You will do well to begin by examining your motivations for pursuing this path. . . . Your motives and needs can work either for you or against you and your future clients." They identify typical needs of helpers that relate to their motivations for wanting to do this type of work. Here we briefly review these motivations by describing productive and counterproductive aspects.

The need to make an impact: Helpers may hope to positively influence their clients. When successful, you may feel satisfaction that others have benefited from your help, yet if unsuccessful, you may feel frustrated and negative.

The need to return a favor occurs when helpers have themseleves been helped by a significant person in their lives and they wish to do the same with their clients. Although rewarding when successful, sometimes it is not possible to emulate the specific role model who once helped you. Helpers are unique and each possesses different strengths and qualities that may not be duplicated.

The need to care for others is one of the most common motivators. Many helpers have fallen into this role in both their family of origin and in their current lives. As caretakers, they are often available to help others. What is potentially dangerous about this role, however, is when helpers care for others to the exclusion of taking care of themselves. This behavior can easily lead to burnout and would not be good modeling for clients.

The need for self-help often occurs when helpers enter the helping field as a way to further explore their own unresolved issues. "It is the wounded healer more so than the well adjusted helper who can be authentically present for others searching to find themselves. If you have struggled successfully with a problem, you're able to identify and empathize with clients who come to you with similar concerns" (Corey & Corey, 2003, p. 5). Although it can be productive for helpers to engage in this self-exploration, care must be taken to not allow one's unresolved issues to negatively influence the helping process. Helpers who attempt to push clients to resolve their issues, or are unaware or unconscious of how their own struggles impact their interventions with clients, tend have difficulties.

The need to be needed can be "psychologically rewarding" for helpers who feel positive about their role. However, if helpers depend exclusively on this "need to be needed" for self-satisfaction, they will likely feel disappointed. Not all clients are appreciative and not all agencies use positive reinforcement to encourage helpers to do their best.

The need for money is an acceptable motivation for wanting to enter the helping profession. Helpers have often invested many years in college, graduate school, and training, and thus it is understandable to expect reasonable compensation. This need can become counterproductive when helpers have unrealistic expectations over the amount of money they will receive for their services. The helping profession is a noble one, but not one where a great deal of money is likely to be made. When helpers' primary motivation is money, they may consciously or unconsciously foster dependency in their clients to ensure a continuous flow of income.

The need for prestige and status is similar to the need for money: helpers may enjoy feeling recognized and rewarded for their efforts. Often, however, helpers are not recognized for their contributions by clients, the agencies they work for, or the community at large. In fact, many helpers may end up feeling isolated and unappreciated. Helpers who are "placed on a pedestal" have only one direction to go—down.

The need to provide answers may be a lifelong quest for many helpers who enjoy the caretaker role. Clients, friends, and family often reinforce this need by seeking advice from a helper. Ideally, helpers are encouraged to help clients find their own answers, despite how time consuming this may be or how emotionally painful. Unfortunately, it is often more tempting to simply provide clients with the answers they request. When helpers succumb to giving advice they may be fostering dependency on their clients. Also, clients may either not follow the advice given or become angry or upset if things turn out negatively as a result. For these reasons, helpers are encouraged to continually discuss this temptation with their supervisors.

The need to be in control is related to the need to provide answers because some helpers enjoy being in control. As with most of these motivations, when this need for control overpowers the best interest of the client, then it becomes counterproductive. Helpers who need to control others will ultimately not benefit their clients and may need to reconsider their professional choice.

The need for variety and flexibility is what often motivates helpers who enjoy the diversity available in human services. Helpers are frequently exposed to a variety of cultures, settings, and client populations; being accepting and flexible are key in appreciating this aspect of helping.

These typical needs or motivators have both positive and negative aspects. To be successful in your fieldwork placement and your overall professional development, you need to acknowledge both aspects.

In our fieldwork courses, when we ask students to consider their motivations and needs, the two most common responses are "the need to make an impact" and "the need to care for others."

SCENARIO: The Need to Make an Impact

During a policy class, you are asked to choose a social policy to research. You choose to explore child-welfare policies; in doing so, you become aware of the widespread existence of child abuse. At this point, you decide that you need to make an impact by working directly with abused children, so you choose to volunteer at a group home for abused children. A positive aspect of this need to help is the support, encouragement, and guidance you are likely to provide many abused children. However, a counterproductive aspect might be the disappointment that is likely when policy changes do not materialize or when you realize that no matter how hard you try, you cannot erase the abusive histories these children have experienced.

What might be some other positive and counterproductive aspects of the need to make an impact?

SCENARIO: The Need to Care for Others

Growing up, you were considered one of the primary caretakers in your family. Siblings, friends, and relatives sought you to help them with a problem or a pressing matter. You naturally cared for others before giving any thought to your own needs. You found yourself enjoying this role, though sometimes feeling overwhelmed. As a helper, you continue to feel this need to care for others. One of the qualities you are most proud of is your great capacity for empathy and caring. Certainly, your ability to be empathic in caring for others is positive—it provides a strong foundation on which to build additional helping skills. But among the counterproductive aspects might be having difficulty recognizing and acting on your own feelings, unclear boundaries, not knowing how to ask for your own needs to be met, and feelings of inadequacy at not being able to "fix" or take care of everyone.

Additional negative motivators identified by Guy (2000) and discussed by Gladding (2004) merit mention here because they are likely to be counterproductive to becoming a helper. These include emotional distress, vicarious coping, loneliness and isolation, a desire for power, a need for love, and vicarious rebellion. To be effective in our work and to prevent unethical behavior, it is essential to assess our true motivations for entering into the field. "To the degree to which we are unaware of our needs and personal dynamics, we will probably use our work primarily to satisfy our own unmet needs or will steer clients away from exploring conflicts that we ourselves are unwilling to acknowledge" (Corey, Corey, & Callanan, 2003, p. 37).

THINKING THINGS THROUGH

From Corey and Corey's list of typical needs of helpers, first select those you most identify with. Second, consider both the positive and negative aspects of each. Lastly, explore ways that you might prevent a negative outcome.

As another exercise, review the negative motivators listed earlier and determine if any exist for you and how you might combat them.

Because understanding oneself is vital to the world of helping, we always start our beginning practicum courses by asking our students to write a paper on personal awareness. Besides helping students learn about themselves and their underlying motives, this assignment can also encourage introspection, which can help them become aware of their personality patterns, strengths, and weaknesses. Often, beginning helpers also gain an appreciation for the helping process and increased respect for themselves and others working in this field.

▨ SELF-AWARENESS EXERCISE

Here are the questions we offer as guidelines for the paper on personal awareness. How would you answer them? Remember, be honest with yourself—this is the only way you will truly learn who you are and what motivates you.

> Discuss what needs and motivations are leading you to pursue a career in the human services. Consider the following aspects: What are some of your needs and motivations? How might these work for you? How might they become a counterproductive force in your professional life?

> Discuss your personal history and significant turning points that have contributed to your desire to become a helper. Review your family of origin for the roles that you were expected to play and the significant relationships that influenced you, and determine what effect these have had on your motivation to become a helper. How did you deal with emotions in your family of origin? How did you handle your feelings with each of your parents or parental figures? Discuss other important events that may have affected your desire to become a helper.

> Provide an honest assessment of your own strengths and limitations, with reference to a career in the helping professions. What attitudes and qualities do you consider counterproductive for you? What attitudes and qualities do you consider strengths? What insecurities are you facing at your current stage of professional development? How will you choose to work on them?

This assignment is helpful not only when it is first given but it can also be a convenient tool for measuring your personal and professional growth later in your career. When our students are completing their education, we ask them to reflect back on their early paper on personal awareness and compare it to their current stage of development. They are often pleasantly surprised to find how much they have grown.

CONCLUDING EXERCISES

1. Describe how you would like to see yourself professionally in ten years' time. Imagine your ideal professional position and working situation. What population do you see yourself working with? What kind of agency do you see yourself working in? What degrees or licenses do you expect to have?

2. Now imagine yourself having retired from the profession and looking back on your accomplishments and personal history. What would you most want to be able to say about yourself professionally at that point? What would you want to be remembered for? What would you most want people to say about you as a person? What do you imagine yourself doing as you grow older?

3. Consider the following scenario: You have been chosen to be on a panel of professionals in behavioral and social sciences for a career day at a local university. Each panelist has been allotted five minutes to present relevant information about his or her profession and give reasons for this career choice. Along with details of the event, the coordinator has included several questions that each presenter must be prepared to answer.

    What does a career in human services entail?

    Is there a fieldwork component? If so, please explain its purpose and significance.

    What are the values of this profession? How do you align with them? Why might they be important in your fieldwork placement?

    Why did you choose this career? What led you to a helping profession?

    Why would you encourage someone else to pursue a career in this field? Why not consider other fields?

CHAPTER TWO

Agency Systems and Policies

AIM OF THE CHAPTER

An integral part of fieldwork is understanding the agency in which you are placed. Knowledge of the agency setting can guide you in developing your skills as a helper. Furthermore, understanding your agency setting is crucial if you are to be successful in working cooperatively with others and fitting in with the agency's mission and goals.

Because working in an agency setting requires a major emotional commitment, it is advantageous to understand the nuances and intricacies of your agency's system. Knowing its policies, practices, strengths, and limitations helps you create an environment where you can maximize your learning. More often than not, agencies seem large, cold, complicated, and overly bureaucratic to the beginning professional. In this chapter, we hope to give a framework for understanding your current agency setting. At times we will refer to a social agency as *a human service organization* because this is consistent with the literature. We will define agencies, describe common types of human service organizations, and identify their function(s) and characteristics. We will review the basic tenets of systems theory as a means to understanding the dynamics of agency functioning, especially in regard to the impact of change on a system. Our focus will then turn to formal and informal agency policies, procedures, and structure. Within this framework, we will discuss sexual harassment to illustrate formal policy and institutional discrimination to illuminate informal norms.

These perspectives can be useful to you as a professional who is beginning to work in agency systems. Once you have a basic sense of what your agency is all about, you will find it easier to develop adaptive coping strategies to deal with various facets of agency life, including rules and regulations, clientele, and personnel. You might also learn better ways of interacting with other professionals at various levels—including management—that will allow you to make the most out of your agency experience. As you make your way, we encourage you to develop your own insights into how agencies function, to maintain your curiosity and fascination, and above all to remain committed to the helping process.

UNDERSTANDING YOUR SOCIAL SERVICE AGENCY

Bureaucracies

Often when one thinks of an agency, the term *bureaucracy* comes to mind, usually conjuring up the image of a big, impersonal, uncaring, rule-driven institution. This is unfortunate. We have found that human service organizations possess both strengths and limitations. In the literature, there is much agreement that bureaucracies are misunderstood and unfairly stereotyped. According to Fairchild (1964, p. 29) bureaucracies are typically described in a negative light and criticized for their devotion to routine, inflexible rules, red tape, procrastination, and reluctance to assume responsibility or to experiment.

Neugeboren (1985) and Woodside and McClam (1994) highlight the typical problems that beginning professionals are likely to face when entering a bureaucratic system.

Tension between the workers and the administration Though working toward the same goals, they have different perspectives. The worker may be focused on the client's point of view, whereas the administration may make decisions from a broader institutional perspective. These differences often create a tense environment, prone to conflict.

SCENARIO

Because of cutbacks and limited availability of resources, many agencies now advocate seeing clients in groups rather than one to one because it is more cost-effective and allows more rapid access. How might this affect a worker who feels strongly that certain clients would benefit more from individual treatment? What if these were your beliefs, but you were told to refrain from individual treatment? How might you handle this?

The impersonality of the bureaucratic structure Almost always, policies and procedures must be adhered to before services can actually be provided.

SCENARIO

A common complaint about large systems is that the participant feels like a number rather than a person with individual needs and concerns. This was evident one day while Lupe was awaiting an appointment at an ophthalmologist's office. Lupe saw an elderly woman who had bandages over her eyes being wheeled in by an attendant. The attendant informed the front desk that she was there and left her alone in the waiting room. A few minutes later, she began to ask for the attendant. When no one responded, she asked more loudly and was becoming visibly upset. Lupe was amazed at how everyone, including the receptionist, ignored her pleas. At that point, Lupe decided to ask her what she needed—she had to use the restroom. Lupe then approached the front desk, and a nurse was sent out to help her. Human needs did not seem as important as keeping the line moving and checking clients in.

How would you feel if you were in that situation? Have you experienced a similar situation?

The bureaucracy's resistance to change Often change is slow and requires persistent efforts. Frustration and apathy are likely to surface if expectations are too high and immediate change is sought.

SCENARIO

As a new intern at a large child guidance center, you are constantly struggling for office space, which is very limited. You and the other interns share frustration about this space issue and decide to voice your concerns at the next staff meeting. You expect that this will resolve the problem, but you find that office space is still limited weeks later. When you broach your frustration with your supervisor, he informs you that diligent efforts are being made to revamp clinicians' schedules so that there will not be so much overlap. He explains that this takes time, because it involves many schedules, and must be approached carefully, because clients' appointments will also have to be rearranged in some instances. After this discussion, you realize that perhaps you expected change too quickly and did not understand the underlying issues at hand.

How might you feel in this situation? Have you ever experienced a situation where you were frustrated by change taking longer than you expected? Explain.

According to Netting, Kettner, and McMurtry (1993), "[t]he term *bureaucracy* has taken on a number of mostly negative connotations" that may not always be accurate. These writers and numerous others (Hansenfeld, 1983; Johnson, 1986; Schmolling, Youkeles, & Burger, 1993) have written about the positive aspects of organizations, not just the negatives. We would also like to focus on the positive traits of agencies rather than just target the stereotypically negative aspects.

There are several reasons for the prevailing pessimistic outlook on bureaucracies. Because of the impersonal nature of many organizations, an individual may feel lost and uncared for. Johnson (1986, p. 7) writes:

> From a human service point of view, one of the most bothersome and problematic aspects of bureaucracy is the insensitivity it may bring on the part of some personnel who are supposed to be serving others and facilitating the meeting of human needs.

SCENARIO

Interns placed within the juvenile correctional system have sometimes reported that they feel frustrated by the impersonal nature of their role. Because many teens are served, and the turnover rate in this population is so high, there is little opportunity for individualized treatment where one-to-one interactions are essential. Also, helpers feel ineffective in that they are unable to examine the underlying issues at the core of a juvenile's delinquency.

If you were placed in this type of organizational setting, how would you cope with this impersonal outlook? How might you reframe your role to avoid such frustration?

In addition, a bureaucracy can often invoke a sense of apathy in its workers, who may feel as if they don't count. Schmolling, Youkeles, and Burger (1993) agree that workers may sometimes feel as if they do not make a difference—that they are just cogs in a huge machine.

SCENARIO

With the advent of managed care in all aspects of health care, many professionals working in medical, rehabilitative, and mental health settings find it difficult to justify the benefit of the limited services they provide. For instance, mental health workers often find it quite difficult to help clients who may require long-term treatment but are allowed only limited visits. Such workers may come to feel ineffective and apathetic about their role.

Think about a situation where you have observed a worker who was apathetic about his or her role.

Because of the rules, regulations, and red tape inherent in many bureaucracies, the organization may seem inefficient and ineffective. Although this may be true, the system in question may be the best or the only provider available. Many organizations have to furnish services on a large scale, which complicates the delivery of service considerably. For example, you might enjoy baking or cooking for your family. Yet if you were to open a bakery or restaurant, your enjoyment might cease; you would have to change your approach and gear your meal preparation to large quantities. The same may be true for large human service organizations, which have had to enlarge their scope of services and therefore change the delivery process.

SCENARIO

Interns who are placed in child welfare agencies as child protective service workers often feel overwhelmed by the bureaucratic structure. One recent graduate, just beginning her career in this type of setting, was frustrated and upset by the incongruence between her values and what she could accomplish. Her primary intent was to help protect children from further abuse and neglect, yet she often struggled with court decisions that defined *protection* differently than she did. "Sometimes I feel like I'm selling myself to the devil," she commented, with reference to having to remove children from their home or return them to a potentially abusive home situation.

If you were in a similar situation, where your values were often discounted by the system, how do you imagine you would feel? How might you cope?

Finally, organizations may have had past difficulties that result in negative perceptions. People develop positive or negative associations with a particular organization, based on past experience. Such negativity could also develop indirectly, through contact with others who are critical of an agency because of their own biases and prior experiences.

SCENARIO

Many people, clients and professionals alike, have a stereotypically negative view of government agencies. Such perceptions, perhaps based on inaccurate information, may limit both the use of these services by clients who could truly benefit from them and the employment of qualified professionals who could be an asset to the system. In fact, government agencies are often teaching facilities that employ motivated, eager, and inquisitive interns, residents, and supervisors.

Can you think of any examples where negative views of an organization directly affect service delivery? How might this be changed? What recommendations do you have?

We hope to engender a more positive, holistic outlook on organizations and the services they provide. We are in accord with Hansenfeld's (1983, p. 1) statement that highlights the importance of organizations in our lives: "The hallmark of modern society, particularly of its advanced states, is the pervasiveness of bureaucratic organizations explicitly designed to manage and promote the personal welfare of its citizens. Our entire life cycle, from birth to death, is mediated by formal organizations that define, shape, and alter our personal status and behavior." He continues by offering his view of human service organizations as ones whose principal function is to protect, maintain, or enhance the personal well-being of individuals by defining, shaping, or altering their personal attributes. Let's consider other definitions of organizations and note their similarities and differences.

Organizations Defined

Lauffer defines organizations as "purposeful social units; that is, they are deliberately constructed to achieve certain goals or to perform tasks and conduct programs that might not be as effectively or efficiently performed by individuals or informal groups" (1984, p. 14). Similarly, Brager and Holloway (1978) state that human service organizations are "the vast array of formal organizations that have as their stated purpose enhancement of the social, emotional, physical, and/or intellectual well being for some component of the population." This definition suggests the vast range of human services that organizations provide. Although we may complain about the deficiencies of organizations, we should take note that many other countries lack such organizations entirely, and many social needs therefore go unmet. Perhaps this thought can reframe our thinking, much as we help our clients do when they focus only on the negative and neglect to acknowledge the positive.

Overall, organizations consist of many individuals who perform specified roles in an effort to provide needed services to certain populations in the community. Certain positions such as manager, line worker, and support staff organize the functioning of the agency. These employees make a coordinated effort to work together and integrate services in a way that is supportive to all.

THINKING THINGS THROUGH

Take a moment to list as many organizations as you can. Freely associate how you view each of these, and then tabulate how many you regard as negative, positive, or in between. Next, take each organization and consider what problems might develop if the organization did not exist.

As you begin your career, it may be helpful to consider how to successfully navigate your way in agency settings. Woodside and McClam (1994, p. 172) offer the following recommendations: First, devote time and energy to studying the structure of the agency system. Consider the strengths and barriers that would enhance or deter optimal service delivery. You can learn more by reading available information about the organization, studying organizational charts, and identifying the people in the system who are resourceful and open to sharing their experiences. Second, be creative and take the initiative to learn to work within the system rather than being paralyzed by it. If you feel a sense of belonging and maintain a belief in your ability to influence the system, you can remain active and make a more positive impact.

Characteristics and Functions of Human Service Organizations

Understanding the functions of an organization can be helpful in various ways to the student in a fieldwork placement or to the beginning human service professional. First, it will help you determine which referrals to make for clients who need a particular service. Also, it may be of assistance in determining the scope of services available in your organization and others, which could help you evaluate whether the agency is fulfilling its mission or not. Further, it is useful to develop a deeper appreciation for the complex nature of organizational systems. This should enhance your regard for the arduous efforts of those who dedicate their lives to working in human service organizations.

Miles (1975) identified these functions of an organization:

- Direction and leadership
- Organizational structure and job design
- Selection, training, appraisal, and development
- Communication and control
- Motivation and reward systems

Schmolling, Youkeles, and Burger (1993) compiled characteristics of human service organizations. We have listed these in paraphrased form and included examples of each.

A separate identity Every organization has an identity—its own territory, program, rules and procedures, history, and vocabulary.

SCENARIO

Consider working at juvenile hall, where the lingo involves such terms as *juveniles, lockdowns, AWOLS,* and *violated probation.* If new to this setting, you may find the atmosphere cold and rigid, yet once you are there for a while, you may come to understand the reasons for this. Similarly, if you work in a day treatment unit for people with chronic mental illness, you may hear such terms as *schizophrenia, undifferentiated type, loose associations,* and *psychotic features.* You are likely to find that the staff are nonchalant about the behaviors of the clients. After a time, you too will probably become accustomed to certain oddities of behavior in the clients and will not react so strongly to them. Also, you may realize that many of the staff are very committed to their clients' well-being and really do care.

Own set of advocates Every organization has supporters and antagonists in the community and in other agencies.

SCENARIO

Working in agency settings, you will learn that many agencies have certain allies and adversaries in the community. By remaining an observant and active participant in the work of the agency, you will gradually learn which other entities are the allies and adversaries. Some are formally delineated, and others are informally spoken about in passing or in staff discussions. For instance, a large organization is likely to have some former clients in the community who have discontinued treatment because of negative experiences. In contrast, many others will have continued with this organization because of positive experiences and their awareness that other agencies may not offer as broad a range of services and may have their own problems.

Established purpose and goals Every organization was originally established to achieve definite purposes and goals in providing human services.

SCENARIO

The purpose of community mental health centers is to provide affordable mental health treatment to people in the community who are experiencing emotional problems. If funds are in abundance, the agency may take on additional functions, such as providing parenting classes and support groups. It may also be open to seeing clients who are experiencing a phase-of-life problem but do not have a serious mental illness. Examples include helping a preteen who is entering adolescence and overwhelmed by biological changes and social pressures. Also, consider a middle-aged adult who is struggling with mid-life crisis issues or an older person who has just retired and struggling with increased free time and role loss. However, when funds are limited, often only the primary goals can be met: direct treatment of mentally ill clients. Also, in such instances, only clients who qualify as having a "mental condition" can be treated within this setting. This may require many outside referrals to make sure that people seeking help are successfully treated.

Subsystems often have different purposes Most organizations have subsystems whose interests are not always compatible with those of the larger agency.

SCENARIO

In many agencies, there may be a division between administration and the line staff who work directly with clients and the public. The management staff is likely to focus on budgetary constraints and thus emphasize limited treatment and increased productivity. The line staff, on the other hand, is committed to the clientele they work with on a daily basis and may not feel it is appropriate to limit their services. They may also find it difficult to do more with less assistance. As you can see, both have valid perceptions. Neither is wrong, yet their views conflict. In such cases, it is beneficial to have cohesion, open communication, negotiation, and the ability to see others' viewpoints.

Growth or decline A given organization is either growing, stable, or declining, and this tendency affects all parts of the organization.

SCENARIO

A large health center, which once was predominant in the community and for many years the largest and most respected, is now experiencing budget constraints due to the changing economy, increased competition, and loss of membership. Professionals who have worked there for a long time will likely have a more difficult time adjusting to the change than those who are new to the organization and only know it as it currently stands.

Can you think of an agency that has seen growth and decline? How has this affected service delivery? How have employees adjusted to the changes inherent in this increase or decrease?

Crises Crises may develop based on conflicts about policy. Such crises have the potential to destroy the agency or bring about its reorganization in positive ways.

SCENARIO

Faced with massive budget cuts, a family service agency is considering the closure of several sites. The staff come up with several ideas that might help save money and still provide needed services. One idea is to recruit graduate interns and unlicensed professionals who need hours, and set up one staff member to concentrate solely on supervising and coordinating this division. Ultimately, this new program could have a dual purpose of continuing to provide services to clients via staff and interns and helping beginning human service professionals receive necessary training and supervision.

Organizational Classifications

Today's society requires organizations to supply services dictated by unmet needs or prevalent societal problems. Numerous human services organizations have evolved to meet those needs. Although not perfect by any means, or responsive to all the issues at hand, they nonetheless make valuable contributions. Netting, Kettner, and McMurtry (1993) provide a helpful view for categorizing human service organizations, distinguishing among nonprofit, public, and for-profit agencies.

Nonprofit (voluntary) Kramer (1981) defines this type of agency as "essentially bureaucratic in structure, governed by an elected volunteer board of directors, employing professional or volunteer staff to provide a continuing service to a clientele in the community." Funding may be from private donations, fundraisers, grants, corporations and government sources. According to Berg-Weger and Birkenmaier (2000), the fact that funds come from a variety of sources can be a source of stress because administrators must be accountable to a variety of entities, or it can be viewed positively because the organization does not depend completely on one source of income. Examples include family service agencies that provide counseling, crisis intervention, parenting classes, respite care for the elderly, teen programs, and anger-management groups. They use professional, paraprofessional, nonprofessional, and volunteer staff to furnish such services to the community.

Public agencies (governmental) Governmental entities (federal, state, regional, county, and municipal) have the main purpose of providing services to designated target populations. Their funding comes from local, state, and federal tax revenues or other governmental funding sources. These agencies are typically accountable to the specified legislative bodies that determine their budget. In comparison to other types of agencies, public agencies are usually large, have a vertical authority structure, and are usually complex, rule-driven entities. Often they handle a great deal of paperwork and have numerous procedures to follow (Swietzer & King, 2004; Berg-Weger & Birkenamaier, 2000). Public departments of mental or behavioral health, public health, corrections, education, aging, and child welfare are some examples.

SCENARIO

A County Mental Health Center provides crisis and short-term counseling to local clients who are suffering from emotional disorders. Individual, conjoint, family, and group treatment are offered as well as parenting and anger-management classes. Also, psychotropic medication is provided along with day treatment services for people with chronic mental illness. The center uses federal, state, and county funds and therefore must abide by government policies and procedures. Clerical, para-professional, and professional staff are county employees.

Although government agencies have a somewhat negative image, our experience has been otherwise. The two of us have worked in numerous agencies that are government funded, and our experiences were favorable. We were able to learn how to work effectively within the system, recognizing the positives instead of focusing on the negatives.

Developing a support system was perhaps what contributed the most to our positive experience. Our colleagues were sources of rich information and constructive feedback, which made a world of difference.

For-profit (commercial/private) For-profit organizations have proliferated because of reduced funding given to public agencies, limited resources for voluntary agencies, and changing political and economic times. These proprietary agencies have a dual function: first, to provide a service, and second, to make a profit in doing so (Kiser, 2000). Because of competition with public and nonprofit agencies, there is much controversy about for-profit agencies (Netting, Kettner, & McMurtry, 1993). Some say that such agencies often become primarily focused on the profit motive rather than the needs of the clientele they are supposed to be serving. Sweitzer & King (2004, p. 139) note that "the issue of funding is especially important because the source of funding has great power over and influence on the organization."

In the 1990s, the issues of privatization and managed care gained much attention (Goodman, Brown, & Dietz, 1992; Kettner & Martin, 1994). Privatization is the shifting of government responsibilities to private entities, with the expectation that market forces will improve productivity, efficiency, and effectiveness in the provision of services (Kettner & Martin, 1994). Privatization in human services includes assigning responsibility to nonpublic entities for any of the following tasks:

- Needs assessment
- Funding
- Policy making
- Program development
- Service provision
- Monitoring
- Evaluation

According to Kettner and Martin (1994), most states devote more than half of their service dollars to private, not-for-profit, and for-profit organizations. Examples are managed care organizations, health maintenance organizations, private corporations, and service agencies.

Understanding the Background Information of Your Agency

To successfully maneuver your way through the organization you work in, you need to understand certain background information. "In both internal and external analyses, organizational theory can help you look at, make sense of, and navigate your placement site" (Sweitzer & King, 2004, p. 135). The following concepts are important in this understanding.

- History: Learning how and when the organization was founded and how it is organized is helpful in understanding current day issues. Knowledge about original mission, goals, objectives, and changes can help you understand the agency.
- Mission: An agency's mission is usually a formally written statement that offers insight into the purpose and priorities of the agency.

Box 2.1 Sample Journal Form

History

How many years has your agency existed? Have there been any major changes during that time? Describe them.

Budget

What is your agency's annual budget? Where does the money come from? Do clients have to pay? If not, does some other agency pay for them? If the services are free of charge, where does the agency get money for payroll and operating expenses?

Staff

How many people work at your agency? How are they organized? Include an organizational chart; if the agency doesn't have one, create one on your own.

Policies

Is there a manual that describes people's duties and responsibilities? If there is no written policy, how do messages about expected behaviors get passed along?

Clients

What kinds of people are served by the agency? What kinds of problems do they have? Are there restrictions on age, gender, or other factors? Does your agency explicitly state that it will not serve certain problems or clients?

Supervisor

Who is your supervisor? What is his or her position in the agency? How have you felt about his or her supervision? Does your supervisor meet with you on a regular basis to provide feedback and promote understanding of the population served?

Source: From Sweitzer and King, (1995). Reprinted by permission.

- Goals and Objectives: Most agencies have delineated specific goals and objectives that often help guide staff to accomplish the mission.
- Values: Most agencies have certain formal and informal values that they try to adhere to. The empowerment of clients, respecting and maintaining confidentiality, and holding to strict ethical and legal guidelines are examples of values.
- Strategies: A strategy is the basic approach an organization takes in working to accomplish its mission. Strategies are specific steps and methods developed to reach agency goals.
- Funding: Information on how much and where the money to operate the organization comes from is also important to understand. Because limits and reductions in funding are common, being aware of these will help you be more sensitive to organizational demands.

According to Bolman and Deal (1991), an agency's structure provides the road map of how the goals will be accomplished. Being proactive in understanding one's agency system is clearly advantageous. Additionally, taking into account all the positive aspects of working in this field is helpful in preventing apathy and burnout. Here we outline several of the reasons we enjoy working with people and consider it our life's work.

- We enjoy working with others in a helping capacity.
- We appreciate the camaraderie of working closely with others who also are devoting their careers to human services.
- We believe in the teamwork approach because it accommodates different perspectives and cooperative effort in providing services to those in need.
- There is often a special connection among human service workers. In contrast, working alone can be very isolating.
- There is never a dull moment, given the varied personalities of clients and coworkers.
- Ultimately, working with others gives rich learning.

Box 2.1 contains a sample journal form that we have used in the beginning of students' placement to encourage them to become more knowledgeable about their particular agency.

▓ SYSTEMS THEORY

Years ago, L. Von Bertalanffy (1934, 1968) wrote about how "general systems theory developed to explain the complex interactions of all types of systems including living systems, family systems, and community systems" (Neukrug, 2004, p. 159). Systems theory is useful for understanding interrelationships within fieldwork agencies and the impact of change on the "system"—the agency setting. Gaining such an understanding as you begin your fieldwork experience can be quite valuable in orienting you to the agency system in which you are placed and recognizing the consequences of change.

Neukrug (2004, p. 159) explains the concept of systems by comparing it to the amoeba:

> Although knowledge of the amoeba may seem like a far cry from our understanding of systems, in actuality there is much we can learn from this one-celled animal. The amoeba has semi-permeable boundaries that allow it to take in nutrition from the environment. This delicate animal could not survive if its boundaries were too rigid or too permeable. Boundaries that are too loose would not allow the amoeba to maintain and digest the food it has found.

We have found this analogy to be a helpful for recognizing the importance of the different parts of a system to the functioning of the system as a whole. Brill and Levine (2002) relate how systems theory originated in biology and has gradually been adopted in other domains such as family therapy, organizational theory, and business. They define a system as:

> A whole made up of interrelated parts. The parts exist in a state of balance, and when changes take place within one, there is a compensatory change within the others. Systems become more complex and effective by constant exchange of both energy and information with their environment. When this exchange does not take place, they tend to be ineffectual. A system is not only made up of interrelated parts, but is itself an interrelated part of a larger system [p. 74].

From an organizational point of view, Netting, Kettner, and McMurtry (1993) use the systems theory approach to discuss large-scale change in organizations. They label the parts of the system as *critical participants*—each individual within the system is viewed as an important part. Therefore, "participants should be viewed as more than simply a collection of individuals who happen to have some common interests and characteristics."

Systems normally exist in some state of balance called *homeostasis*. When change is introduced in any one part of the system, there is likely to be a resultant change in other parts of the system. This can lead to a state of disequilibrium. Often the system is most comfortable when functioning in its original state, so systems may strive to return to their previous state of equilibrium. At times, this is not optimal; it can reinforce maladaptive behaviors. Among our roles as human service workers is to help clients and systems aspire to a more healthy level and thereby form a new homeostatic state.

SCENARIO

At the Women's Center, you are working with a family in which the father is alcoholic and the mother co-dependent. When the father is intoxicated, he often becomes irrational and violent, battering his wife as the children look on. Although there has not been any actual physical abuse of the children, the mother is concerned because they are often present during these violent episodes. She arrives for guidance and is assigned to you, a new caseworker. Through your work with this family, the father seeks help by entering a chemical dependency program, and you give treatment to the mother and children at the center. The mother begins to recognize her co-dependent patterns and starts to make changes with her children, spouse, and extended family. Although no one wants to continue the violent, chaotic home environment, there is a great temptation to regress into old behaviors that are comfortable and unchallenging. You caution this family about its reluctance to follow through with difficult change. You encourage family members to talk about their struggles in adjusting to the changes and to focus on the positives.

Systems can exist in many forms: an individual, a family, an agency, a neighborhood, a community, a government, a society, and even the universe. Individuals themselves contain various systems—their emotions, biology, spirituality, social roles, and intellectual capacity. Families are systems made up of parents, children, and other relatives. In fact, family system theories have been extensively written about. A community is a type of system, consisting of such entities as its residents, police, government agencies, medical facilities, fire department, schools, pharmacies, grocery stores, restaurants, and so on. Of course, the agency setting you work in is a system made up of all its employees, support staff, professionals, and managerial personnel. According to Neukrug (2004, p. 174), "[f]or human service professionals to work effectively with clients, they must have effective client skills and have a clear understanding of the boundaries, rules, hierarchy, and information flow within the system." We have found it very helpful to use this concept of systems; in the next section, we make comparisons between a family and an agency when discussing change.

Eco-Maps

Seeing how the parts react to each other and how the system strives to maintain equilibrium is helpful in understanding the underlying dynamics of your agency. This approach can be used to map the subdivisions within an agency. By using lines to connect different parts, the various relationships between members of the system can be indicated. This is similar to the *eco-maps* used to help practitioners assess specific difficulties in client systems so that appropriate interventions can be developed. Zastrow and Kirst-Ashman (1994, p. 502) write: "The eco-map is a drawing of the client/family in its social environment. . . . It helps both the worker and the client achieve a holistic or ecological view of the client's family life and the nature of the family's relationships with groups, associations, organizations, other families, and individuals." Gladding (2005, p. 160) describes an eco-map as "a visual tool used to map the 'ecology,' or the client's relationship to social systems in his or her life, including relationships, resources, and systems that are strong, stressful, and unavailable or tenuous."

Eco-maps typically consist of a family diagram, surrounded by a "set of circles and lines used to describe the family within an environmental context" (Zastrow & Kirst-Ashman, 1994, p. 523). To bring this to life, we have included a sample eco-map of a family (Figure 2.1) and of an agency (Figure 2.2).

THINKING THINGS THROUGH

Using the two eco-maps in Figures 2.1 and 2.2, draw an eco-map of your immediate family and one of the agency in which you are currently placed. Use solid lines to connect the parts of the system that interact closely, dotted lines between those that are more tentatively tied, and no lines between those that appear to have little to do with each other. Draw triangles or circles around those parts of the system that seem to interact as a triad or close twosome.

What can you learn from these charts? How can they be helpful in planning needed change?

Engaging in such an exercise with regard to an agency can be very helpful, for various reasons. First, it may help you recognize the overall picture of the agency and its functioning. Second, it may help you assess how individuals interrelate, and you can then hypothesize how internal or external change may affect them. Third, you may be able to identify how and where the introduction of change could be least disruptive and most effective. Within the world of human services, change is inevitable, so it is advantageous to be prepared for it and to minimize its potentially negative impact on the system. Here we outline two systems, by way of illustration: first, in their ideal state; second, as they are affected by change; and finally, how preparation and development of useful tactics can be helpful.

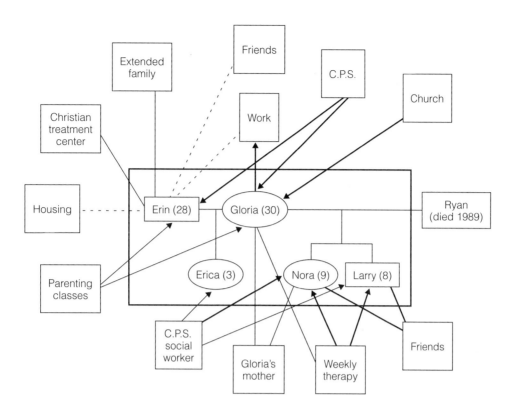

- - - - - - Tenuous connection

───────── Strong connection

━━━━━━━ Stressful connection

───────▶ Arrow signifies flow of energy

Figure 2.1
Eco-Map of a Family

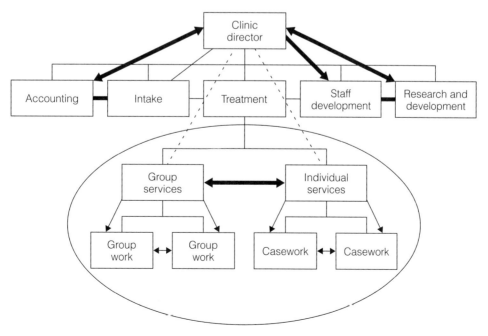

Figure 2.2
Eco-Map of an Agency

SCENARIO: An Individual

1. Ideal State
   The individual is a 35-year-old married Latino male who has two children—a daughter, age 12, and a son, age 9. He is gainfully employed as the X-ray technician at a local hospital. He is active in sports, playing basketball on weekends and acting as an assistant coach for his son's soccer team. He enjoys hiking and camping with the family.

2. As Affected by Change
   One day on his way home from work, he had a serious accident, suffering injury to the spinal cord. The client became quadriplegic, resulting in the following changes.

   Physical He is unable to walk and is wheelchair bound. Thus, he is less mobile and is limited physically. He is more prone to bedsores and atrophy of the muscles.

   Financial Due to the client's hospitalization and long-term rehabilitation, he has had to go on disability, which is half his previous income. Because of the nature of his injury and the physical demands of his job, he is unlikely to return to his previous employment, so he will be retrained, which will also create financial strain. Lastly, insurance has not paid all the massive financial expenses incurred, leaving the client in extreme debt.

   Marital There has been a role change. The wife has had to become the primary provider for the family, and the client is now the one who is at home. The wife often feels overwhelmed by the added responsibility and care of her husband. Also, due to the spinal injury, there has been a major change in their sexual relationship. There seems to be more tension and conflict in the home since the client returned from his stay in the rehabilitation facility.

   Family There has been considerable stress on the entire family. As mentioned, the marital relationship has suffered. Both the children have also struggled to adjust to the changes in their father's role. They particularly miss his participation in their sports and in the outdoor lifestyle they were accustomed to. The client is reluctant even to contemplate a camping trip because his wheelchair would be a great obstacle. He is also more irritable and easily frustrated by the children, who are also not used to him being home as often.

   Psychological The client has experienced an array of emotions since he was first injured. He has felt shock, denial, depression, anger, and anxiety. More often than not, he seems depressed and pessimistic. He has even contemplated suicide as a way of coping with his major change.

   Social Because of the client's lack of mobility and outside outlets, and with his depressed state, he has become increasingly withdrawn and isolated. Although friends have visited and encouraged him to get out, he has refused so often that friends stop by less and less.

3. Use of Interventions
   Several interventions could potentially help this client develop healthier coping and overcome his current state of depression:

   Individual counseling to help him deal with the many emotions secondary to the major loss of mobility he has experienced. Encouraging him to vent. Providing much-needed support and validation. Instilling hope and encouraging exploration of alternative ways of coping.
   Involvement in physical and occupational therapy along with vocational rehabilitation.
   Family therapy to help the client and family deal with the changes.
   Involvement in a support group with others who have experienced a physical trauma and subsequent changes in various roles and functions.

SCENARIO: An Agency

1. Ideal State
   An agency whose primary role is to provide emergency assistance is well known and respected in the community for providing low-income families with needed aid that often

prevents hunger and homelessness. Federal, state, and county funds are necessary for the ongoing functioning of this agency.

2. As Affected by Change

Due to massive budget cuts at all the governmental levels, this agency is no longer able to provide the same level of emergency assistance. Staff have been cut, resources limited, and many applicants turned away without any aid. Some people who have previously relied on this help to survive are now in a precarious position. The affected children exhibit poor school performance, increased truancy, and shoplifting and antisocial behaviors. Families may end up living in homeless shelters, where they can at least be maintained. Increased depression, apathy, and anger are likely emotional reactions.

3. Use of Interventions

Providing families with helpful referrals and suggestions on how to cope with reduced assistance is crucial. For instance, encouraging parents to register their children in all school meal programs is often helpful; then the children will have at least one complete meal daily. Education regarding meal preparation and nutritional needs is also important in helping families make the most out of their limited resources. Assisting clients to find gainful employment and pursue job training and educational opportunities serves long-term goals. In the present, efforts must be made to give them hope and emotional support, as well as to allow them to express their appropriate frustration. Lastly, professionals must advocate that resources be restored and funding increased. This may be through fund raising, public speaking, grant writing, and correspondence with entities that might offer some assistance through donations, waivers, or actual relief funding. Volunteers are necessary, and a support network among staff is important to help them cope with the changes resulting from the cutbacks.

THINKING THINGS THROUGH

Using the individual and agency systems just depicted as guides in your eco-maps, describe the ideal state, how it has been (or might be) affected by change, and which interventions would be most helpful. Are you surprised by any of your conclusions? Did you realize how change could affect the system?

AGENCY POLICIES AND PRACTICES

In your agency or fieldwork setting, consider the formal and informal lines of command that might be useful in effecting change. It is essential to understand agency policies and practices and be aware of the formal and informal aspects of organizational functions.

Formal Organization

The human service field is so broad that it would be impossible for one organization to provide comprehensive services to everyone in need. Different types of agencies have developed to meet the divergent needs of varying populations. You can certainly attest to this from your own involvement in fieldwork. In such agencies, formal structures and policies have been developed to ensure uniformity and continuity in the delivery of services. In most agencies, policies and procedures are spelled out in some formal manner. Many agencies have a manual that outlines rules, regulations, and procedures. Similarly, most agencies have formal organizational structures that specify the lines of command. Hansenfeld (1983) states, "Every organization establishes an internal structure which defines the authority of each person and the mechanisms of coordination among them." Such a structure is developed to promote cohesiveness and efficiency, but in reality, this is not always the result. Max Weber (1946), the founder of the sociological analysis of organizations, wrote extensively about organizational structure. The Weberian model describes the hierarchical distribution of power and authority, according to which responsibilities and operating decisions are delegated to different members of the organization—for example, typing to a secretary, child-abuse reports to a social welfare worker, suicide crisis calls to a clinician, and the budget to a manager. Roles and positions become specialized and the organization's activities formalized and standardized.

Example: Hierarchical Distribution of Power

Board of directors
Administrative director
Clinical Director
Supervisors
Clinical staff
Support staff

Lauffer's (1984) writings illustrate how an organization chart can represent an organization's formal structure. (See Figures 2.3 and 2.4.) Sweitzer & King (2004) discuss how the chain of command, visible in the organizational chart, is defined by the authority structure. They note that "vertical coordination and control" refer "to the ways in which the work of the agency is assigned, controlled, and supported. This dimension consists of structures for authority (the chain of command), supervision, communication, decision making, policies, and rules. The authority structure of an organization defines the chain of command, and this chain is visible in the organizational chart" (p. 141). To better understand agency systems, we highly recommend the use of organizational charting, a valuable tool for examining the following organizational issues:

- Formal hierarchical dimensions
- Formal relationship between positions
- Issues of independence
- Areas for potential conflicts
- Communication and work flow
- Organizational setup

THINKING THINGS THROUGH

Devise or acquire a formal organization chart of your agency. Does this chart reflect the reality of the agency structure? Is it representative of the actual agency setting? How so? If not, consider some reasons why.

Informal Agency Structure

Understanding the informal structure of the agency and its norms is also important to those beginning their fieldwork experience. *Informal agency structure* includes all the policies, rules, and norms that are unspoken and unwritten, yet clearly govern the behavior of workers. Bolman and Deal (1991) describe these less formal and more flexible structures as "lateral coordination and control" as opposed to "vertical coordination and control" discussed earlier. Although it is important to learn the formal setup of the agency, it is perhaps more significant to recognize the informal aspects because this is where actual agency life takes place. To survive in an agency, it is wise to learn informal policies (to guide you in procedural issues) and the informal structure (to develop relationships with key people in the agency). Unlike the formal structure, informal structure is not delineated in any written manuals or charts. This vital information is learned from daily interactions, information seeking, and keen observational skills. You will be amazed at how much you can learn about an agency by paying attention to the manner in which policies are actually implemented.

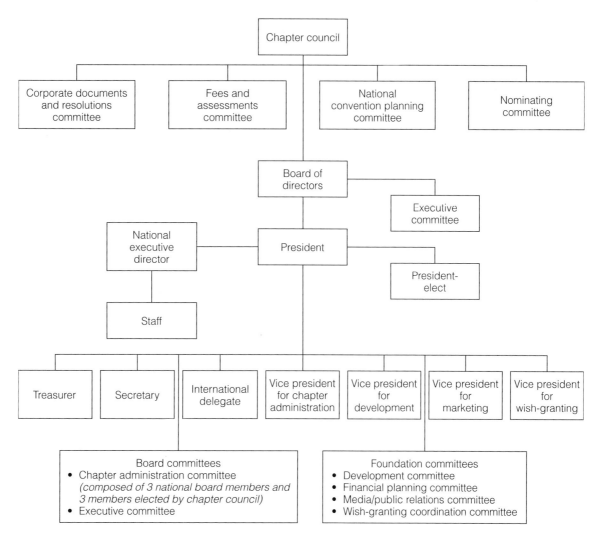

Figure 2.3
Traditional Organization Chart

There also exist informal groups. You will find it immensely helpful to identify such groups and understand their functions and personalities. This could prevent you from experiencing conflict later on. Both the literature (Lauffer, 1984; Bolman & Deal, 1991; Sweitzer & King, 2004) and our actual experience confirm that informal groups are inevitable in any organization. Among people's reasons for joining such groups are that it helps the individual member establish an identity, it may increase task completion, and it can foster emotional support. Understanding the "actual' or "real" agency structure in terms of communication skills, norms, informal roles, cliques, management style, and staff development is advised by many professionals who write about internships, practicum, or fieldwork (Baird, 2002; Faiver, Eisengart, & Collona, 2004; Royse, Dhooper, & Rompf, 1999; Sweitzer & King, 2004). In addition to becoming aware of the informal structure of an agency, understanding the politics that may be at play is equally important. Bolman and Deal (1991) encourage viewing the agency through a lens they call the "political frame." This involves becoming aware of how agencies are made up of different coalitions with differing agendas. Decisions about funding and the realities about the actual power and influence within an agency are other aspects of organizational politics (Bolman & Deal, 1991; Homan, 2004; Sweitzer & King, 2004).

Observing how workers interact with each other, as well as with the administration and upper management, can give you insights into the functioning of the organization.

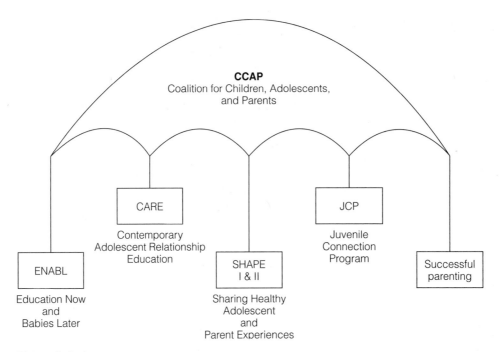

Figure 2.4
Unconventional, Innovative Organizational Chart

Is there a definite hierarchy with well-defined, rigid roles? Or are the roles less formal, allowing more flexibility? Is there room for negotiation in the system, or is it closed to new ideas? Is there camaraderie and cohesion among staff members, or are there strong divisions and distance? Is the system open to accepting newcomers and their ideas, or are outsiders considered threats to the functioning of the organization? These are just a few of the questions that you might consider exploring once you have settled into your fieldwork or new human service position. The search for answers will expose you to the structure (formal and informal) of the agency and help you recognize early on whom you can trust, learn from, and develop a supportive relationship with. You may also identify individuals with whom you want little contact, based on their style of interacting and patterns of personality. All of this information can be useful when thinking about how to get the most from your fieldwork experience.

THINKING THINGS THROUGH

Using your agency or school as an example, begin to explore the informal structure of the organization. Notice who talks to whom and how the individuals interact. Write in journal form all of your perceptions, thoughts, and feelings about your interactions with certain staff members. Don't censor your reactions; this journal is only for you to examine.

Go back over the journal later—even read it out loud. What do you notice about the observations you made? Do they say something about the informal structure of the organization? Could this reveal anything about you as well?

THINKING THINGS THROUGH: Designing an Informal Organization Chart

Using the information derived from the preceding exercise, attempt to chart the informal structure. Begin by identifying those who seem most powerful in the agency, as well as the least influential participants.

IDENTIFYING CRITICAL LOCATIONS

Show where in the agency there are strong bonds, tension and conflict, and cohesion and support. Based on your observations, what changes would you consider making to improve the agency's functioning? What reactions or repercussions might you expect if such changes were implemented?

Cases in Point

We will now examine two very important issues, relevant to the preceding material, that you will undoubtedly face during your fieldwork experience: sexual harassment and institutionalized discrimination.

Sexual Harassment and Formal Organizational Policy

Recognizing sexual harassment in the workplace is essential for anyone beginning work in the human services field. Early recognition can lead to the implementation of formal policies specifically designed to address such unethical behavior. The prevalence of sexual harassment is widely documented in the literature (Farley, 1980; Fellin, 1995; Horton, Leslie, & Larson, 1988; MacKinnon, 1979; Maypole & Skaine, 1983; Saltsman, 1988; Zastrow, 1995). Legally speaking, it is a form of sexual discrimination. Widespread recognition that it is a serious problem, coupled with the increase in reported cases of sexual harassment, has led to policies, regulations, and legal mandates prohibiting such behavior. This is a recent phenomenon, however. It was not until 1976, in the legal case of *Williams* v. *Saxbe* (413 F. Supp. 654, D.D.C. 1976), that the crime became the subject of a court decision. Now, organizations are required to develop policies and procedures to ensure prevention and proper management of actual or alleged sexual harassment. Working in agency settings and with others (coworkers, superiors, support staff, and clients), you or someone you know will very likely experience some form of sexual harassment, covert or overt. Thus, you must be knowledgeable about the many facets of this problem behavior. Maypole and Skaine (1983, p. 385) write:

> Knowledge about sexual harassment and the legal protections against it is important for workers in human service organizations when their clients encounter this behavior at their work-place. Workers need to be cognizant of sexual harassment policies in relation to their own work conditions and in order to work at the community level in helping to improve work settings, especially as social environments for women.

Most agencies require supervisory staff to take a course on this subject to familiarize them with all aspects of this unethical and illegal behavior. Let's start with some definitions.

Zastrow (1990) defines sexual harassment as "repeated and unwanted sexual advances." Others emphasize that it occurs in situations where there is an imbalance of power between the parties involved (MacKinnon, 1979). The National Organization for Women is more specific in its definition, stating that sexual harassment is:

> Any repeated or unwarranted verbal or physical sexual advances, sexually explicit derogatory statements, or sexually discriminatory remarks made by someone in the workplace, which is offensive or objectionable to the recipient or which causes the recipient discomfort or humiliation or which interferes with the recipient's job performance.

Although much more extensive than the others, this definition may still seem vague. Note that sexual harassment can be seen on a continuum from mild to severe. For instance, Sandler (1993) writes, "It may range from sexual innuendos made at inappropriate times, perhaps in the guise of humor, to coerced sexual relations." It must be further stated that sexual harassment is different from flirting, although there may sometimes be a very fine line between the two. The Project on the Status and

Education of Women (1986) and Sandler (1993) both cite all the following as examples of actual harassment:

- Verbal harassment or abuse
- Subtle pressure for sexual activity
- Sexist remarks about a woman's clothing, body, or sexual activities
- Unnecessary touching, patting, or pinching
- Leering or ogling a woman's body
- Demanding sexual favors accompanied by implied or overt threats concerning one's job, grades, letters of recommendation, and so on
- Physical assault

Sandler (1993) goes on to describe the most extreme case, wherein a perpetrator uses his position of authority to coerce a woman into sexual relations or to punish her refusal. What makes this most difficult to deal with is the fact that the male is usually in a position to "control, influence, or affect" a woman's job, career, or grades. More often than not, the woman remains quiet about the harassment, assuming nothing can or will be done (Sandler, 1993). With established policies, we hope that more women will feel safe in seeking appropriate resolution.

Sexual harassment is illegal. Victims have recourse through various channels (Fellin, 1995):

- Sex discrimination laws
- Titles VII and IX of the Civil Rights Act of 1964
- Criminal, tort, and state employment laws

Although most of the literature refers to males as the perpetrators and women as the victims, the reverse is also possible. Men are victimized in a similar fashion when their superior is a woman who engages in behaviors such as those described. "Sexual harassment is not, of course, solely a women's issue. Both men and women suffer under a system that fails to provide established remedies" (Sandler, 1993, p. 228). Victimized men and women have similar reasons for not disclosing the harassment: shame, low self-esteem, the possibility of losing the job, and concern that they will not be taken seriously or will be teased. Women may additionally fear being blamed and further victimized, and men may fear excessive ridicule and be concerned that promotional opportunities will not materialize.

Types of sexually harassing behavior, from mild to severe, include:

- Sexual teasing, jokes, remarks, or gestures
- Pressure for dates
- Letters, phone calls, or published material of a sexual nature
- Sexually suggestive looks or gestures
- Deliberate touching, leaning over, cornering, or pinching
- Pressure for sexual favors
- Actual or attempted rape or sexual assault

Legal aspects Victims of sexual harassment have legal recourse, which may differ from jurisdiction to jurisdiction. In considering the legal aspects, sexually harassing behaviors may be categorized into those involving conditional liability compared to those with strict liability, as follows:

> Conditional Liability: An intimidating, hostile and/or offensive work environment is created, as noted in the following behaviors: verbal acts, physical acts, and graphic displays. These behaviors tend to interfere with the affected person's job performance.
> Strict Liability: Employment rewards offered for sexual favors, and employment punishment threatened if sexual favors are not given.

After reading this, you may recall an incident you have witnessed or experienced that constituted sexual harassment, although you may never have recognized it as such.

Working in a professional capacity, you can educate and empower your clients and yourself to respond appropriately to such behavior. The development of formal policies in this regard by no means ensures the elimination of sexual harassment. Nonetheless, such dictates may discourage sexual harassment and give victims a means of gaining justice.

SCENARIO

You are newly employed in your first professional position as a caseworker for a foster-care agency. In the process of meeting the staff, you become aware that a high level of sexual joking and innuendo permeates the agency atmosphere. You learn to tolerate this, as it seems an informal norm in this particular agency. One day, however, you are directly affected by what you believe is sexual harassment. A supervisor from another unit corners you in the staff lounge and begins to make direct sexual comments that are offensive and make you quite uncomfortable. You manage to escape any further harassment but are left feeling distraught and fearful of any further encounters with this supervisor. At staff functions, you catch him looking at you leeringly and feel a certain embarrassment and trepidation. You are hesitant to discuss this with anyone else because you are concerned about your reputation and position.

How might you feel if this happened to you? What would you think? How might you handle this situation?

THINKING THINGS THROUGH

Using the previous scenario and the knowledge you have gained thus far, what steps would you take if you were the victim of this type of harassment?

Institutional Discrimination and Informal Agency Norms

Without question, institutionalized discrimination is alive and well in agency settings. You will probably be exposed to it in some manner, even in your fieldwork placement. Because of the unspoken, unintentional, or unconscious nature of discrimination, it is often an informal norm that directly or indirectly influences policies and regulations and ultimately colors the way services are provided.

What exactly is institutionalized discrimination? Cherry (1993, p. 134) says institutionalized discrimination occurs "when individuals and institutions may use decision making procedures that inadvertently discriminate and reinforce inequalities." Brill (1990, p. 57) defines discrimination as "unequal or preferential treatment, injustice toward particular individuals, groups or peoples." It may be personal, as expressed by individuals or informal groups, or institutionalized, as expressed by policies and laws. Discriminatory behavior may be directed toward clients or against human service workers themselves. Whether conscious or not, such discrimination threatens those who are most oppressed in our society. Kinds of discrimination include racism, sexism, ageism, and ableism. Let's look at these forms of discrimination and examine how they might directly or indirectly affect your work in human services. We will expand further on issues of diversity in Chapter 5.

Institutionalized racism
Racism has been defined as "the belief that race determines human traits and capabilities and that particular races are superior to others" (Brill, 1990, p. 249). The domination of one social or ethnic group by another is a logical consequence. According to Lum (1996, p. 57), racism is used as an "ideological system to justify the institutional discrimination of certain racial groups against others." Lum sees racism and its effects on a continuum. Racism is considered to be the ideological belief that leads to an attitude of prejudice, which in turn results in discrimination, which leads to the "expression" of discrimination: oppression, powerlessness, exploitation, acculturation, and stereotyping. Figure 2.5 illustrates this process. Our view is that institutionalized discrimination can be informal and is often difficult to

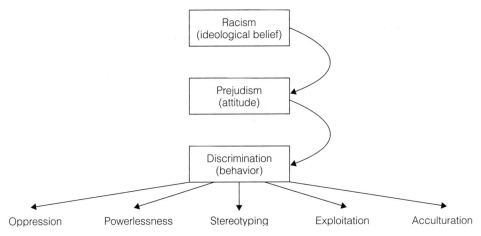

Figure 2.5
Institutional Racism

confront when disguised. Therefore, we must be concerned not only about overtly prejudiced behavior but also about hidden forms of racism.

Although we believe we live in liberated times, recent years have seen a steady rise in racism. According to Zastrow (1990), in the 1970s and 1980s institutionalized racism came to be regarded as a major problem. Sue and Sue (1990, p. 4) have also noted an increase in individual and institutional racism: "Racism is alive, well, and thriving in the United States. Indeed, the 1980s have seen a historic rise in the incidents of overt racism throughout the country." In the 1990s, hate crimes increased and efforts to reverse affirmative action are well documented.

What implications does racial discrimination have for the workplace? In particular, how might human services be affected by institutionalized discrimination?

SCENARIO

You are a Latino male who has worked at the same human service agency for ten years. You have gained much experience and feel quite confident in your skills and abilities. For several years now, you have been applying for various promotions within the agency, only to be overlooked in favor of others who have less experience and are hired from outside. You are trying to maintain a positive attitude and not personalize this situation, yet it is becoming increasingly difficult to view the agency's hiring policy as nondiscriminatory.

Over time, how might you feel in this situation? What might you consider doing to cope? What are some appropriate measures that you could take?

THINKING THINGS THROUGH

Have you ever witnessed institutionalized racism? If so, please provide an example and discuss it in light of the various issues outlined here. How might you react today if you were to see some form of racism acted out in your agency setting? What measures might you take?

Institutionalized sexism Sexism is the subordination of an individual or group of women and the assumption of the superiority of an individual man or group of men, based solely on sex. As with racism, sexism is reflected in both individual and institutional acts, decisions, habits, procedures, and policies that neglect, overlook, exploit, subjugate, or maintain the subordination of an individual woman or all women (Cyrus, 1993, p. 219).

Clearly, women can also be oppressed in an informal, subtle manner by society in general and institutions in particular. Further, as stated by Cole (1992, p. 148), "Patriarchal oppression is not limited to women of one race or of one particular ethnic group, women in one class, women of one age group or sexual preference, women who live in one part of the country, women of any one religion, or women with certain physical abilities or disabilities." Sexism is pervasive and has effects on the workplace. It has also been established (Cole, 1992) that a woman who belongs to a minority is often doubly discriminated against, further increasing her sense of power-lessness and oppression.

According to Schmolling, Youkeles, and Burger (1993, p. 33), sexism has its roots in prehistory. They present a historical perspective:

> In early societies, the physical strength of males determined their dominant position as hunters, warriors, and leaders. Females were relegated to food preparation, child care, and other domestic chores. Modern sex stereotypes, which are over simple ideas about the differences between the sexes, still reflect this early division of labor. Men are seen as "naturally" more aggressive, venturesome, and dominating. Women are supposed to be relatively more expressive, "intuitive" (irrational), nurturing, and submissive. Although these qualities have some inherited basis, they are largely produced by differential patterns of upbringing for boys and girls.

Given that sexism is so pervasive and so rooted in the fabric of our gender roles, how does it affect how we function in the workplace? This needs to be considered not only as it relates to female clients who are victims of sexist behavior but also to female human service professionals. Also, sexism can be discovered in many other situations because of narrow gender definitions. Note that men can also be discriminated against by women.

SCENARIO

Female clinicians working in a male-dominated environment often face certain prevailing sexist behaviors. For instance, males may hold all administrative and supervisory positions while the support staff consists solely of females. Requests for educational leave and opportunities for promotion seem more common for the men within the agency. Input and decision making by the women within the agency seem to take a backseat to those by the men.

Have you ever personally experienced this type of discrimination based on your gender? If so, how did you feel? How did you respond? Looking back, would you respond differently today?

SCENARIO

In certain situations, there is sexism against males. For instance, in such female-dominated agencies as children's centers and battered women's shelters, male staff can be subtly discriminated against. Fewer positions exist for males within these settings. At times, added pressure is placed on male staff because of traditional male stereotypes. Although not overt, an underlying attitude may cast them as potentially unworthy, so they must prove themselves differently. There is often fear that a male is more likely to engage in sexually inappropriate behaviors with children or to act out impulsively.

Have you personally witnessed this type of sexism or known someone who has? How did you (or they) feel, think, and respond?

THINKING THINGS THROUGH

Consider whether you have witnessed any form of sexism within your agency setting. If so, how did you feel? How did you react? If this were to happen today, would you respond any differently?

Institutionalized heterosexism Related to the issue of sexism is that of sexual orientation—whether individuals are heterosexual (physically and emotionally attracted to members of the other sex), or homosexual (attracted to the same sex), or bisexual (attracted to both). *Heterosexism is* "the belief that heterosexuality is or should be the only acceptable sexual orientation" (Morales & Sheafor, 1995). Heterosexism can also be viewed as a "cultural assumption that heterosexuality is natural and the only proper sexual behavior" (Cyrus, 1993). The fear of being labeled as a homosexual, and the accompanying hatred of homosexuals, is known as *homophobia.* Such fear and hatred contribute to prejudice, discrimination, harassment, and acts of violence against homosexuals—both gay men and lesbian women (Brownsworth, 1993).

In relationship to the workplace, Blood, Tuttle, and Lackey (1992, p. 160) write:

> Gay oppression is one of the ways the potential unity of all workers is prevented. . . . Everyone is hurt by gay oppression. The fear of being considered gay limits and distorts everyone's life choices and relationships. An essential prop for sexism, in keeping people within their accustomed roles, is this fear of homosexuality, or homophobia. Because of this, women's liberation and men's liberation depend partly on gay liberation.

Fellin (1995) also emphasizes the prevalence of discrimination against gays and lesbians in many organizations. "Some occupational groups, such as the military, schoolteachers, government workers, and clergy, appear to be most negative toward gay and lesbian individuals" (p. 193).

Taking into account the common feelings of fear and even hatred for individuals who do not fit the norm of heterosexuality, what are the implications for human service employees?

SCENARIO

Although you would think that people working in human services would be more liberal and accepting of differences in sexual identity, this is not always the case. In fact, homophobia often permeates agency settings. Discrimination directed at gay and lesbian staff is often subtle and indirect. It may be that these staff members are criticized or ridiculed when they are not present and that their credibility and skills are questioned because of their sexual orientation.

How might you feel if you were gay or lesbian and were discriminated against at work for this reason? How might you react? What might you do to deal with this issue?

THINKING THINGS THROUGH

In your agency or in any other organization, have you witnessed institutionalized heterosexism? Take a moment to describe the circumstances and imagine how this situation could best have been handled. What sorts of agencies would you guess informally endorse heterosexist beliefs? Why?

Institutionalized ageism According to Butler (1997), there exists a widespread personal and institutional prejudice against the elderly. He contends that these prejudices stem from people's basic fear of aging and eventual death. Zastrow (1990) also considers that ageism may have at its core the fear of death (thanatophobia). He finds it ironic that our society is so caught up in preserving youth, and thus denying the beauty of the aging process, that we neglect the elderly whose lives have been extended. People are living longer and growing older. This reality has increased our awareness of ageism. Blood, Tuttle, and Lackey (1992, p. 154) state, "People are just beginning to have a glimpse of what oppression based on age involves. The fact is that our society is almost totally blind to the dignity and capacities of the very young and the very old." Robert Butler further discusses ageism as it relates to our myths and stereotypes regarding older persons. The words of Butler (1975, p. 6) clearly describe the nature of such exaggerated, overgeneralized, or false beliefs:

> An older person thinks and moves slowly. He does not think as he used to or as creatively. He is bound to himself and can no longer change or grow. He can learn neither well nor swiftly and, even if he could, he would not wish to. Tied to his personal traditions and growing

conservatism, he dislikes innovations and is not disposed to new ideas. Not only can he not move forward, he often moves backward. He enters a second childhood, caught up in increasing ego-centricity and demanding more from his environment than he is willing to give to it. Feeble, uninteresting, he awaits his death, a burden to society, to his family, to himself.

Just imagine what the world would be like if we all surrendered to the many negative and pessimistic views presented by Butler. Most of us would not want to even live long enough to be victims of the type of old age presented. Sadly, many people still adhere to these beliefs and therefore reinforce such myths and stereotypes.

Recognizing that older people are subject to discrimination and exploitation, you should consider the implications for your work with older persons in your agency. This applies to older staff members, approaching or past retirement, or elder clients you might come into contact with.

SCENARIO

In agencies, there is often an elder who continues to work despite advanced age. Is he kept on out of respect for his previous endeavors? Perhaps he was a pioneer in the field, or responsible for developing the agency or department itself. But is he still capable of doing the work required of others? Or has the impact of age left him less capable?

We have witnessed this situation, as well as the other side—an older person remains quite capable, yet is thought to be inadequate because his pace is slower and his views and interventions more conservative. What if he continues to execute his work appropriately? Should he be penalized because his productivity is a little less than that of a young, energetic, new professional? What about the wisdom and life experience he has to offer? Doesn't this count for something?

THINKING THINGS THROUGH

Consider family, friends, acquaintances, or any others you know that have been discriminated against because of their age. Please describe one of these situations in detail and discuss how such discrimination affected the individual. Consider older clients. Has there been any type of discrimination toward them? As a human service worker, how might you handle this issue?

Institutionalized ableism   The term *ableism* means discrimination against individuals who are in some way disabled. This could apply to people who are impaired physically, mentally, developmentally, socially, or economically. *Disability* is a diagnosed condition such as blindness or deafness; *handicap* refers to the consequences of the disability (Schmolling, Youkeles, & Burger, 1993). Our focus here is broader; by *disability* we do not necessarily mean a diagnosed condition. For instance, people who are of a lower socioeconomic class may be considered disabled—they are oppressed in ways that have major consequences. The literature substantiates what many of us have experienced: negative attitudes toward disabled individuals. With reference to physically or mentally disabled people, Asch (1984) discusses the strong negative emotions these individuals evoke in others who are "able-bodied" and "emotionally healthy." Often, our own anxieties about loss, vulnerability, and weakness are sparked when we encounter a disabled person. The literature (Ascha, 1984; Schmolling, Youkles, & Burger, 1993) also indicates that most people who are unaffected by a disability react in one of two ways: by pretending the disability doesn't exist or matter, or by being excessively helpful and overly reassuring.

SCENARIO

You work in a clinical setting. Several coworkers have become disabled by illnesses and progressive conditions. Some, as a result of their disabilities, require the use of canes, walkers, or wheelchairs. Despite their diminished physical capacity, their mental state is still acute and thus their therapeutic abilities still intact. Nonetheless, some staff members question their capacity and tend to treat them differently. Sometimes there is a tendency to infantilize and overprotect. The fact that they are treated differently because of the disability indicates discriminatory practice.

THINKING THINGS THROUGH

Consider which disabilities you have some reactions to and why. How do you think your feelings will affect your work when you encounter someone with a certain type of disability? Do you feel there is sometimes discrimination?

Additional forms of institutional discrimination include classism and lookism, which can be equally oppressive and victimizing. In these, individuals are discriminated against because of the class they belong to (e.g., lower class) or their physical appearance (e.g., unattractive, heavy, not svelte and model-like).

In this section on institutionalized discrimination, we hope that you have become more cognizant of how many people suffer from discrimination and how it is manifested in the human service arena. We trust that you will be more sensitive to these issues, personally and professionally.

CONCLUDING EXERCISES

1. If you were to think of your placement agency as a family, draw a picture of how you see yourself fitting into the system. Draw whatever comes to your mind, without censoring any thoughts or ideas. Let yourself be creative in your representation of the agency and your position within it. If the drawing is one you are willing to share, consider taking it to your supervisor and discussing it with him or her.

2. Draw a picture representing your family system. It can be of any family you see yourself as part of—your family of origin or your own current family. Be creative, without censoring your thoughts or feelings. Once it is completed, compare this family drawing with the drawing of your agency above. Do you notice any similarities? Any differences? How might you explain these?

3. If you had the luxury of developing an ideal agency committed to providing quality human services (you specify the type), what would it be like? Imagine you had complete latitude in planning and implementing this vision. Funding, support, staff, and clientele would all be readily available. Describe specific aspects of this agency, taking into account the issues you have read about in this chapter.

CHAPTER THREE

How to Make the Most of Your Agency Experience

AIM OF THE CHAPTER

In the helping professions, be it counseling, psychology, social work, human services, or child development, fieldwork is an integral part of the educational process. In fact, fieldwork training is among the most valuable practical learning experiences a beginning helper will have. It enables a beginner to transfer theoretical knowledge to real-life situations when practicing in an agency setting. Beginning human service professionals learn so much more about themselves and about the helping process when they are challenged to work directly with clients, or indirectly through community organization and administrative tasks. Somehow the intricacies and subtleties of actually helping a person can never be learned solely from a theoretical perspective. The heart of true learning lies in actual experiences—seeing a theory or technique come to life via actual application or expressed as a part of program development or grant implementation. Thus, fieldwork experience is not only exciting but also crucial in the development of a professional helper. This chapter describes the importance of fieldwork and making the most of this experience. Effective communication skills (including assertiveness) and time-management skills will be discussed, as well as the value of supervision. We conclude with a discussion of pathways to a human service position.

FIELDWORK

As a student or beginning professional, you will probably begin your career in some form of internship or training. Here, we will use the term *fieldwork* interchangeably with such terms as *internship, practicum placement,* or *job training.* Although our main focus will be on students in a formal educational program, this material is just as useful to beginning helpers who have already completed their education.

The Value of Fieldwork

The value of engaging in some form of fieldwork experience goes without saying. In fieldwork, helpers begin to integrate all that they have learned during their formal education. In fact, most helping professions require that their members participate in several fieldwork placement experiences as part of their professional development. Some undergraduate programs incorporate a fieldwork component in the curriculum. Also, fieldwork practicum is an integral part of most graduate programs in helping fields. Royse, Dhooper, and Rompf (1999, p. 1), who are involved in social work programs, note, "As professionals in the making, social work students attend classes to learn practice principles, values and ethical behaviors, a body of specialized knowledge, and the scientific basis for practice. In field instruction, the preparation to become a social worker is composed of formal learning as well as practical experience." In both undergraduate and graduate programs, the fieldwork component provides students with activities that help bridge theory and practice. It offers them an opportunity to learn "first-hand about paperwork, agency policies and procedures, and the challenge of working with a wide range of client populations and problems" (Corey & Corey, 2003, p. 45). Students are able to interact professionally with individuals, groups, and organizations.

Fieldwork is a collaborative partnership among the university, social service agencies, human service professionals, and students. Generally, placements are made at social service agencies throughout the region where the university is located. Such agencies are usually carefully screened on the basis of the quality of their professional practice, their dedication to addressing human service issues, and their interest in human service education. Such a partnership represents two important principles. First, it symbolizes a commitment by all concerned to providing high-quality educational experiences for students. Second, it renders a valuable service to the community and often motivates interns to return to the site of their field placement to work once they have graduated.

Every undergraduate or graduate program has its own set of requirements, guidelines, and goals. General fieldwork goals described by Corey and Corey (2003), include:

- To provide students with knowledge of the varied approaches and methods used in human service programs
- To help students extend self-awareness and achieve a sense of professional identity
- To broaden students' sociocultural understanding of the individual, the family, and relevant social systems
- To help students recognize and respect cultural diversity and to offer ways of applying this understanding in practice
- To help students expand their awareness of professional roles and relationships in the organization, as well as the agency's role in the community

These goals reflect the extensive learning that takes place during an internship. Engaging in a fieldwork placement can be one of the most important vehicles for a beginning helper to cross the bridge from theoretical knowledge to real-life practice.

Our own experiences during our undergraduate and graduate internships, as well as our work as field supervisors, instructors, and liaisons, have allowed us to witness the benefits of engaging in a fieldwork placement. As the culmination of one's academic learning, internship can be an exciting but anxiety-producing time. Faiver, Eisengart, and Colonna (2004, p. 1) write, "Internship students often find themselves facing unfamiliar situations, engaging in intense encounters, and processing powerful feelings, which all lead to increased introspection, personal reassessment, change and growth." Clearly, your fieldwork placement can furnish a wealth of opportunities to learn about yourself and others, to further your skill development, and to enhance your knowledge base. But opportunities alone are not enough to ensure an optimal learning experience. You must also be willing to take risks, be open, and allow yourself to be vulnerable. Specific characteristics agency supervisors tend to look for in interns include a strong desire to help, interest and ability to function in a particular setting, emotional maturity, and honesty (Royse, Dhooper, & Rompf (1999, p. 16). Above all, you must take fieldwork seriously.

Take Fieldwork Seriously

Taking fieldwork seriously is vital if the beginning professional is to gain the skills and awareness essential to becoming an effective helper. If, from the outset, helpers value all of their learning experiences, both classroom and fieldwork, they will be able to integrate their knowledge and also gain respect for the process of working in community settings. Fieldwork should also entail an appreciation for the ongoing process of learning.

Individuals who do not regard fieldwork as important will learn less and probably carry this negative attitude into their professional careers. One must recognize that learning can occur at many different levels. We believe that learning results from every task performed or interaction encountered—whether positive or negative, significant or insignificant. Through the years, many of our students who initially disliked some aspects of their field-work placement have later said this experience was meaningful to their learning.

SCENARIO

Undergraduates in fieldwork were asked to help out in their agencies by answering phones, filing, or typing. They not only learned to appreciate that these tasks are essential in the functioning of an agency but also gained respect for the clerical and support staff who routinely perform these functions.

Have you been in a similar situation? Were you able to learn from the situation? If so, how? If not, why not?

SCENARIO

Many students enter their fieldwork placements with little experience and are fearful of working directly with certain client populations (e.g., abused children, perpetrators, chemically dependent teens, battered women, juvenile offenders, or the frail elderly). In discussing their anxieties with their supervisors, they were encouraged to challenge their perceived limitations and discovered that they could effectively help these populations. Some were amazed at their heightened interest and natural ability to work with such clients. Others realized, however, that they were neither suited for nor interested in working with such clientele.

Can you identify with either of these situations? Do you feel learning is possible here?

From these examples, it is evident that students who have a positive attitude regarding their learning and are willing to take risks are more likely to benefit from the overall fieldwork experience. Collins, Thomlinson, and Grinnell (1992, p. 135) stress the point that in fieldwork "[e]mphasis is placed on the student's responsibility to create and manage a positive learning environment."

Although fieldwork settings are considered ideal for student learning, this is not always the case. The agency placement may not be appropriate, supervision may be lacking, or shortage of funding may jeopardize the functioning of the agency. In such instances, the student cannot be considered responsible for the unfortunate outcomes. However, certain students are primarily accountable for their negative fieldwork experiences. So that you won't be among them, the next section details our recommendations for student involvement in fieldwork. Above all, take an active stance. Engage in effective communication, learn to be assertive, and apply effective time-management skills. If you add these to your repertoire of skills, your learning will be maximized and negative situations lessened.

TAKING AN ACTIVE STANCE

In our personal and professional experiences, we have noted the profound impact of being active in one's own learning. We believe that beginning helpers are likely to gain the most from their professional experiences when they get involved actively. Active involvement is present when students freely ask questions and seek feedback, both in class and in their placements; participate willingly in class discussions, role playing, group supervision, and agency meetings; and remain open to understanding them-selves in an effort to become more effective helpers. Corey and Corey (2003) cite the following guidelines to becoming an active learner: preparing, dealing with anxieties, taking risks, establishing trust, self-disclosing, listening, and practicing outside class.

By contrast, many beginning helpers who remain passive in their learning subsequently become negative and bitter about their educational experiences. Corey and Corey (2003, p. 38) counsel students against "passively complaining about their department, their professors, and everything and everybody but themselves." Such blaming is useless and sets up a negative pattern. Deciding to focus on the positive aspects of one's education and training, rather than concentrating on agency shortcomings and imperfections, is essential to making the most of your fieldwork experience. Learn to work creatively within a system, instead of allowing yourself to be demoralized by its limitations.

SCENARIO

A student was placed in a group home for abused adolescents. Early in her internship, she felt frustrated by a lack of support and input from the staff. She began to adopt a blaming stance: She was critical of the agency and its personnel. While discussing this in class, it became clear that this attitude was related to her own fear of being assertive. The instructor and fellow classmates challenged her to be more direct and assertive about her needs. Despite her fears, she spoke up in a staff meeting and was surprised by the positive response. This opened several doors for her. She learned to see this agency in a different light; she was able to have her needs more readily met; and most important, she learned a valuable lesson about her own fears and abilities.

Can you identify with anything presented in this example? Have you ever experienced a similar situation? If so, how did you handle it? Do you believe you took an active stance in your learning?

SCENARIO

A student was placed at a crisis hotline, where he felt overwhelmed by the suicidal callers and the lack of supervision. Because the agency was underfunded and understaffed, it relied on volunteers and interns to maintain services to the community. The student attempted to seek guidance from his assigned supervisor, only to find it difficult even to make a connection. He became bitter and wasted energy bemoaning the lack of appropriate supervision.

In a fieldwork seminar, he was asked if anyone else in the agency was capable of supervising him. He was surprised at the question; he had never even considered this to be an alternative. When he returned to the agency the next week, he explored other sources of supervision. When he actively sought supervision by another staff member on his shift, he found her willing and receptive, and he himself benefited greatly from their interaction. His anger toward the agency and his former supervisor decreased, leaving him with more energy to use in working on the hotline.

Have you ever experienced a similar situation, where you used more energy on the problem than on working on the solution? Explain how.

Taking an active stance implies being vigorous and energetic in seeking out learning experiences. Those who adopt this dynamic style can stretch their learning beyond the limitations that are often inherent in agency systems. Furthermore, by demonstrating their genuine desire and enthusiasm for learning, they are likely to affect those around them in a constructive manner and thereby increase the likelihood of attaining personal and agency goals.

Effective Communication

Being active in one's learning also means being able to communicate effectively. Effective communication is at the core of successful human interaction, and it is certainly crucial for anyone hoping to be an effective helper. By contrast, poor communication can result in gross misinterpretations, misleading information, and erroneous perceptions. In the human services, both individuals and agencies are harmed by poor communication. From an organizational perspective, Travers (1993, p. 189) highlights the dangers of ineffective communication: "Without good communications, a number of serious problems can plague the organization. There can be misdirected effort, duplicated effort, or no effort at all toward organizational goals. . . . All these conditions negatively affect productivity and morale." A key to preventing poor communication is understanding the principles of communication itself. In this section, we review these basic concepts in hopes that they will make you a more effective communicator, in both personal and professional life.

Communication is defined as "the process of passing along information and understanding from one person to another person or group" (Travers, 1993, p. 188). Hackney and Cormier (2005) believe that communication is a skill that is essential for helpers to master. In regard to counseling, they state, "Counselors and clients alike continually transmit and receive verbal and nonverbal messages during the interview process. Therefore, awareness of and sensitivity to the kinds of messages being communicated is an important prerequisite for counselor effectiveness" (p. 7). There are many different types of communication. Active listening and nonverbal communication will be discussed here because they are most conducive to developing effective communication skills.

Effective communication is instrumental in achieving goals and facilitating self-understanding. According to Long (1996, p. 155), "[t]he process of communication has three essential aspects: listening, responding, and expressing." She develops a five-stage communication model that stresses such skills as listening, centering, empathizing, focusing, and providing directional support. In Chapter 5, we will elaborate further on the importance of communication.

Active Listening

We believe that active listening is extremely important if one is to become an effective communicator and helper. This skill is the "ability to listen actively and attentively to a client" (Cormier & Hackney, 2005). Hutchins and Cole (1992, p. 57) elaborate on the concept of active listening:

> In attending to the client, the effective helper usually:
>
> - Listens carefully to what the client says and how it is said
> - Avoids interrupting and allows the client to complete sentences and ideas
> - Uses silence to encourage the client to talk when ready
> - Uses reflection to clarify the client's meaning
> - Asks questions to ascertain important details, including answers to the questions of what, where, when, how, etc.
> - Helps the client systematically explore relevant content
> - Reflects similarities and discrepancies among what the client says (thoughts), how the client says it (feelings), and what the client does (actions)
> - Elicits feedback from the client to determine and ensure accuracy of the helper's perceptions of the client

This extensive list highlights the importance of attentive or active listening as an integral part of effective communication.

Attentiveness is communicated through four channels: facial expressions, eye contact, bodily positions and movement, and verbal responses (Cormier & Nurius, 2003). Egan (2002) also believes in the importance of active or attentive listening, which includes "tuning in, listening, processing, sharing highlights, and probing" (p. 134).

Even though most of us who enter the helping professions are caring and empathic, we may not have mastered the art of attentive listening. "Although attending to clients on the surface appears relatively simple, it is easier said than done" (Cormier & Hackney, 2005, p. 37). Unlike personal qualities that many of us naturally possess, active or attentive listening is a skill that requires continual practice. Ongoing self-awareness regarding our ability to listen to others attentively is important, both at a professional and personal level. Part of this process involves being aware of the possible obstacles that may interfere with listening. Egan (2002) cites the following as such barriers to the attending process: being judgmental, having biases, pigeonholing clients, attending to facts, sympathizing, and interrupting.

SCENARIO: Demonstrating Barriers to Listening

In a counseling setting, a middle-age female client enters treatment because of relationship issues. As the counselor learns more about her past, he finds that she has a history of being passive and often is involved in abusive relationships with men. In seeing the dynamics of her situations with men, the counselor is quick to encourage separation and independence from these abusive-type relationships. Although well intended, the counselor may not have attended to the client's deeper feelings, which needed further exploration before change could be possible. Taking time to have the client work through her emotions and unresolved past issues may seem cumbersome, yet this is essential to helping the client successfully make long-lasting changes in her life. Allowing clients to go at their pace and make changes only if they wish or when they are ready to is ultimately what attending to the client's needs is all about.

SCENARIO: Successful Listening

In working with the elderly, the art of active listening is essential. Many students placed in the Alzheimer's Association, senior citizen centers, and caregiver support services have commented on how active listening skills were most valuable and instrumental in helping both the elderly client and his or her family. Allowing seniors to reminisce and engage in life review was a vital aspect of their helping. Similarly, caregivers benefited when family members were able to vent their pent-up frustrations and emotions to someone who was sensitive and listened with compassion and empathy. Listening carefully seemed to promote healing much more readily than giving advice or any other types of active intervention.

Can you think of another situation where active listening is essential to the helping process?

Nonverbal Language

Nonverbal communication is considered the most prevalent form of communication. According to Young (2001, p. 79), "As much as 80 percent of communication takes place on the nonverbal level. Where emotions are concerned, nonverbals are even more important. . . . Nonverbals can be compared to the musical score in a movie. They can affect us tremendously, but we rarely notice their presence." Mehrabian (1971) offers the following statistics about communication:

- Messages from *what* people say (their words) account for 7 percent of communication
- Messages from *vocal aspects* of their communication (tone of voice, volume, etc., but not actual words used) make up 38 percent of communication
- Messages from *nonverbal aspects* of their communication (i.e., body language) account for 55 percent of communication

Egan's (2002, p. 67) statement that "nonverbal behaviors regulate conversations, communicate emotions, modify verbal messages, provide important messages about the helping relationships, give insights into self-perceptions, and provide clues that clients (or counselors) are not saying what they are thinking" underscores our belief in the importance of nonverbal communication.

Hutchins and Cole (1992, p. 58) cite the following as aspects of nonverbal behavior and communication: vocal or subvocal sounds, vocal tones, or pitch, speed of talk, body language (including eye contact), facial expressions, gestures, posture, and overall body movement. Similarly, Egan (2002) identifies bodily behaviors, eye behavior, facial expressions, voice-related behaviors, observable autonomic physiological responses, and space as major categories of nonverbal communication. We elaborate further on these:

- Bodily behavior can include posture, body movements, and gestures. Cormier and Nurius (2003) have found that "body positions and movement regulate space or distance between the counselor [helper] and a client, greeting of a client, termination of a session, and turn taking." Young (2001, p. 80) echoes a commonly heard statement: "Actions speak louder than words." He adds that because "posture may be the most often noticed aspect of body language it becomes important for the helper to have a 'posture of involvement'." Similarly body gestures are important communication tools that often convey emotions and emphasize important points.

- Eye behavior can involve eye contact, staring, and eye movement. Research into interpersonal interactions according to Hackney and Cormier (2005, p. 38) demonstrates that "eye contact has more than one effect. . . . It may signal a need for affiliation, involvement, or inclusion; it may reflect the quality of an existing relationship; or it may enhance the communication of complex messages." Young (2001, p. 80) adds that "eye-to-eye contact is the first and most important indicator of listening" and "conveys the helper's confidence and involvement."

- Facial expressions can consist of smiles, frowns, raised eyebrows, twisted lips, and so on. The face is the principal way in which individuals communicate information about their emotional states (Knapp & Hall, 1997). Such basic emotions as anger, disgust, fear, sadness, and happiness are all typically expressed by facial expressions. It is important for helpers to pay close attention to how closely clients' facial expression match their words. When there are noticeable incongruencies, this may signal the need for further exploration of deeper issues or may be "clues to lack of self-awareness and conflict" (Young, 2001, p. 81).

- Voice-related behavior refers to tone of voice, pitch, volume, intensity, inflection, spacing of words, emphases, pauses, silences, and fluency. Cormier and Hackney (2005) and Egan (2002) contend that what the helper says can have an immediate and significant impact on the client. Similarly, clients can communicate a wealth of information regarding their emotions through their voice. Young (2001, p. 81) says, "We can tell from the client's voice which issues are the most painful and the source of the greatest motivation and excitement. Similarly, clients respond to the helper's voice tone."

- Silence, depending on when it occurs and who uses it, can signal important information about a client's feelings, a helper's skills, or the client-helper relationship. When clients become silent, they may be in deep thought about what has been discussed, may have nothing further to say, or may be uncomfortable. It is important to check comprehension before making assumptions about a client's silence. Many researchers (Cormier & Hackney 2005; Hutchins & Cole, 1997; Kleinke, 1994; Neukrug, 2004; Young, 2001) consider the appropriate use of silence to be a powerful tool in the helping relationship. Silence can be beneficial to client growth by allowing clients to reflect on what they have been saying, and it lets the helper process this information and prepare for a therapeutic response or further questioning. Silence can be powerful because it may raise enough anxiety within a client to encourage further disclosure or may create so much discomfort that a client no longer continues in treatment. Cormier and Hackney (2005) specify the following types of silence: helper-induced silence, client-induced silence, therapeutic silence, pacing the interview, silent focusing, and responding to defenses. To use silence appropriately in your work, you need to be well versed in these types of silence. With experience, you will better understand the many intricacies of silence.

- Observable autonomic physiological responses such as quickened breath, blushing, paleness, and pupil dilation also provide invaluable information about a person's emotional state, mood, and beliefs. A seasoned helper is able to pick up on these typically subtle changes and can decide when and how to use them in a therapeutic way in the helping process.

- Space relates to how close or far a person chooses to stand or sit during a conversation. Also referred to as physical distance, space varies considerably among cultures, with gender, and depending on a client's expression of feelings (Cormier & Hackeny, 2005; Young, 2001). Awareness of this issue of space is equally important in better understanding the client and providing a safe and therapeutic environment for the client to flourish.

■ **TABLE 3.1 Some Nonverbal Cues and Their Meanings**

Cue	Meaning
Lifting an eyebrow	Disbelief
Lowering eyebrows	Uneasiness, suspicion
Lifting both eyebrows	Surprise
Eye contact	Interest, confidence
Winking an eye	Intimacy
Slapping forehead	Forgetfulness
Rubbing the nose	Puzzlement
Turning up corners of mouth	Happiness
Cocking head to one side	Friendly, human
Little or no head and/or hand movement	Cool, no emotion
Raised shoulders	Fear
Retracted shoulders	Fear
Square shoulders	Responsibility
Bowed shoulders	Carry a burden
Shrugging shoulders	Indifference
Sitting with arms and legs crossed	Withdrawal, resistance
Clasping arms	Protection
Tapping fingers	Impatience
Fondling and touching inanimate objects	Loneliness
Leaning forward in a chair	Interest
Learning back, arms uncrossed	Open to suggestion
Leaning back, arms crossed	Not open to suggestion
Postural change—facing away	Finished with interaction

Source: Woodside and McClam (1994, p. 197).

- Touch, when used appropriately, can have a positive impact on clients. Touch can convey caring and concern, especially during times of crisis; foster communication and bonding; show support to clients who are grieving, depressed, or traumatized; be used as a greeting or ending; and emphasize or underline important points (Driscoll, Newman, & Seals, 1988; Willison & Masson, 1986; Holroyd & Brodsky, 1980; Older, 1982; Young, 2001). Because touch can arouse powerful sexual and transference reactions in clients, it is important to be cautious in the use of touch. With clients who have been physically or sexually abuse, battered, or who come from a different culture, touch may be inappropriate. Young (2001) suggests that helpers know the client well before initiating even the safest forms of touch. A guideline regarding touch that we follow is to initially use it only sparingly to communicate encouragement and concern. Being aware of individual client issues and needs will determine when the additional use of touch is deemed appropriate.

As we have noted, nonverbal behavior offers invaluable cues to a person's feelings. Table 3.1 presents some nonverbal cues and their possible meanings.

SCENARIO: Clients' Nonverbal Communication

- A helper asks a client how he or she is doing, and the response is "Fine" or "OK," yet there are tears in the client's eyes.
- Individuals may show that they are upset or angry by crossing their arms or giving angry grimaces or dirty looks.
- A person's speech may seem pressured, possibly indicating anxiety or fear. Raising one's voice or lowering the tone can likewise indicate anger or being upset.
- A client who is grieving the loss of a child welcomes hugs at the end of each session.
- A client from a different culture may sit far away from the helper and at an angle where eye contact is obscured or wear a veil covering the face. Conversely, a client may sit very close to the helper.

If these subtle aspects of communication are not attended to, helpers may be missing very important clues to the client's emotional state and may therefore choose an inappropriate approach.

SCENARIO: Helpers' Nonverbal Communication

Helpers may communicate nonverbally by continually glancing at the clock, possibly conveying a desire for the session to end. Sighs or lack of eye contact may also communicate discomfort, lack of interest, or boredom. Although often unconscious and unintentional, such communication can be detrimental to the working relationship. It must be recognized and examined by the helper.

Oral communication should be used when one wishes to communicate emotions or when information is especially important. Certain feelings, if not communicated effectively, can interfere with your relationship with others. Angry, sad, and anxious feelings are common emotions that we all share. To be an effective communicator, recognize what you feel and then assertively communicate these feelings when necessary. Learning to be assertive can help you express your feelings and attitudes in the ways that they are most likely to be heard.

Multicultural and Gender Issues and Communication

Patterns and styles of verbal and nonverbal communication vary from culture to culture and between males and females in each culture (Alle-Corliss & Alle-Corliss, 1999). Different communication styles exist among different generations even within the same ethnic group. Attending and listening patterns of Euro-North Americans differ from those of other cultures (Evans, Hearn, Ulhemann, & Ivey, 1993). Corey and Corey (2003, p. 201) share that "personal space requirements, eye contact, handshaking, dress, formality of greeting, perspective on time, and so forth all vary among cultural groups." It is important to be aware that "different cultural groups ascribed varied meanings to certain nonverbal behavior" (Hepworth, Rooney, & Larsen, 2002, p. 172). Clearly, both verbal and nonverbal communication "must be understood in a cultural context" (Lum, 2004, p. 175). Because there are likely to be cultural-, generational-, and gender-based variations in communication, helpers are encouraged to be alert to these differences and to be aware of their own stereotypes, values, and cultural assumptions. Cormier and Hackney (2005, p. 98) assert that "it is very important not to assume the meaning of nonverbal affect cue without checking it out with the client." Clients are often the best resource to learn about these difference in communication patterns and styles. While listening skills are foundational and important in all cultures, you must recognize the cultural differences in basic listening skills. Here we address some well-documented differences:

Multicultural Differences

Eye contact   The direct eye contact pattern of mainstream Americans is considered intrusive and rude in some cultures (Evans et al., 1993). "In many societies (e.g., Native American and Latino) less eye contact is preferred; in some, direct eye contact is avoided altogether" (Alle-Corliss & Alle-Corliss, 1999, p, 48). Cormier and Hackney (2005, p. 98) add that "for some American Indian clients and for some Asian American clients, lack of eye contact may be consistent with their cultural norms and values and does not suggest lack of attention or disrespect. Corey and Corey (2003) highlight research that indicates African Americans may look more when talking and slightly less when listening. Furthermore, they have found that some Hispanic and Native American groups may view direct eye contact by the young as disrespectful. When serious or private subject matter is being discussed, some cultural groups avoid eye contact (Ivey & Ivey, 1999). When discussing private matters, many clients may feel more comfortable if the helper avoids direct eye contact. In sum, helpers need to modify their use and expectations of eye contact to meet clients' individual and cultural needs.

Body language   The meaning of body language varies widely from culture to culture. For instance, Russians may shake their heads up and down to signify no; Vietnamese may consider motioning to a person with one's fingers to "come here" a rude and derogatory gesture; or for a Japanese person, nodding the head does not necessarily indicate agreement, only attentiveness (Alle-Corliss & Alle-Corliss, 1999; Lum, 2004). Value judgments can also be made through one's gestures. Lum (2004, p. 175) and Franklin (1983) discuss how some streetwise African-American youth may share their antisocial exploits to engender a reaction and/or value judgment from the helper in an effort to "scare the worker off, test sincerity, and measure empathy for ghetto conditions." Such body language as "raised eyebrows, furrowed forehead, and shifting in one's seat at sensational stories about drugs, sex, alcohol, and delinquency are nonverbal signs to these youths that the worker made a value judgment."

Space   There are many cultural variations regarding physical distance or space. Young (2001) in his research found that Northern Europeans require more space during a conversation than do Southern Europeans or Middle Easterners, while Italians may be completely at ease conversing almost nose to nose. Alle-Corliss and Alle-Corliss (1999, p. 48) add that "some North Americans consider an arm's length between people to be a comfortable conversational distance. Others, especially those of Arabic descent, may prefer 6 to 12 inches. Still others, such as Australian aboriginal people, are not comfortable unless there is great distance. For many, maintaining a lengthy distance is desirable until trust and rapport are established." Once again, helpers need to consider the client when determining proper physical distance. Stone and Morder (1976) have found that five feet is an optimal distance between clients and helpers; we suggest using this as a guide in arranging office space. Young (2001, p. 82) advocates "allowing the client to arrange the chair in a comfortable way if it feels too close or too distant."

Silence   Silence on the part of the helper or client may be somewhat culturally determined (Neukrug, 2004). Tafoya (1996) has found that pause time for some cultural groups is longer than for others and can therefore affect how helpers interpret client responses. For instance, Native Americans whose pause time is longer are often labeled as reticent to talk and resistant to treatment when they may just be reacting more slowly than the helper might expect. Lum (2004) adds that in many cultures silence is often accepted and regarded as a sign of respect. "To talk back may be misconstrued as being rude, disagreeable, or argumentative. Silence may be an unwritten message that a person is listening, is open to what is being said, and is willing to go along with the speaker. However, silence may make a person feel uncomfortable, particularly when we are accustomed to feedback" (Lum, 2004, p. 175). Because silence can mean different things to different groups, helpers are encouraged to research the meaning of silence to the specific group they are working with.

Touch In some cultures, touch is encouraged and is viewed as a sign of trust and caring. However, in other cultures touch may be viewed as very private and personal. Problems may arise when the helper and client greatly differ in their views. For instance, if a helper who is uncomfortable with touch works with a client who sees touch more positively, the client may feel disliked or believe the helper to be cold and uncaring, neither which may be true.

Verbal following Helpers from North American tend to listen to clients by following their words as directly and literally as possible. However, Chinese and Japanese Americans may prefer a "more subtle, indirect style of listening for generalities, rather than specifics" (Alle-Corliss & Alle-Corliss, 1999, p. 48).

Listen first, then act Most American helpers believe in listening to a client before reaching a conclusion. In some cultures, as with Asian clients, trust is heightened only when there is sign of direct helper action. Because helpers are viewed as authorities who can solve client's problems by direct advice, clients may not feel comfortable in sharing deeper emotions until they sense that the helper is committed to being direct with them. "This points to the importance of the timing between listening and acting that may vary among cultures" (Alle-Corliss & Alle-Corliss, 1999, p. 48).

Gender Differences

Along with differences related to culture, communication styles differ according to gender. Males and females often differ in their responses, which are further complicated by their culture. For instance North American men may intellectualize their feelings whereas women may react more emotionally. Similarly, American men may be more geared to focus on thinking and problem solving, while American women are taught to focus more on feelings and process. Asian women, on the other hand, may be more reserved and hesitant to show emotions, unlike African American females, who may be more open and free in their expression of feelings.

Helpers should modify their style and approach to listening and acting when necessary because of gender, culture/race, affectional orientation, physical issues the client may be facing, age of the client, and a multitude of other factors. Professional helpers must be committed to continue learning about gender and cultural differences in the diverse population they will be working with (Alle-Corliss & Alle-Corliss, 1999). Lum (2004) emphasizes that "social workers [helpers] should enhance their communication style with ethnic clients and should be aware of these cultural dimensions of communication. Multicultural practice calls for highly developed communication skills" (p. 175).

The Value of Assertiveness

Being active, learning to communicate effectively, and becoming assertive will all contribute to your making the most of your fieldwork experience. As a beginning helper, you must learn the value of assertiveness. Corey and Corey (2003, p. 53) stress its importance in your relationship with your supervisor: "The assertion skills you practice in getting adequate supervision will be useful in your relationships with both clients and colleagues."

Any relationship—personal, professional, or social—is enhanced when assertive communication takes place. In fact, many of the problems in the field of human services seem to be rooted in inadequate communication by clients, staff, coworkers, and professional colleagues. "Many of the relationship problems helpers face are linked to difficulties in being assertive in their personal and professional lives" (Alle-Corliss & Alle-Corliss, 1999, p. 70). Passive (nonassertive), aggressive, and passive-aggressive communication abounds. Why is assertive communication not more widely used? This section will explore the underlying dynamics of assertiveness to clarify why it is such a difficult skill to attain.

What Is Assertiveness?

"The term 'assertiveness training' became popular in the 1970s and refers to a broad set of social skills used to enhance self-esteem and deal effectively with emotions" (Young, 2001, p. 248). According to Lange and Jakubowski (1976, p. 7), *assertiveness* involves standing up for personal rights and expressing thoughts, feelings, and beliefs in direct, honest, and appropriate ways that do not violate another person's rights. The basic message is:

- This is what I think.
- This is what I feel.
- This is how I see the situation.

The message expresses who the person is.

Assertion skills are helpful personally and in work with clients, colleagues, supervisors, administrators, and other professionals in outside organizations. "It has been used successfully with a wide variety of client concerns, including marriage problems, depression, sexual dysfunction, aggressive behavior, substance abuse, and dependency, as well as raising low self-esteem" (Young, 2001, p. 248). Alberti and Emmons (1995) characterize assertive behavior not only as self-affirmation but also as an expression of respect for other people. Lange and Jakubowski (1976, p. 7) agree that assertiveness involves two types of respect: (1) respect for oneself—that is, expressing one's own needs and defending one's rights; and (2) respect for the other's needs and rights—for example, accepting that another person may refuse or not condone one's request.

With regard to being respectful of our own needs, Alberti and Emmons (1995, p. 19) state, "Each of us has the right to be and to express ourselves, and to feel good (not powerless or guilty) about doing so, as long as we do not hurt others in the process."

Thus, assertiveness is a respectful communication technique that encourages mutuality and compromise and preserves a person's basic integrity. However, in some people's minds, the term *assertive* is associated with individuals who simply want to get their needs met without concern for those around them. Such lack of respect for others signals selfish (even narcissistic) behavior, not true assertiveness. The term *responsible assertiveness* is more in line with our view of the concept. According to Lange and Jakubowski (1976, p. 8), this involves "mutuality, asking for fair play, and using one's greater assertive power to help others become more able to stand up for themselves." In assertive communication, both verbal and nonverbal skills are used in helping to maintain respect for oneself and others, satisfying one's needs, and defending one's rights, without dominating, manipulating, abusing, or controlling others (Alle-Corliss & Alle-Corliss, 1999; Bolton, 1979). Such behavior encourages cooperation; most people respond favorably to a person who is acting in an assertive way that is responsible.

No doubt, practicing assertion skills can be beneficial in all areas of one's life (Alle-Corliss & Alle-Corliss, 1999).

SCENARIO: Responsible Assertive Behavior

As a new intern, you are excited about the many learning opportunities you will have. During your initial interview with your supervisor, some learning objectives were addressed. However, after a few weeks, you are not anywhere near these objectives. You have mentioned it in passing to your supervisor, and she promised that you would soon begin to interact with clients and become involved in group work. In an effort to act assertively, you approach the supervisor and say you need to meet soon to discuss this situation. When the meeting finally takes place, you directly, clearly, and specifically address your desire for increased responsibility and your concern that this has not yet taken place. By respecting your own need to engage in active learning and showing a willingness to express your emotions, you are in fact acting assertively.

TABLE 3.2

Assertive	Nonassertive	Aggressive
Definition: When people express their feelings and thinking in ways in which they show respect for the right to personal worth and to equal self-expression of those with whom they are interacting. This can be constructive for both parties.	*Definition:* When people deny their own right to be assertive and do not express their own thinking and feeling, which contributes nothing to enhancing the situation and can destroy self-confidence and self-respect.	*Definition:* When people exercise their right to express their own feeling and thinking but in ways that attack those with whom they are interacting. This mobilizes aggression in others.
Reasons for acting: Often people fear that acting assertively will result in losing other people's approval, becoming too vulnerable, or losing control. A major reason for acting assertively is that it increases one's own self-respect. It also increases one's control over oneself.	*5 reasons identified:* Mistaking firm assertiveness for aggression; mistaking nonassertiveness for politeness; failure to accept one's own personal rights; anxiety about negative consequences; mistaking nonassertiveness for being helpful.	*5 behavioral patterns:* Powerless; prior nonassertiveness; overreaction due to past emotional experiences; beliefs about aggression; reinforcement and skills deficits.
Consequences: Increased likelihood of having one's needs met and of feeling self-respect, respect from others, and enhanced self-esteem and confidence. Personal rights are valued, and one has the opportunity to express one's feelings.	*Consequences:* In an effort to avoid conflict, one risks loss of integrity and eventually loss of self-esteem. Also, many who remain nonassertive wind up feeling hurt, angry, and taken advantage of.	*Consequences:* Immediate consequences may be positive: emotional release, sense of power, and having needs met. Yet in the long run, those who continue to act aggressively risk alienating others, losing jobs or promotions, and feeling misunderstood.

THINKING THINGS THROUGH

Can you think of a situation in your current fieldwork placement where you have either acted assertively or should have done so? If so, did you take into consideration the question of mutual respect? In that situation, think of one central assertive statement that you could make. Attempt to develop this statement in a way that would be mutually respectful to you and the other person involved.

Assertive, Nonassertive, and Aggressive Behavior
These three behaviors are often confused with one another. To clarify the distinctions, we have adapted (in Table 3.2) definitions given by Alberti and Emmons (1995), Brill (1995), Lange and Jakubowski (1976), and Hepworth, Rooney, and Larsen (2002).

Basic assertiveness
This involves *a* single *expression,* by which an individual stands *up* for his or her personal rights, beliefs, feelings, or opinions. It can also be a way to express affection and appreciation toward others.

SCENARIO

Your supervisor asks you to consider coming in on your day off. You simply state, "Thank you for the invitation, but I have plans already and will not be able to work that day." You are appreciative of all the time and energy your supervisor has spent in orienting you to the agency and working with you on some difficult cases. At your next visit, you tell him, "I have really appreciated your guidance and support. I feel fortunate to have a supervisor like you."

Can you think of an example of basic assertiveness? Please elaborate.

Empathetic assertiveness This involves "making a statement that conveys recognition of the other person's situation or feelings and is followed by another statement which stands up for the speaker's rights" (Lange & Jakubowski, 1979, p. 15).

SCENARIO

In an educational meeting at your agency, a coworker interrupts loudly while an invited speaker is presenting a topic that is vital to your work. You feel irritated and annoyed and decide to assert yourself. You state, "I know that what you are discussing may be important to you, but your talking is making it difficult for me to concentrate on the presentation. I would appreciate it if you could save your remarks until the speaker has finished talking."

Consider how you might respond in this type of situation. Would you use empathetic assertiveness? Why or why not?

Escalating assertiveness This involves using gradually stronger assertive statements, of increasing firmness. It is useful when numerous attempts at basic and empathetic assertiveness have yielded minimal results.

SCENARIO

You are beginning to work with a client who insists that you can provide her with "a written off-work order" because she is feeling "too anxious and depressed to go to work." You have told her numerous times that this is not something you are able to do and that a referral to a physician is necessary. She still persists. You state, "I am sorry that I cannot help you with an 'off-work order' as you desire. However, as I have already explained, this is not something I am able to do. I would be more than glad to ask the clinic administrator to explain this policy and also to make an appointment with a physician."

Can you think of a nonclinical example where escalating assertiveness is most appropriate?

Confrontational assertiveness This type of assertiveness is used when there are incongruencies in others' words and actions. The assertive response involves describing the incongruencies directly and objectively.

SCENARIO

You are a newly hired child-care worker at a children's center. When oriented, you were explicitly told that everyone must work one Saturday per month as part of the job—no exceptions. You would rather not work on weekends but did agree to the one Saturday a month. Now that you have been on the job for over a month, you notice that at least half of the staff who are in the same position as you do not work Saturdays. Moreover, you have been scheduled several Saturdays per month. You approach your supervisor and say, "I'm a little confused about the Saturday work schedule. It was my understanding that all the line staff were to work one Saturday a month only. In the two months since I began, various staff members have not worked Saturdays and have indicated that they will not. Also, I've been asked to work several Saturdays per month when my agreement was to work only one per month. I am feeling frustrated and would like to clarify the policy."

How would you handle such a situation?

SCENARIO

You are an intern at a high school. Your job title is task counselor, yet you have not seen a single student. Instead, you have been involved in filing and computer work. Each day that goes by, you feel more frustrated and less optimistic about your experience. You decide to discuss this at your next meeting with the supervisor. "I realize that with the beginning of school, there have been many administrative areas requiring attention. However, now that we are past that phase, I'm beginning to feel frustrated at my lack of direct contact with students. As you recall, my learning agreement specifies that I will be working directly with students as a task counselor, and I hope this will begin to occur soon. Can you help me with this matter?"

Would this be your approach? If not, how would you proceed?

"I" assertiveness   Thomas Gordon (1970, p. 18) says that "I" assertiveness often involves a four-part statement: When . . . , the effects are. . . . I feel. . . . I'd prefer. . . .

SCENARIO

You have given your supervisor an evaluation form to complete, three weeks prior to the due date. Now, three weeks later, your supervisor informs you that she will get to it as soon as possible. At your next meeting later that week, the evaluation is still not completed. You use an "I" statement to assert yourself. "When I gave you my evaluation form to complete, I recall letting you know exactly when it would be due. Now that it is past that date, I'm feeling anxious about it being late. Also, I am feeling frustrated by this process. I'd prefer that we focus on completing the evaluation today rather than discussing any other issues."

Can you see how "I" assertiveness was used? Would you respond similarly? If not, how might you respond in this situation?

Being aware of your feelings and needs is only the first step in meeting your goals. The second and most important step is being active and assertive in expressing your thoughts, feelings, and desires. The following self-awareness exercise is based on our work with clients, students, colleagues, and administrators in human services. Take a moment to answer the questions and see how you rate with regard to your communication skills.

SELF-AWARENESS EXERCISE

Assessment of Assertiveness

Assess your assertiveness on a scale from one to ten (one being passive and ten being optimally assertive). Rate yourself in the following four categories: interpersonal, academic, agency, and social. Are they all consistent, or are there differences? If so, what do you suppose the reasons are? If your score is on the lower end (one to four), how might you work on developing your assertiveness skills?

Similar to this exercise is another we have used in our fieldwork courses. This exercise asks that beginning helpers assess their communication skills by determining if they are aggressive, assertive, or nonassertive (passive).

SELF-AWARENESS EXERCISE

On the scale below, circle the number that most clearly corresponds to your assertiveness skills at this point.

−10 −9 −8 −7 −6 −5 −4 −3 −2 −1 <u>Optimal</u> +1 +2 +3 +4 +5 +6 +7 +8 +9 +10
    Nonassertive   Assertive   Aggressive

Discuss how you might improve your assertiveness skills.

Effective Time Management

Although there are vast learning opportunities in many fieldwork placements, using your time wisely is essential to maximizing your experience. Being aware of the limited time you have can help motivate you to plan and manage your time more constructively. For instance, if you are allotted ten hours a week at your placement, explore how you can optimize your learning during this time rather than waiting until your supervisor directs your activities. Relying on others to manage your time can result in gaps in your schedule and frustration in your work. Avoid this by taking ownership of your time and recognizing that it is precious. Of course, everyone has other responsibilities and commitments that make demands on their time, but don't think that you have no control over your time. If you take a little time to plan your schedule wisely, you will ultimately conserve time and feel less overwhelmed and more productive. Many beginning helpers are hesitant to discuss their schedules or activities with their supervisors, but once they do, they may find their work more worthwhile. Corey & Corey (2003, pp. 373–374) offer various suggestions for effective time management that include identifying your long-term goals, establishing subgoals, prioritizing your activities, thinning out your schedule, evaluating potential new commitments, avoiding procrastination, identifying time wasters, learning to delegate, trying to do a task only once, and striving to live in the present moment. In our own practice, we have found that attempting to follow as many of these tips as possible is well worth the effort.

SCENARIO: Time Management

You are an administrative intern in a hospital. Part of your role is to develop a newsletter for the employees. Your supervisor is very glad to have you at her facility and welcomes you warmly. She gives you an overview of what is to be incorporated into this newsletter and then leaves you on your own to follow through. At first you feel overwhelmed and immobilized by all that you have to accomplish. You discuss your struggles in your practicum/fieldwork seminar, where you are encouraged to set a time line and outline the steps needed to write the newsletter. When you return to your placement, you attempt to follow through on these suggestions and find that your job begins to go more smoothly. In a few weeks, you can actually see results; the newsletter is beginning to take shape. What time management skills would you have applied in this situation? Please elaborate.

Consider experimenting to learn to manage your time more effectively and tailor your schedule to your specific needs.

Journaling and Process Recordings

Several additional ways of remaining active in your learning are to engage in journal writing and process recordings. expressing themselves in writing helps workers clarify and focus their efforts. As a beginning professional, you may be asked to "journal" your experiences or record your work in a process recording. Both of these experiential forms of documentation can enhance learning.

Journaling

In "journaling," thoughts and feelings are put in writing, which increases insight into the writer's thoughts, behaviors, and emotions. Basically, journaling is free-form writing about various topics within your field placement setting and/or even within your life. It can be a way to access your feelings and thoughts in regard to the work and relationships you are forming within your agency. At times your instructors may ask you to write about specific topics to encourage your critical thinking on the various practice and process issues that evolve out of your work in the agency. Sometimes you may just want to keep a diary of your reactions from day to day in your work to help you become more focused and evaluate your growing self-awareness.

According to Collins, Thomlison, and Grinnell (1992, p. 155), "[a] journal should be a record of your learning during your practicum experience," reflecting the two aspects of learning: the information acquired and your interpretation of that information's meaning.

SCENARIO

Randy remembers that it was extremely helpful in his beginning years in the field to keep a journal and just write about what was happening in his thoughts and feelings at the time. Years later he could return to these writings to see some of the progress he had made in his ability to recognize and deal with his feelings and thoughts.

Process Recordings

A process recording is a written account of interactions you have had with clients, staff, supervisors, and/or community personnel. Usually, the helper documents the content of the interview, along with his or her feelings and impressions. Some recording forms contain a place for supervisory feedback. According to Wilson (1980, p. 18), process (narrative) recordings are a "specialized and highly detailed form of recording that is a very valuable tool for enhancing student learning if it is properly used." Nelson (1984, p. 13) considers process recordings "the chief learning vehicle of field instruction, sorting out feelings, making inferences, developing critical thinking, and selecting alternatives for action." Process recordings were developed by social work educators who envisioned this type of learning to be a catalyst for professional growth whereby individual experiences, feelings, and values could be integrated into a professional self (Wilson, 1980). Today, many disciplines within the helping professions use process recordings. They are primarily used during a student's internship or fieldwork experience and by field supervisors whose purpose is to help students apply theory or empirical knowledge to practice. In sum, process recordings serve as a method for both professional/personal growth and acquisition/integration of knowledge (Alle-Corliss & Alle-Corliss, 1999). Specifically, the written process recording facilitates student learning by helping students do the following:

- Rethink or process the interview content
  - Deal with their feelings about the content and the client
  - Analyze interventions
  - Draw conclusions
- Understand the helper's impact on the client system
  - Demonstrate the helper's understanding of information and theory gained in the classroom
  - Build clinical skills
  - Participate in a growth relationship with the field instructor (Nelson, 1994)

Process recordings can have additional purposes:

- Help students learn interviewing skills
- Assist supervisor in evaluating student's performance on an ongoing basis as well as in the review during termination; comparing process recordings written at the beginning of the internship with those written later often demonstrates student growth
- Help students enhance their self-awareness by encouraging them to separate factual data from gut-level reactions of what they have observed
- Serve as a learning experience for the entire group when process recording are shared in a peer group

Typically, a process recording contains the following information:

Supervisory comments
Content

Dialogue
Student's feelings
Analysis of the situation

The literature describes many other forms of process recordings, which we have included in Appendixes A, B, and C for further review.

■ SUPERVISION

The Importance of Supervision

In all fieldwork placements, and in most organizations in general, some form of supervision exists. Often it is a formal relationship, as mandated by the curriculum, the organization's policies, or the state licensing boards. However, supervision can be informal, as when a professional seeks guidance from a higher-ranking colleague. According to Royse, Dhooper, and Rompf (1999, p. 45), "[g]etting the most out of supervision requires that you know what supervision is, what the dynamics of the supervisory relationship are, and what specific actions facilitate effective supervision."

Corey, Corey, and Callanan (2003, p. 321) define supervision as "an integral part of training helping professionals and is one of the ways in which trainees can acquire the competence needed to fulfill their professional responsibilities." The primary purpose of supervision is to enhance the personal and professional development of the student (Gordon, McBride, & Hage, 2001). It represents one of the most important experiences in the development of professional competence (Gladding, 2004). Although its purpose is didactic, supervision involves dynamic and emotional factors. The supervisory relationship is therefore a complex one. A unique link between preparation and skilled practice, supervision is critical to your career development.

Learning to appreciate the value of the supervisory relationship is often a challenge for the beginning helper. But it's usually worth the effort; having a positive working relationship with your supervisor can lead to greater satisfaction in your fieldwork. Recognize the importance of this relationship early in your training. It will enrich your learning and perhaps inspire you to become a supervisor yourself one day.

Supervisory Roles and Responsibilities

Supervisors are as diverse as those they oversee, and they play a multitude of roles. Ideally, your supervisor should be an experienced, competent, well-trained helper with a sound background in his or her area of expertise. Faiver, Eisengart, and Colonna (2004) emphasize that the "SUPERvisor" is a professional who provides necessary encouragement and constructive feedback and tailors it to meet each student's individual needs.

Supervisors perform varied roles, depending on the professional status of the supervisee, the particular style and philosophy of the supervisor, and the nature of the agency. From our experience and a review of the literature (Corey, Corey & Callanan, 2003; Faiver, Eisengart, & Colonna, 2004; Martin & Moore, 1995; Travers, 1993), we have compiled a list of typical supervisory roles:

- Teacher
- Model
- Evaluator
- Mentor
- Counselor
- Advisor

We concur with many who believe that what role the supervisor performs is not as important as the relationship between supervisor and supervisee. Effective supervision is most likely to occur when there is a match between the learning needs of the supervisee and the approach of the supervisor. With your own supervisor, find out what you have in common and how you can work together productively to achieve mutual goals.

THINKING THINGS THROUGH

Take a moment to assess which of the roles in the preceding list are met by your supervisor. For those that are unmet, consider how you might use assertiveness skills to address these issues. Then consider any commonality between you and your supervisor that might help create a positive working relationship.

To understand the supervisory relationship, be aware of three functions of a supervisor: administrative, supportive, and educational (Alle-Corliss & Alle-Corliss, 1999; Baird, 2002).

Functions of Supervisors

Administrative Regardless of the setting, supervisors are responsible for directing and evaluating the work of the beginning helper so that the organization can deliver services in the most effective and efficient manner. Included in this function are the ethical and legal responsibilities inherent in the supervisory role. According to Borders and Leddick (1987), supervisors need to be cognizant of the following concepts to observe their professional codes:

- Informed consent
- Evaluation
- Due process
- Confidentiality
- Dual relationships
- Legal liability

Administrative leadership is among the major roles of a supervisor in an agency setting. The supervisor's job responsibilities and requirements are designated by the agency, taking into account the levels of experience available and the particular needs of the supervisees. In supervising students and interns, the supervisor is likely to play a primarily educational role, whereas supervision of employees may take on a more administrative slant.

Supportive Supervisors who offer reassurance, encouragement, and positive reinforcement foster trust and rapport with their supervisees, just as a counselor does with a client. Such trust can be furthered by providing approval, constructive criticism, catharsis, and attentive listening. According to Martin and Moore (1995), supportiveness is a primary characteristic of effective supervisors. Supervisors work supportively through coaching, modeling skills, and offering suggestions regarding practice issues. On an emotional level, they can offer support, reassurance, encouragement, and affirmation. Acknowledging the beginning helper's uncertainties is also supportive. Neukrug (2004, p. 82) adds that "like an effective helper, the good supervisor is empathic, flexible, genuine, and open."

Educational Supervisors are often educators in their own way, serving to extend your formal academic training (Brashears, 1995; Faiver, Eisengart, & Colonna, 2004). The educational function includes:

- Knowledge of professional roles that will foster professional development and harmony with other agency staff
- Clinical knowledge and skills to model for students, allowing them to sharpen their clinical abilities
- Case conceptualization skills, including teaching the student about social, cultural, and behavioral patterns

- Self-awareness skills—being aware of one's own interpersonal behavior, to model appropriately for supervisees
- Effective teaching techniques, possibly including one-to-one (traditional) supervision, the use of audiotapes or videotapes and process recordings, one-way mirror interviews, live supervision, and group supervision

The ideal supervisor would be able to incorporate all three of these functions perfectly. Real-life supervisors will be stronger in one of these areas than in the others, but you can still obtain great benefit from such a relationship. If forced to choose just one aspect, we believe that supportiveness is the most valuable. Berger and Graff (1995) also regard the supportive component as the most important for effective clinical supervision. Approachability, style of feedback, and ability as a therapist are other important characteristics of effective supervisors, in their view.

Supervisory Responsibilities

Human service supervisors must maintain various areas of accountability. From the literature (Brashears, 1995; Berger & Graff, 1995; Corey, Corey, & Callanan, 2003; Faiver, Eisengart, & Colonna, 2004; Morales & Sheafor, 1995; Nelson, 1993), we have assembled a comprehensive list of supervisory responsibilities, applicable to human service disciplines.

Ethical and legal Supervisors are ultimately responsible for the actions of those whom they supervise, both ethically and legally (Corey, Corey, & Callanan, 2003). Depending on the setting and the clientele served, this can be an overwhelming responsibility for the supervisor. Issues of child abuse, battery, suicide, homicide, AIDS, and sexual misconduct are among ethical and legal concerns a supervisor must tackle.

Sherry (1991) stresses that supervisors are "ethically vulnerable" because there is a power differential between the participants, the supervisory relationship often has "therapylike" qualities, and there may be conflicting roles between the supervisor and supervisees. Ultimately, the supervisory role is a complex one in which "supervisors are responsible for ensuring compliance with relevant legal, ethical, and professional standards of clinical practice" (Corey, Corey, & Callanan, 2003, p. 322).

SCENARIO

As a new intern in a battered women's shelter, you encounter a case where child abuse is suspected. Although you have completed an orientation on domestic violence, you feel overwhelmed and anxious. Your supervisor asks about your level of comfort in handling this case. When you share your discomfort, she kindly suggests a seasoned helper who might be more appropriate. She encourages you to consider transferring the client to that worker and remaining involved by observing the treatment process. She also recommends that Child Protective Services be contacted to determine whether a child-abuse report should be filed.

This supervisor responded to both ethical and legal responsibilities. Do you agree with the manner in which she intervened? Would you have responded differently? How might your own supervisor have reacted?

Evaluative A supervisor must provide ongoing feedback to supervisees on their performance and evaluate their acquisition of knowledge, skills, and values (Corey, Corey, & Callanan, 2003; Nelson, 1993; Neukrug, 2004). This helps the supervisee gain an awareness of his or her strengths and limitations and enhances skill development. The process can be simple or difficult, depending on the supervisor's skill, confidence level, and relationship with the supervisee and the supervisee's own personality and level of defensiveness. (Some beginning helpers tend to take certain feedback personally.)

SCENARIO

For the first time, you have just acted as leader of a parenting class at a group home. As you meet with your supervisor to discuss the class, she begins by giving you positive feedback regarding your performance. She is clear about both your strengths and limitations. She conveys the message that you are there to learn, and she gives you constructive feedback and direction.

How would you feel if you were supervised in this manner? Would your supervisor have responded differently?

Educational "A supervisor is an educator, and as such he or she enhances your formal academic training" (Faiver, Eisengart, & Collona, 2004, p. 19). Supervisors often instruct as part of their supervisory role; they may focus on treatment issues, agency policies, or interpersonal aspects. Sometimes there is a mismatch between a particular supervisor and supervisee. It is expected that the supervisor will be more knowledgeable and experienced, but this may not always be the case. When the supervisee is more knowledgeable and skilled than the supervisor, the situation is delicate.

SCENARIO

In your placement at a child-care center, you meet with your supervisor weekly. Having had no experience with young children, you are often unclear about the appropriateness of their behavior. Your supervisor is very knowledgeable in this area. She consistently provides you with pertinent information (verbal and written) about normal child development. When maladaptive behavior is observed, she gives you the needed guidance and support to intervene appropriately.

In contrast, let's imagine that you have been placed in a group home for abused children and are being supervised by a person who recently attained an MSW degree. You are an undergraduate seeking a BS degree, but for ten years previously you worked in Child Protective Services as a social work assistant. Many of the cases at the group home are similar to those you worked with while employed. On the other hand, your supervisor is new to the area of child welfare and is less informed than you are about child-abuse issues. You often feel frustrated because you are more skilled and experienced than your supervisor.

How might you handle this situation?

Protection Sherry (1991) considers the issue of protection to be all-important for supervisors: "Supervisors are faced with the responsibility of protecting the welfare of the client, the supervisee, the public, and the profession." Although this responsibility may appear clear cut, helpers may sometimes genuinely believe that they are placing the client first, but their own needs are actually being given precedence. When this occurs, supervisors have their work cut out for them and must be assertive.

SCENARIO

As a beginning helper at a family service agency, you are assigned a client who is volatile and potentially dangerous. You are able to bond with the client and feel quite comfortable in your work with him. However, others in the agency are not very comfortable with the presence of this intimidating individual. Your supervisor shares her concern with you; there have been numerous reports of the client's inappropriate behavior while in the lobby. Because young children are frequently exposed to this person's anger, she is concerned about their emotional and physical welfare. She asks that you consider a referral to a more appropriate setting where the client's needs can be better met and the potential danger minimized. Initially, you become upset and take her recommendation as a direct insult to your work with this client. As you review the case more closely, however, you come to understand your supervisor's concern for others, for the client, and for you.

Do you agree with the manner in which the supervisor responded to this potentially violent situation? Would you have approached it differently? Would there have been other ways to be protective yet not have to refer the client elsewhere?

Compliance Supervisors are responsible for ensuring the competence of those they supervise. According to ACES (Association for Counselor Education and Supervision, 1995), supervisors must consider relevant legal, ethical, and professional standards of practice and evaluate if and how these are being met.

SCENARIO

Due to various unexpected circumstances, you have not been able to attend your placement during the times originally agreed upon with your supervisor. You have been there only for a limited time and cannot foresee completing your hours in the time designated or fulfilling the requirements of your contract. Your supervisor discusses his concerns about your compliance and encourages you to reconsider your options.

Do you agree that this is a compliance issue? How do you think it could be handled?

Individual supervisors approach the task of supervision in their own unique ways. Common approaches and styles of supervision are discussed next.

Approaches and Styles of Supervision

Berger and Graff (1995), Corey and Corey (2003), and Corey, Corey, and Callanan (2003) identify several main approaches to supervision:

- Sharing information
- Content-oriented supervision
- Process-oriented supervision
- Emphasis on helper's dynamics
- Emphasis on client's dynamics

Supervisory styles are also as varied as supervisors themselves, based on:

- The supervisor's own personality
- The level of desire and willingness to supervise
- The supervisor's abilities
- The supervisor's time availability
- The personality match with the supervisee

Some supervisors are extremely positive and supportive, others quite distant and critical, and most fall somewhere in between. An effective supervisor can highlight your strengths as well as give you constructive feedback on your limitations. Avoiding defensiveness and allowing yourself to accept feedback in a positive manner can lead to a beneficial relationship with your supervisor, no matter what his or her style may be. You have little control over who your supervisor is, or his or her personality style, but you can choose to make the most of the experience.

Aspects of the Supervisory Relationship

The supervisory relationship is unequal in terms of relative power, status, and expertise between the supervisor and the trainee.

The process of supervision possesses many therapylike qualities, but it is not a therapy experience. Dual relationships can develop if the supervisor crosses the line between remaining a supervisor and becoming a therapist. Faiver, Eisengart, and Colonna (2004) write that it is often difficult to keep your involvement with your supervisor strictly professional. An ethical supervisor should always bear in mind the need to maintain proper boundaries. The supervisor has specific ethical responsibilities toward the trainee.

SCENARIO

You are being supervised by a well-trained and qualified clinical social worker. She is open, honest, and amenable to working with you. She often helps you deal with countertransference issues with clients, yet she does not actually engage in therapy with you. When she feels you may need to seek further guidance regarding a particular issue, she conveys this with care, pointing out that many beginning helpers engage in professional therapy of their own. Although you have become very close, you are aware that the relationship is not an equal one, especially when it comes time for formal evaluation. Also, you are cognizant of her ethical and legal responsibilities: She has had to co-sign several child-abuse reports and has been alert to the ethical issues involved.

Do you agree with the ethical approach taken by this supervisor? In what other ways can she fulfill her ethical responsibilities?

Effective Involvement in Supervision

We have found the following guidelines useful in working with students in fieldwork experiences.

Be Clear About What You Expect
In this regard, a learning contract can be helpful. Collins, Thomlinson, and Grinnell (1992, p. 111) describe the *learning environment* and *learning contract* as follows:

> Creating a learning environment is like setting the stage for a play. The stage is the agency and the community is the wings. In your roles as writer, director, and major actor, you are free to set up the props as you wish, guide the action, and establish the script. Remember, you are a beginning director. You will find that the other actors and directors have strong opinions, and the play itself will limit your options by its very nature. The script sets out the story line. Your learning contract and field evaluation form are the script for your practicum.

This creative approach stresses the importance of using your learning contract as an avenue to discuss the process with your supervisor.

According to Nelson (1993), a *learning agreement plan* formalizes expectations for student performance in achieving the learning objectives of the field practicum. Of course, such learning agreements may need to be modified later, and they must reflect the differing needs and skill levels of individual trainees.

The tasks involved in developing this learning contract include defining learning objectives, identifying activities for achieving these objectives, and delineating the procedure for evaluating the student's progress. Specific learning objectives may cover the basic areas of practice: individual, family, group, administrative, and community. The learning plan may also include development of certain competencies:

- Awareness of and ability to use individual learning patterns
- Effective interaction with the agency
- Demonstration of the beliefs and ethics of the profession
- Skill as a practitioner
- Sensitivity to cultural diversity

Developing this agreement in writing about your learning plan at the outset of your fieldwork helps provide structure and direction. This written plan can help prevent misunderstandings or confusion about your role and expectations. It can also help begin a solid working relationship with your supervisor as you assess your areas of strengths and limitations. This process can also help alert students early on about what objectives are realistic and which are not. Early interaction with supervisors gives students insight about their styles and approaches that will be useful as they work together.

It is important to take your role in developing this learning plan seriously. Your learning is at stake. When students do not actively participate in their learning agreement, they often feel frustrated later when their needs have not been met or the learning goals and objectives delineated are not truly theirs. It's acceptable to ask your supervisor to renegotiate the plan as needed. If the contract does not fit what the agency can offer the student, then it may need to be revised as well.

In essence, the learning plan can be helpful throughout your entire fieldwork experience. At the beginning, it sets the tone for the work that is to be accomplished; in the middle, it is a guide that keeps learning focused and continual; at the end, it is an essential evaluation tool (Alle-Corliss & Alle-Corliss, 1999).

SCENARIO

As a first-time intern at a newly opened teen chemical dependency shelter, you have your initial meeting with your supervisor. You show her your formal learning agreement and let her know that your primary interest is in working directly with the teens. She indicates that a large part of the job will entail coordination and administration, yet there will be room for individual and group work. You make sure that this is incorporated into your learning objectives. You also make clear that you will not be available to do "on call" work because of your parenting responsibilities.

Can you offer an example of how you have used or might use your learning contract to clarify your role and learning objectives?

Be Assertive About Your Learning

Assertiveness is especially necessary if supervision has started out on the wrong track or is inadequate or if the role you expected to play is not an option. Because of the power differential between you and your supervisor, acting assertively may be difficult. Beginning helpers, as well as seasoned professionals, often fear adverse repercussions (such as negative evaluations) if they speak up. Remember our earlier discussion of responsible assertiveness: Assertive behavior includes respect for the possible refusal of one's request. Even if you are direct and open about your needs and desires, there are no guarantees that they will all be met, yet the odds are certainly greater. By contrast, remaining passive and withholding your views can only serve to alienate you from your supervisor and distance you from the learning opportunities your workplace has to offer.

SCENARIO

You are employed in a hospital. You originally were told that you would be leading groups with both staff and clients. After several months of waiting for permission to begin a group on stress management, you are beginning to feel frustrated with all the red tape. You tell your supervisor about your eagerness to begin a group. "I realize that there have been a lot of unforeseen changes in our work demands since I started. However, I would like to let you know that as we agreed upon at my interview, I am very interested in being involved in group work. I am still very committed to this and am ready to start as soon as is possible. Can you help me with this?"

How would you have handled this situation? Based on the assertiveness skills you've learned about, how else might you respond assertively?

Take Advantage of Your Role as Learner

By admitting your inadequacies and acknowledging your limitations, you can freely seek out learning opportunities that might not be available otherwise. For instance, if you give the impression that you already know how to perform a certain task, your supervisor may not offer you such an experience or may expect you to perform it without difficulty. Also by remaining open to learning and cognizant of your limitations, you show the supervisor your willingness and motivation, thus contributing to a climate of openness, trust, and two-way communication.

SCENARIO

Your second-year graduate placement is at a community mental health center. Your supervisor has a great deal of faith in your skills and has given you much latitude in how you conduct your work. You feel very appreciative of this. As the year progresses, however, you find that some of the referrals you are given are quite complicated and perhaps a little over your head. At first, you hesitate to let your supervisor know, for fear she will think you are not capable enough. After some soul searching and encouragement from peers, you tell your supervisor that as a beginning counseling student, you are sometimes unsure about how to work with certain complicated cases. She listens carefully and asks about the specific cases at hand. Once she hears about the complexities, she admits that they are clearly difficult. She is sensitive to your struggles and agrees to be more cautious about assigning you such cases.

In the same situation, what if the supervisor had not been so open and understanding? Would it have still been appropriate to let her know of your concerns, referring to your role as a learner? Might you have been more assertive, stating, "I'm very eager to learn all the intricacies of working with difficult cases such as these, but as a student I need some help to start"? How would you handle this type of situation if it took place in your agency? Would you consider using your role as learner to your benefit?

A Certain Amount of Anxiety Is Appropriate

If you give yourself permission to feel some anxiety and apprehension, you will find it easier to confront fears and worries. If you deny your feelings, you will not be as likely to challenge yourself or explore the underlying dynamics; your feelings will probably spill out in other ways if you do not recognize and deal with them.

SCENARIO

It is your first time working in an inpatient chemical dependency unit. In your first group experience, group members expressed much anger and resentment about the program. You left feeling tense, anxious, and upset. Your supervisor tells you that this is a normal reaction, one almost anyone would have. She says that you should seek support and validation and give yourself permission to have these feelings from time to time, as they are a natural process of working in the field.

At your fieldwork placement, have you sometimes felt anxious, insecure, or fearful? Can you elaborate? Do you feel that your emotions were valid?

Take an Active Stance

Take an active stance in your learning and be assertive. When employed effectively, these skills can help your fieldwork become an optimal learning experience. Corey and Corey (2003), Royse, Dhooper, and Rompf (1999), and many others offer great suggestions for students in getting the most of supervision. They recommend:

- Planning for the supervisory conference or meeting by writing down questions you have, problems you want to discuss, or observations you want to make (i.e., maximizing your time in supervision). Creating an agenda for each meeting with their supervisor helps students stay on track by planning and organizing their supervisory hour. These agendas are most helpful to supervisors if submitted at least three days before scheduled meetings, giving the supervisor some time to prepare.

- Informing your supervisor which assignments you've completed and which are still works in progress.
- Being accountable for your time.
- Asking for new assignments and additional responsibilities if you are not satisfied with your tasks or role.
- Being willing to discuss problem areas, to learn from your mistakes, and to say "I don't know."
  - Being open to constructive feedback.
  - Inquiring about things you want to learn.
  - Expressing your reactions.
  - Learning without copying.

SCENARIO

As a new intern, you find that you are able to complete your assigned tasks efficiently and have time left over. You approach your supervisor about engaging in additional projects when you have completed your work. He is thrilled by this possibility but says that you will have to take responsibility for deciding which projects because he is too wrapped up in an upcoming audit. He asks that you brainstorm and return with some ideas at your next meeting. You gladly agree. Your mind is ticking away with new ideas for projects.

Can you see how this is an example of an active stance? Can you think of other situations where you have been active in your learning? At times, might you have benefited from being more active?

Evaluation and Termination Process

Be aware that an internship has a beginning, middle, and ending. Being active during all of these phases is important. During the termination process, it will be essential to meet with your supervisor to review your learning and evaluate your progress. Reviewing the learning contract is often part of this process. Many internship sites have specific evaluation forms that the supervisor must complete. It is ideal if the student is a part of the evaluation; during this time, the supervisor can emphasize particular strengths in the student and identify limitations.

If students are active in their own termination and evaluation process, it can facilitate learning about the ending phase in relationships and about areas for continued growth. In contrast, students who are passive during this process may leave the agency feeling frustrated, resentful, and/or unfulfilled. During termination and evaluation, interns also have the opportunity to offer their own feedback regarding their learning experiences and areas within the agency needing improvement.

Because termination can bring about certain feelings in the student and/or supervisor, it is important to remain active in processing them. It is sometimes easier to overlook or dismiss one's emotions; however, in doing so, valuable learning may be missed. We encourage you to take the opportunity to explore and discuss the issues that are brought about in endings.

SCENARIO

Julie was placed at a chemical dependency outpatient program for her first internship. She had never worked with this population and at first felt overwhelmed. Her supervisor was older and serious in demeanor. Initially, Julie felt anxious and inadequate, especially during supervision. As the semester progressed Julie had even contemplated quitting. Despite her struggles, she eventually seemed to find her niche; she developed strong therapeutic relationships with several clients and was surprisingly even beginning to look forward to her supervision hour. As the end of the semester was approaching, Julie suddenly felt upset and confused; the placement she had initially wanted to leave had now become a comfortable learning environment. She found herself avoiding discussing termination with her supervisor and called in sick on several occasions.

Based on this scenario, consider how a passive student would handle the termination and evaluation process differently from an active student.

Challenges in Supervisory Relationships

In closing, we explore several areas where the supervisory relationship can present challenges, along with advice on how you might handle them.

Poor match between supervisor and supervisee Accept the fact that you will have to learn to work with different personalities in a cooperative fashion. Rather than being angry or resentful, make an effort to concentrate on your work and recognize that everyone has a unique personality. If it seems that your supervisor wants you to be exactly like him or her, you must deal with this directly by letting the supervisor know that you appreciate his or her help but have your own style.

Frequent change in supervisors Recognize that it is normal to be disappointed and upset at a change in supervisors. Remain open and assertive about your needs being met. Attempt to use your people skills to adapt as well as you can to the changes. If you feel the problem to be serious, you may want to talk to your school fieldwork instructor about the instability of your supervision.

The absent supervisor Try to deal with your concern as soon as you notice a potential problem. Be clear about your need for direct and regular supervision. Use assertiveness skills. If appropriate, use the leverage of your academic requirements to substantiate your requests and needs. Certain supervisors may not be able to provide the supervision you need; perhaps they will refer you to another person. However, if you have not mentioned your concerns, how can you expect them to know about your needs and frustrations?

The inadequate supervisor First of all, it would be most appropriate to share your concerns as they arise and be assertive about your needs not being met. This does not always work, so you may need to seek out others whom you can learn from as mentors and role models. This may also be discussed with your fieldwork instructor.

The critical or difficult supervisor Working with critical or difficult supervisors is a trying experience. Make an effort to understand where they may be coming from and recognize that they may be displacing some of their own frustrations onto you. Remaining assertive is crucial, but it is also important not to personalize the situation. Develop a support system where safe venting is acceptable and where you can receive validation.

The unethical supervisor Depending on the nature of the unethical behavior, you may have to report it or at least consult with others—fieldwork instructors, faculty members, or advisors. This is a difficult situation, especially for beginning human service workers, who may be surprised that unethical behavior does exist. Try to learn from the experience and contemplate how you might handle a similar situation differently.

The overloaded, overworked, or overwhelmed supervisor Try to be sensitive and supportive, yet remain assertive about having your needs met. You may ask if there are others who could help out with your supervision. Informally, you can seek out others who can provide appropriate feedback.

Dual relationships Make an effort to keep the line between supervisor and supervisee clear. Be aware of the complications and potential negative consequences of dual relationships. If you are still interested in becoming friends with your supervisor, wait until the supervisor/supervisee relationship has ended. This may not seem necessary, but in the long run it may be the most desirable position.

Maintaining a positive supervisory relationship is valuable during your fieldwork placement, and it can also help when it comes to exploring pathways to a career position. Networking could begin with your supervisor, since he or she may be the person who is most aware of your skills. Neukrug (2004) concurs: "There is nothing better than a good supervisory relationship that is based on trust, mutual respect, and understanding to help us take a good look at what we are doing" (p. 81).

■ PATHWAYS TO A HUMAN SERVICE POSITION

When you are in your field placement, you are in an excellent position to network. Developing positive relationships within your agency, and with community agencies and individuals you come into contact with, can be the bridge you need to help you cross over into the world of employment. Many who have become successful in their careers became involved in networking early on.

Successful networking involves using many resources—family, friends, professional colleagues, community contacts, and social and political organizations. Finding a job requires your active involvement, creativity, research skills, and above all your motivation and determination. We examine several areas where networking can be instrumental in obtaining a suitable placement or job.

Pursuing an Excellent Fieldwork Experience

Various authors emphasize the importance of taking an active part in the selection of an internship site or fieldwork placement (Baird, 2002; Collins, Thomlinson, & Grinnell, 1992; Royse, Dhooper, & Rompf, 1999; Faiver, Eisengart, & Colonna, 2004). Your role in this process may be major or minor, depending on your academic program and the availability of specialized agencies that best fit your needs. In any case, your active participation in the process will increase the likelihood that your placement will be rewarding. Faiver, Eisengart, and Colonna (2004, p. 2) identify four basic factors as essential to this selection process:

- Your own needs
- The field placement guidelines of your university's human service program
- The state requirements (if any) for on-the-job experience for the professional position you are seeking
- The opportunities for training and experience at the placement site

Fieldwork is a most valuable aspect of your overall education, and taking time to participate in the selection process can help make your experience the most meaningful possible. Those who delay or take this process lightly are often the ones who are most dissatisfied with their placements and eventually the most disillusioned with the helping process. Just as it is the client's responsibility to change, it is the student's responsibility to participate actively in the educational program.

Obtaining a Community Position

Through the years, we have marveled at how many of our students have obtained employment through their fieldwork experiences. These students were generally those who took their fieldwork seriously. We remember them as motivated, eager, and committed to learning. They were not reluctant to look at themselves and their personality patterns, and they weren't afraid of saying, "I don't know." These students were willing to ask for help when they needed it and to learn from their mistakes. In their placements, they demonstrated their dedication to the helping professions and treated their fieldwork as a job.

Some students who developed positive working relationships with staff members and community contacts while at their placement also readily found positions in the field upon graduation. This occurred either directly—from connections developed during their placement—or through recommendations from their supervisors.

Returning to Your Fieldwork Placement to Begin Your Career

If you take fieldwork seriously and maintain a professional stance, your chances of being considered for a potential job opening are greatly increased. An agency may prefer to hire an intern who is a known entity rather than an individual whom they know only through interviews. Also, hiring an intern reduces the need for costly training and orientation. Every semester some of the interns in our classes are hired by the agencies where they did their fieldwork.

SCENARIO

A student in a beginning practicum class successfully completed an internship at a university research center. He demonstrated dedication, honesty, and motivation. His keen ability in research, coupled with his willingness to work cooperatively with others, led the center to consider him for a position.

SCENARIO

A graduate student who interned at a child guidance agency was quite successful in developing several support and didactic groups within the organization. The clinical and administrative personnel were struck by her initiative and organizational skills. When the new budget was being formulated, they were able to create a tailor-made position for her to continue her work.

SCENARIO: Randy's Experience

I remember an internship experience that has greatly influenced my career. After I completed the program, I began applying for employment, with a certain amount of apprehension and anxiety. Although I experienced some discouragement at not immediately securing a position, I persisted. Through the connections that I had made with professionals at my internship, I learned of a new project at this agency. When I applied, I was pleased to find that I had made a positive impact as an intern. I was soon hired and was eventually able to tailor this position to fit my interests and clinical skills. This underscores the importance of treating your fieldwork experience with respect, remaining persistent in your pursuit of a job, and continuing to assert yourself.

SCENARIO: Lupe's Experience

In both my undergraduate and graduate internships, I was integrally involved in my learning and developed positive working relationships with my supervisors and colleagues. These experiences and contacts were instrumental in my future job options and led directly to actual employment. My first-year graduate internship led to my first job as a medical social worker upon completion of my graduate studies. There is no doubt in my mind that these opportunities would not have existed if not for my enthusiasm and commitment during my internship. My skills, willingness to learn, and work ethic clearly helped me be considered for job openings.

Obtaining a Position Through Fieldwork Experience and Contacts

Many students find that their experience at their placement has a positive impact on their employment prospects elsewhere—perhaps through a positive letter of recommendation, a favorable reference, or contacts made during work at the fieldwork agency. Do not underestimate the benefits of doing well in your current placement.

You are bound to reap the rewards of having been a responsible, dedicated, and motivated learner. Although not every placement will lead directly to a paid position, the positive connections you make can certainly benefit you.

SCENARIO

A graduate student developed a bereavement group for children at her first graduate internship at a family service agency. During her second year, when she was considering a topic for her master's thesis, she was encouraged to present her work to the local elementary schools. A school principal became very interested in her project and hired her to lead several groups at the school site. Additionally, the information she gathers from her group work will be the source of her research data.

Can you imagine ways in which your current placement can lead to a position, either at this same agency or in a different one? What steps might you take to achieve this?

CONCLUDING EXERCISES

1. If you were to wake up tommorrow morning magically transformed into the most assertive person you could be, how would your life be different? What would you do differently, from the very moment you woke up? Would you act any differently with those closest to you? With those you work with? In your fieldwork placement? Make a list of some of the ways you might change. Now pick at least one item on the list that you believe you could put into practice today.

2. Imagine that you are conducting a fieldwork seminar in your agency for students new to the agency and to fieldwork. How would you orient them? What would be your primary focus? What might you share about your own past experiences, both positive and negative, that could give them added insight into fieldwork? Provide an outline of these and other factors you feel are important to present at the first meeting of the seminar.

CHAPTER FOUR

The Helping Process

▨ AIM OF THE CHAPTER

As a human service provider, your role will involve helping others in some capacity. In your fieldwork, you may already have been involved in some form of helping. Inherent in this role is active participation in the helping process, the essence of human services. This chapter is dedicated to the helping process. We begin with definitions of helping and the characteristics needed to be an effective helper. We address the importance of establishing a helping relationship in which trust and rapport are foremost.

We then discuss the variety of approaches to helping, including "micro" practice with individuals, "mezzo" work with couples, families, and small groups, and "macro" practice in organizations, institutions, and entire communities.

In presenting effective helping approaches, we first focus on the TFA model, which emphasizes the role of thoughts, feelings, and action in human behavior. Also covered are the biopsychosocial approach to helping and the strengths perspective. We then review problem-solving models in an effort to determine common themes. Our own integrated approach to helping—the AGATE model—is elaborated next, followed by case applications that demonstrate the model's effectiveness in a variety of contexts. The chapter concludes by emphasizing the importance of developing your own style and approach to helping.

▨ HELPING

Definition and Dimensions

Helping is a term that has many individualized meanings, varying with a person's role in human services, educational background and experience, and personal motivations for becoming a helper. Helping is also interpreted differently in various subcultures. It can be seen as informal—family and friends assist each other—or as formal—where an individual seeks professional help to cope with life's problems. Along with Brammer (1988), we view it as the process whereby human service workers "help others to grow toward their personal goals and to strengthen their capacities for coping with life." The helping person or human service worker is designated as the *helper* and the client as the *helpee*. Other possible terms for this pair are *counselor–counselee, worker–client, therapist–patient, clinician–client, case manager–client,* and *interviewer–interviewee.* In any type of helping relationship, the importance of self-awareness, knowledge development, and skill attainment remains the same.

Dimensions of Helping

In this text, we will consider helping from a professional perspective, in which the helping process is formal. Hackney and Cormier (1988) consider the process of professional helping to have several dimensions: helping conditions, preconditions for helping, and results or outcomes.

Helping conditions Someone seeks help. Clients (helpees) reach out for help in some form. This describes the voluntary client; there also exist involuntary clients, who are referred for help without specifically requesting it. Involuntary clients may have been ordered to seek help by a court, physician, school, employer, or family members. Others may be indirectly asking for help by "acting out" behavior. Clearly, when clients seek help on their own, the process is likely to go more smoothly; there is less chance of resistance and usually greater motivation. This does not mean that involuntary clients cannot be helped—they too are capable of change.

Someone is willing to give help. Clearly, clients need to know that helpers are willing and motivated to provide services. When clients sense that helpers are uninterested or do not care, the success of the helping process is likely to be severely compromised. We will elaborate further on this in our discussion of essential helping characteristics.

Someone is capable of helping or trained to help. Professional helping is necessary to the success of the helping process. Formal training is crucial for anyone involved in the complexities of helping others.

The setting permits help to be given and received. The physical and emotional surroundings are important if worker and client are to be successful in the helping process. Ideally, each helper should have a personal office or space, comfortable and aesthetically pleasing, in which to work with clients. It would also be great if everyone in the environment were open and receptive to the helping process. In real life, these conditions are not always met. Many human service workers have to share space, work out in the community, or go to the client rather than vice versa. Although such surroundings may not always be conducive, helping can still take place. Also, realistically, not everyone in the client's environment may be thrilled with the helper's presence; some may, in fact, attempt to sabotage the treatment process. Nonetheless, effective intervention is possible; this may require more skill, willingness, and determination on the part of the worker.

THINKING THINGS THROUGH

Take a moment to assess your fieldwork agency and your own role to determine whether all these conditions for helping are met. Explain which are met and which are not.

Preconditions for Helping
Characteristics of helper and client. Clients bring to the helping process their history of coping skills, their problem-solving abilities, their view of themselves, their personality, and their interpersonal experiences. As pointed out by Hackney and Cormier (1988, p. 3), "[a] deficiency in any one of these areas can produce obstacles to the helping process." The degree of motivation and ability to change differ from client to client. Similarly, helpers bring a great deal to the helping process: their experience as helpers, their view of themselves, their professional demeanor, their interpersonal experiences and skills, and their self-awareness. Each of these contributes to the success or ineffectiveness of the process.

The clients' intentions in seeking help, the types of change they desire, and their level of motivation and commitment. Clients seek help for different reasons and have different levels of expectation of themselves and the helper. Some clients may seek help because they are forced to and are basically unmotivated. Others seek help in hopes that the helper will give them direction on how to change others in their lives or give them magic answers or "cures." Some clients may expect change overnight, whereas others may be well aware that change is a process. Certain clients may expect the helper to do all the work; they may struggle when asked to participate actively in the helping process.

Being able to assess these aspects early on will be crucial in developing appropriate interventions.

THINKING THINGS THROUGH

Consider a client you are currently working with or one you have worked with in the past. Assess the existence of these conditions. Where any aspect is not conducive to positive change, explore ways it can be altered for the better. Address the impact of the conditions on the client's overall success in the helping process.

Results or outcomes In some cases, clients feel that they have been greatly helped, while the helper may feel ineffective. Conversely, the client may believe the process has been ineffectual, yet the helper feels positive about the outcome. It would be great if every helping encounter pleased both parties and proved to be effective in achieving the desired goals. Most outcomes fall somewhere in between, but being aware, knowledgeable, and skilled in working with many different types of clients will certainly enhance your success rate. Certain helping characteristics are essential to the helping process; we discuss them in the next subsection.

THINKING THINGS THROUGH

Consider a case where you and a client have had differing views regarding the results or outcome of the process. How did you handle this? How did you feel? What might you do differently in the future?

Essential Helping Characteristics

In Chapter 1, we introduced characteristics of effective helpers, including empathy, respect and positive regard, genuineness, congruence, trustworthiness, acceptance, warmth, relationship building, being open/open-mindedness, immediacy, objectivity, self-awareness and self-understanding, competence, cognitive complexity, being an experiencer of life, concreteness, psychological health, and effective "personal qualities" of effective helpers. Here, we discuss additional characteristics and therapeutic factors that have been found to be essential to successful helping. Determining which characteristics are the most helpful is not a simple task because the helping process is an individualized endeavor that is experienced differently for each client–helper interaction. Each instance of helping is a unique process that cannot be repeated exactly. Feedback from clients and helpers has given human service academicians insight regarding the value of certain helping characteristics. For instance, Neukrug (1994, p. 96) states, "An effective human service worker has learned to integrate listening skills, empathetic responses, questioning, confronting, probing, advice giving, and offering alternatives, information giving, and self-esteem building to help problem solve for the client." Hackney and Cormier (1988) consider Carl Rogers's person-centered approach to be at the core of the helping process. They encourage helpers to develop skills in empathy, positive regard, genuineness, appropriate self-disclosure, and intentionality. Let's explore issues of appropriate self-disclosure and intentionality further.

Appropriate Self-Disclosure

In many instances, it is appropriate to engage in self-disclosure by expressing one's own thoughts, ideas, or feelings. It is an important way to let clients know that helpers are also real people, not just playing a role. Simply put, "[h]elper self-disclosure occurs when the helper relates personal information verbally or nonverbally to a client" (Young, 2001, p. 55). Two different types of self-disclosure identified in the literature are content self-disclosure, where the helper shares information about himself or herself; and process self-disclosure, which occurs when the helpers shares how he or she feels toward the client in the moment (Kleinke, 1994).

Clearly, it is essential that self-disclosure be appropriate rather than indiscriminate. Also, it is more successful with higher-functioning clients. "Self-disclosure needs to be done sparingly, at the right time, and only as a means for client growth, not to satisfy the helper's needs" (Neukrug, 2002, p. 78). A helpful way to determine the appropriateness of any self-disclosure is to ask whose needs it would meet—the client's or the helper's? If the helper's needs would be taking precedence, the self-disclosure could be more harmful than helpful.

Intentionality

This involves being able to generate alternative behaviors in a given situation and approach a problem from different vantage points. An intentional helper is flexible and continually strives to develop a broad range of skills and strategies.

Overall, these characteristics are ones you can continue to build on. Even the most experienced helper occasionally needs to be reminded of these essential helping skills.

THINKING THINGS THROUGH

Identify which of the preceding characteristics you currently possess. Make a game plan as to how you will work toward incorporating those needing further development into your helping relationships. Consider discussing this plan with your supervisor.

Establishing Helping Relationships

Our experiences have demonstrated time and time again that the stronger the helping relationship is, the more likely a positive outcome becomes. Combs (1982) highlights this concept in her suggestion of "self as an instrument" indicating that we, as helpers, are often the most important tool in our work with clients. Brammer (1988, p. 26) agrees: "Our principal tool is ourselves acting spontaneously in response to the rapidly changing inter-personal demands of the helping relationship."

Brill and Levine (2002, p. 89) consider the helping relationship to be neither personal nor a friendship, but rather "transactional in nature and based on mutual trust and respect. . . . [Positive relationships are fostered by a worker who] makes conscious use of self in establishing a specialized kind of goal-directed connection with another person." Many helpers are naturally adept at beginning a positive interaction with their clients, but the test of an effective helper is whether he or she can remain actively engaged in learning and successfully apply the necessary skills to specific situations. Brill and Levine (2002, pp. 114–120) summarize the necessary characteristics of helping relationships:

> *Accepting:* "Acceptance of the individual's right to existence, importance, and value."
> *Dynamic:* Active participation by both worker and client.
> *Emotional:* "The essence of relationships is emotional rather than intellectual."
> *Purposeful, time-limited, and unequal:* "By definition, a helping relationship is purposeful and goal directed" and not equal because "they are not friendships."
> *Honest, realistic, responsible, and safe:* "In order to be effective and ethical, a helping relationship must be honest, responsible, and safe."
> *Authoritative in appropriate ways:* Two kinds of authority are appropriate in helping relationships: the authority of knowledge and the authority of social sanction.

THINKING THINGS THROUGH

Examine the characteristics just listed in an effort to determine which ones you possess and which you need to develop further. Consider a particular helping relationship at your fieldwork placement and explore ways to improve it by developing these characteristics.

The Importance of Rapport and Trust

The development of rapport is essential for a helping relationship to be effective. Words such as understanding, compatibility, and harmony have been used to describe rapport. "Rapport refers to the psychological climate that emerges from the interpersonal contact between you and your client" (Hackney & Cormier, 2005, p. 25). In fact, such factors as respect, trust, psychological comfort, and mutual purpose tend to develop when positive rapport is established with clients.

Long (1996, p. 81) characterizes rapport as an interactional process that contains three components: (1) "a perceived warmth and caring"; (2) "a perceived safety and acceptance"; and (3) "a perceived trustworthiness and credibility." Trust is thus an ingredient of rapport.

Trust is an affective experience and a desired outcome within the relationship. It grows from rapport and is built on acknowledged cognitive beliefs, particularly the belief that individuals have the right to be themselves and to have their own feelings, thoughts, and actions (Long, 1996, p. 81).

We believe that the development of trust is the key factor in the helping process. It is the underlying force that enables helpers and clients to negotiate additional stages of helping. Trust may be continually tested by the client, so the helper may need to reestablish it continually. Difficulty in trusting may be due to the client's personality, developmental history, or current situation. Also, many clients' trust in the helper may waiver when faced with new challenges that are painful or frightening. Helpers should not only provide reassurance but also be selective in their interventions to match the difficulty of the intervention with the client's level of trust.

Without trust, a client will seldom risk opening up and enlisting help from professionals. Clients differ in their level of initial trust of the helper. For some to develop trust, the worker must show that he or she is reliable and dependable; start and end sessions on time; make certain to pay full attention to the client; honor confidentiality; refrain from talking about the problems of others; be supportive, accepting, and nonjudgmental; respect clients for who they are; and maintain professional boundaries. Other clients freely trust helpers until they detect a reason not to. Maintaining trust requires that helpers remain responsible, committed, and ethical in their practices.

SCENARIO: Limited Rapport and Trust

A male client who has had negative experiences with helpers in the past becomes upset when a new helper is consistently late, does not follow through with phone calls, and has very loose boundaries. The client feels uncomfortable, and little rapport or trust develops in the helping relationship. This creates an environment of distrust and resistance.

SCENARIO: Appropriate Level of Rapport and Trust

In the same situation just cited, suppose that the helper is very sensitive to the client's past negative experiences and attempts to build rapport and trust by demonstrating her commitment through responsible, ethical behavior. As the client sees a continuous pattern in the helper's professionalism, he begins to accept the idea that not all helpers are ineffective or uncaring. He becomes more hopeful and trusting.

THINKING THINGS THROUGH

Consider cases of yours where there has been either minimal trust or rapport or an appropriate level. Why? Did you contribute to the client's reactions and overall level of rapport and trust? If so, how?

The Process

Helping is not an isolated event; it is a process that begins at your first interaction with the client and continues long after. Although many of you will interact with a client on only one occasion, helping may continue well beyond this encounter. What is learned in the helping process often carries through to clients' daily lives. Helpers may plant seeds that later flourish and develop. The helping process evolves over time and may consist of phases—beginning, middle, and ending. Later, when we present an overview of problem-solving models, we will elaborate on the stages of helping. Here, we want to emphasize the need to approach the helping process in a patient and committed manner.

THINKING THINGS THROUGH

Describe your experience with the process aspect of helping. Were you surprised at the ongoing, continuous nature of helping? Or have you experienced situations where the process was time limited?

Therapeutic Factors

In addition to possessing many of the essential helping characteristics we have identified thus far, successful helping occurs when specific *therapeutic factors* are present. Also termed *common curative factors*, therapeutic factors are the activities common to all effective helpers (Young, 2001, p. 30). As identified by Frank (1991), six therapeutic factors cut across all theoretical perspectives:

1. Maintaining a strong helper/client relationship
2. Increasing the client's motivation and expectation of help
3. Enhancing the client's sense of mastery or self-efficacy
4. Providing new learning experiences
5. Raising emotional arousal (or lowering through relaxation)
6. Providing opportunities to practice new behaviors

Keeping these factors in mind when working with clients will no doubt be helpful and increase the likelihood of a productive outcome.

◼ APPROACHES TO WORKING IN HUMAN SERVICES

Working in human services will expose you to many types of client-helper interactions. Many approaches to helping can be employed at various practice levels: "micro," "mezzo," and "macro" practice. Furthermore, "[K]nowledge of and skills in practice methods vary according to the level of the client system served by practitioners" (Hepworth, Rooney, & Larsen, 2002, p. 14).

"Micro" Practice

Micro practice involves working with individuals on a one-to-one basis. According to Barker (1991, p. 144), a micro orientation to practice involves "an emphasis on the individual client's psychological conflicts and on the enhancement of technical skills for use in efficient treatment of these problems." This involves working with individuals to help them resolve conflicts and enhance their personal functioning. Individual work with clients can take many different forms, from intake, to information and referral and case management services, to individual counseling or psychotherapy. Crisis intervention, as well as brief and short-term treatments are quite prevalent, but long-term treatments also continue to be provided in some settings. The opportunity to work with individuals in a variety of helping capacities is part of the beauty of human services. Ideally, services are tailored to the individual needs of the client, but this is not always possible in today's world of budget cuts and program reductions. Nonetheless, we can strive to meet the needs of our clients in whatever manner is most appropriate.

To be effective in working with individuals, one must understand human nature, be diversity sensitive, and be knowledgeable about the range of appropriate interventions.

Designated as direct practice, micro-level interventions typically involve delivering services to clients in face-to-face contact (Hepworth, Rooney, & Larsen, 2002).

SCENARIO: Micro Practice

A social worker, counselor, or case manager works with grieving clients on a one-to-one basis to help them cope with the loss of a loved one to AIDS and also refers clients to appropriate support groups and community services.

"Mezzo" Practice

Mezzo practice involves working with "small groups, including family work groups, and other social groups" (Zastrow & Kirst-Ashman, 1994, p. 16). This is quite common because working with individuals often entails dealing with various systems they may interact with, such as family, school, work, social authorities, the judicial system, and the medical establishment. Mezzo intervention is usually designated to change the systems that directly affect clients, such as family, peer groups, classrooms, and work environment (Hepworth, Rooney, & Larsen, 2002). Mezzo practice therefore consists of two basic areas: working with families and group work.

Working with Families

Here it is appropriate to apply a systems approach that takes into account family norms, structure, and life cycle. The amount of work with families is growing, due to the increase in single parenthood, divorce, blended families, and households where both parents work. Family assessment is important and can take various forms: family systems assessment, cultural family assessment, and family functioning assessment. Treatment orientations include couples/marital therapy, family therapy, multifamily therapy, and family support groups. Often, work with a family begins because one family member is struggling with issues such as parenting difficulties, marital problems, chemical addiction, eating disorders, legal problems, and psychiatric hospitalization. To work with such problems successfully, family involvement is necessary.

SCENARIO: "Mezzo" Practice with Families

A helper in a chemical dependency program conducts family groups. He works on educating the family about the disease aspects of substance use, the triggers, and family members' possible roles as enablers for the addicted person.

Group Work

This can occur either formally or informally. Either way, one must understand the nature and process of groups and have group leadership skills. In formal group work, it is important to determine the purpose of the group, understand group development and stages, and have effective group skills.

Group work in agency settings requires careful planning and implementation. One must be able to determine the need for certain types of groups and to enlist the support of the administration and staff for these groups. Appropriate evaluation and follow-up are vital.

SCENARIO: "Mezzo" Practice with Groups

The number of specialized groups is increasing because of pressure for short-term, cost-effective treatment. Groups are often age related—children, adolescent, or adult groups, for example. Some groups are gender based—for example, women's and men's support groups. Groups are often formed to deal with a certain problem or concern: loss, domestic violence, incest, eating disorders, drug diversion, or divorce adjustment. Many groups are also psychoeducational in nature, with a teaching emphasis: parenting, nutrition, assertiveness training, anger management, and stress reduction.

"Macro" Practice

Macro practice involves working with systems larger than small groups. It often aims at broad-based changes in organizations, institutions, communities, and society at large. Frequently, the emphasis is on seeking changes in laws, regulations, and social policies. According to Barker (1991, p. 136), a macro orientation involves "an emphasis on the sociopolitical, historical, economic, and environmental forces that influence the overall human condition, causing problems for individuals or providing opportunities for their fulfillment and equality . . . bringing about changes in the general society." Hepworth, Rooney, and Larsen (2002, p. 15) also stress that at this level, helpers "serve as change agents who assist community action systems composed of individuals, groups, or organizations to deal with social problems."

In work with organizations or institutions, helpers must understand the nature of the organization and be able to determine the need for change. Then planning and effecting change at an organizational level can take place. Similarly, when working with the community as a whole, one must understand the community and the perceived need for change. Various models for promoting community change have been employed by macro practitioners, including social action, social planning, and locality development. Locality development involves "broad participation of a wide spectrum of people at the local community level in goal determination and action." According to Homan "this process emphasizes economic and social progress. It is intended that "a wide range of community [are] involved in determining their 'felt' needs and solving their own problems—this is a model that has a self-help orientation to it. Administrative policy must also be understood if community change is to be achieved. Macro practice among minority groups is common. Although the focus may be on a large-scale system, assessment, problem-solving, and interpersonal skills are just as necessary as in micro and mezzo practice.

SCENARIO: "Macro" Practice with Communities and Organizations

Human service professionals work as community planners with senior citizens. Their role is to plan and implement senior services within the community, in which the number of seniors is rising.

THINKING THINGS THROUGH

Consider your experience. Which types of practice have you been involved in? Is there an area you feel you need more experience in? If so, consider interviewing helpers who do that work to learn more about aspects of helping at this level.

▒ EFFECTIVE HELPING APPROACHES

As you evolve professionally, each of you will ultimately develop your own style of helping. For many of you, this process has already begun in your fieldwork experience. A prerequisite for developing a personal approach or style is to understand the following factors:

> The interaction among thoughts, feelings, and actions (the TFA model)
> The biopsychosocial approach
> The strengths perspective

The TFA Model

The TFA approach to understanding human behavior focuses on thoughts, feelings, and actions. These are paramount in human services, where we ourselves constitute the most important tool in helping others. Hutchins and Cole (1992, p. 4) maintain that "to be an effective helper, you must first understand major aspects of your own behavior and how you interact with others. 'Know thyself' is a critical admonition that is as important today as it was in the time of Plato and Socrates." Understanding one's behavior is linked to recognizing the role of one's thoughts, feelings, and actions. This is a crucial aspect of the TFA model, which was introduced by Hutchins (1979). Its key points follow (Hutchins & Cole, 1992, p. 16):

- Knowledge of self is essential for working effectively with others.
- Behavior is the interaction of thoughts (T), feelings (F), and actions (A).
- Behavior is dimensional, not static. It may change in different situations.
- Assessing behavior in a specific situation provides useful clues to understanding self and working with others in a helping relationship.

It is important to be congruent in one's presentation; helpers who are positive role models live what they believe in.

SCENARIO: Incongruent Helper

A helper states that she is unbiased about homosexuality and can work with this population. However, when actually attempting to help a lesbian client, the helper tends to focus on sexual orientation rather than the issues for which the client is seeking help.
    Is there congruence or incongruence? How might the client feel?

SCENARIO: Congruent Helper

Another helper is open from the outset about his lack of experience in working with clients who are gay. He takes this into account in choosing his approach, and he tries to learn more about this population from clients and through additional training.
    Is this helper more congruent? How might his clients feel?

In addition to focusing on our own behavior, we want to help clients work on changing theirs. Helping them recognize how one aspect of their behavior (say, their feelings) may affect other aspects, such as their thoughts and actions, can help them determine what changes are necessary.

SCENARIO: Interaction Among T, F, and A

Consider a client who is very negative in his thinking. Wouldn't he be likely to feel depressed and act in ways that reflect his pessimistic outlook on life?

The TFA model integrates three major approaches in the helping professions. Hutchins (1979) outlines these three approaches as follows.

The Cognitive or Rational Approach, Influenced by:
Ellis (rational-emotive therapy)
Meichenbaum (cognitive modification)
Other cognitive therapies

The Affective or Humanistic Approach, Influenced by:
Rogers (client-centered or person-centered therapy)
Perls (Gestalt therapy)
Maslow's hierarchy of needs
Phenomenological, humanistic, and existential writers

The Behavioral Approach, Influenced by:
Skinner and Bandura (behavior modification)
Wolpe and Lazarus (behavior therapy)
Krumboltz and Thorensen (behavioral counseling)
Other behavioral sciences

Many of these pioneers have applied the TFA model in some form. According to Smith (1982) and Ward (1983), the integration of these various approaches is the most noteworthy trend in the field of helping.

SCENARIO: The TFA Model at Work

In working with a child who thinks he is not wanted, feels angry and hurt, and acts out by becoming aggressive toward others, it may be helpful to encourage him to identify and express the underlying feelings of fear and hurt. These feelings have probably not been recognized or verbalized before. If the feelings are expressed, the child may not resort to acting them out, and he may think more positively of himself.

How would you proceed?

THINKING THINGS THROUGH

Can you think of an example from your own experience where an individual's thinking has affected how he or she feels and behaves? How feelings affect thinking and actions? How actions affect feelings and thoughts? Give an example of each.

Used as a foundation for working with people, the TFA model still allows for your own individual approach. Recognizing the unique aspects of thoughts, feelings, and actions can also be a helpful approach to work with others in and of itself.

The Biopsychosocial Approach

Zastrow (1995, p. 24) highlights the importance of understanding clients through an ecological approach that considers all systems within the social environment. This approach emphasizes transactions between people and their physical and social environments. It views human beings as developing and adapting through transactions with all elements of their environments. An ecological model gives attention to both internal and external factors. It does not view people as passive reactors to their environments, but rather as being involved in dynamic and reciprocal interactions with them. According to Lum (2004, p. 90), "[e]cological theory focuses on the reciprocal relationship of person and environment. The emphasis is on the consequences of exchanges between the two entities and the changes or modifications that are made as a result."

This approach takes into consideration all aspects of an individual's life to understand clients within their environment. Three major areas (biological, psychological, and societal) are assessed; when they are viewed together, a complete picture emerges. Subsystems include individual (biophysical, cognitive, emotional, behavioral, motivational); interpersonal (parent-child, marital, family, kin, friends, neighbors, cultural reference groups, spiritual belief systems, and others in people's social networks); organizations, institutions, and communities; and the physical environment (housing, neighborhood environment, buildings, other artificial creations, water, and weather and climate) (Hepworth, Rooney, & Larsen, 2002, p. 19).

Biological Systems

These include all the physiological processes necessary for the overall functioning of an individual. "Sensory capacities, motor responses, and the workings of the respiratory, endocrine, and circulatory systems. . . . They develop and change as a consequence of genetically guided maturation, environmental resources such as nutrition and sunlight, exposure to environmental toxins, encounters with accidents and diseases, and lifestyle patterns of behavior including daily exercise, eating, sleeping, and the use of drugs" (Newman & Newman, 2003, p. 6).

Psychological or Emotional Systems

These include "those mental processes central to the person's ability to make meaning of experiences and take action. Emotion, memory, and perception, our problem-solving, language, and symbolic abilities, and our orientation the to the future—all require the use of psychological processes thinking, reasoning, and making meaning of experiences. . . . [This system] provides the resources for processing information and navigating reality" (Newman & Newman, 2003, p. 6). As in the case of biological aspects, individuals also change psychologically or emotionally as a result of maturation and positive and negative experiences.

Societal Systems

These are defined as "those processes through which a person becomes integrated into society. Societal influences include social roles; social support; culture including rituals, myths, and social expectations; leadership styles; communication patterns, family organization; ethnic and subcultural influences; political and religious ideologies; patterns of economic prosperity or poverty and war or peace; and exposure to racism, sexism, and other forms of discrimination, intolerance, or intergroup hostility" (Newman & Newman, 2003, p. 7). As with biological and psychological systems, societal systems change over the life span as a result of changing roles and relocation into new environments.

SCENARIO

A client's chronic lower back pain affects her emotionally. She feels depressed and becomes increasingly withdrawn and isolated socially.

THINKING THINGS THROUGH

Take the three areas in the biopsychosocial approach and apply them to a client situation, present or past. How did you do? Did you learn any new information about the client? What areas need further exploration?

In the various realms of a person's life, there is bound to be overlap, and certain aspects may prove to be irrelevant or inconsequential. Most often, however, factors in each of the spheres directly or indirectly affect the client and have a significant bearing on the development of the helping relationship and the outcome of the helping process. A biopsychosocial approach is desirable when concern for the client system is of utmost importance. Failure to evaluate the impact of any of the systems can lead to inappropriate interventions and misunderstanding the client.

The biopsychosocial perspective is most commonly incorporated into actual practice through a psychosocial evaluation. This entails exploring many facets of the individual's life, both present and past. Newman and Newman (2003, p. 9) state, "The psychosocial approach seeks to understand the internal experiences that are the products of interactions among biological, psychological, and societal processes. Changes in one of the three systems generally brings about changes it the others." There is continuous interaction of the individual and the social environment. Some of you may be applying a biopsychosocial approach in your fieldwork, and others may apply only certain aspects, depending on the agency setting and your helping role. In any case, grasping the basics of this approach is important to understanding your clients more completely.

A biopsychosocial approach can be quite effective because it is thorough and comprehensive. Faulty perceptions and gross misinterpretations can result if a client's problem is not considered within its overall context. Also, a psychosocial evaluation is an opportunity to establish the important initial bonding between client and helper, which lies at the very foundation of successful practice.

An integral aspect of the biopsychosocial approach is its emphasis on client strengths—known as *the strengths perspective.*

The Strengths Perspective

The strengths perspective, first proposed by Weick, Rapp, Sullivan, and Kisthardt (1989), assumes that helping can proceed successfully from the identification and enhancement of individual strengths. When maximized, these facilitate the helping process and foster an individual's interaction with the community. In this alternative approach to helping, the emphasis is on clients' strengths and their ability to use them, together with resources from family, friends, and the community, to overcome obstacles and proceed on a positive path. Saleeby (1997) has written extensively on the strengths perspective and has outlined several assumptions and principles of strengths-based practice with at-risk populations. DeJong and Miller (1995, p. 729) describe them as follows.

Assumptions of the Strengths Perspective

- "Despite life's problems, all people and environments possess strengths that can be marshaled to improve the quality of clients' lives. Practitioners should respect these strengths and the directions in which clients wish to apply them."
- "Client motivation is fostered by a consistent emphasis on strengths as the client defines these."
- "Discovering strengths requires a process of cooperative exploration between clients and workers; 'expert' practitioners do not have the last word on what clients need."
- "Focusing on strengths turns the practitioner's attention away from the temptation to 'blame the victim' and toward discovering how clients have managed to survive even in the most inhospitable of circumstances."
- "All environments—even the most bleak—contain resources."

Others have substantiated these assumptions. According to Weick, Rapp, Sullivan, and Kisthardt (1989, p. 352), "[a]ll people possess a wide range of talents, abilities, capacities, skills, resources, and aspirations." No matter how little or how much may be expressed at one time, a belief in human potential is tied to the notion that people have untapped, undetermined reservoirs of mental, physical, emotional, social, and spiritual abilities that can be expressed.

Focusing on people's strengths rather than their limitations emphasizes "the already realized capacities of an individual," which encourages continued growth and enhances their confidence and self-esteem.

In addition to these assumptions, Saleeby (1992, 1997) identifies six key concepts related to the worker–client relationship:

Empowerment The belief that individuals can make their own decisions and that they possess the strengths to resolve their own conflicts increases their ability to grow and to contribute to society.

Membership Helpers must foster a sense of membership in alienated clients. Saleeby (1992) states that "[c]ertain things are required of [helpers] at the outset" (p. 9). DeJong and Miller (1992) define these requirements as:

- Working collaboratively with clients
- Affirming client perceptions and stories
- Recognizing the survival efforts and successes of clients
- Fostering links to contexts where client strengths can flourish

Regeneration and healing from within As the strengths perspective helps clients define their own goals and capabilities, they can discover their own resources for better lives.

Synergy As described by DeJong and Miller (1992, p. 734), "[a] synergic relationship is one in which the participants, by virtue of their interaction, are able to create a larger, more beneficial result than either could have created alone using individual resources."

Dialogue and communication Affirmation of the "otherness" of the client is essential, as is collaborating with clients rather than acting as the "expert."

Suspension of disbelief Respecting the client's perceptions and statements rather than distrusting them can enhance acceptance and affirmation by the helper. It can serve to improve clients' view of themselves, strengthen the client–helper relationship, and increase the likelihood that productive work will continue.

Assessment of Client's Strengths

Traditionally, assessment has focused on diagnosing the pathology, shortcomings, and dysfunctions of clients. Many feel this is a holdover from the Freudian view, based on a medical model and geared to identify illness or pathology but having very few ways to identify strengths. Weick et al. (1989, p. 353) state:

> A strengths assessment is necessary to practice according to the strengths perspective. This assessment focuses exclusively on the client's capacity and aspirations in all life domains. In making this assessment, both the client and worker seek to discover the individual and communal resources from which the client can draw in shaping an agenda.

Hepworth, Larsen, and Rooney (2002, p. 191) also highlight the importance of assessing the client's strengths. They believe that concentrating on clients' deficits can impede a worker's "ability to discern clients' potential for growth," reinforce self-doubts and feelings of inadequacy in clients, and induce workers to provide services longer than is necessary.

Cowger (1992, pp. 139–147) proposes 12 practice guidelines that foster a strengths perspective.

- Give preeminence to the client's understanding of the facts.
- Believe the client.
- Discover what the client wants.
- Move the assessment toward personal and environmental strengths.
- Make assessment of strengths multidimensional.
- Use the assessment to discover uniqueness.
- Use language the client can understand.
- Make assessment a joint activity between worker and client.
- Reach a mutual agreement on the assessment.

- Avoid blame and blaming.
- Avoid cause-and-effect thinking.
- Assess; do not diagnose.

Interviewing for client's strengths during assessment In addition to these guidelines, solution-focused interviewing (DeJong & Miller, 1995) can help identify client strengths.

Exception-finding questions are used to discover clients' present and past successes in relation to their goals.

SCENARIO

In working with a client who struggles with being too angry at and critical of her daughter, ask, "Has there ever been a time when you were not critical and you felt supportive and close to your daughter?"

Scaling questions are clever ways to make complex features of a client's life more concrete and accessible for both the client and the worker. With each progressive step, the client is positively reinforced for existing strengths.

SCENARIO

In working with a client who is experiencing marital difficulties, one way to assess exactly where she is at is to ask, "On a scale from one to ten, tell me where you are today in regard to your marriage." The rating one indicates the worst (i.e., wanting a divorce) and ten the best (i.e., as if on a honeymoon again).

Coping questions help clients recognize how they have been able to cope with overwhelming circumstances and feelings. Be sensitive to clients' need to talk about their problems.

SCENARIO

You are working with a client who has just sustained a major loss. It is important to ask how she has dealt with loss before and to point out instances of successful coping, to help her recognize her abilities. If past coping was ineffective, help the client recognize what she learned not to do. If there has been no previous experience with loss, reassure the client: Since she has no experience with loss, it is no wonder that she is having difficulty. In all cases, provide support and reassurance and instill hope.

Use of the Strengths Perspective in Goal Setting and Intervention
Once a client's strengths are assessed, the worker proceeds to the next phase: goal setting that leads to intervention. Well-formed goals are essential to effective practice. Goals must be negotiated between helper and client; by working together, they define achievable goals within the client's frame of reference. Berg and Miller (1992, p. 731) have identified seven characteristics of well-formed goals. Such well-formed goals:

- Are important to the client
- Are small
- Are concrete, specific, and behavioral
- Seek presence rather than absence
- Have beginnings rather than endings (e.g., visualize first steps toward a happy marriage)
- Are realistic in the context of the client's life
- Are perceived by the client as involving hard work

In interviewing for well-formed goals, workers must "listen to clients' concerns, assess to determine that there is not an emergency, and then turn the focus on developing goals" (DeJong & Miller, 1995, p. 731). Using the "miracle question" is a good way to begin this process.

SCENARIO

Suppose that while you are sleeping tonight, a miracle happens. The miracle is that the problem that has you here talking to me is somehow solved. Only you don't know that because you are asleep. What will you notice tomorrow morning that will tell you that a miracle has happened? (DeJong & Miller, 1995, p. 731).

Once clients can imagine a future in which the problem is solved, specific goals can be developed to achieve this in reality. It is important that workers and clients recognize that well-formed goals are the presence of something rather than the absence.

SCENARIO

If a client responds to the miracle question in the previous scenario by stating, "I would be less irritable or depressed," the worker might respond, "What might there be instead of the irritability and depression?"

In conclusion, we believe the strengths perspective has many valuable aspects that can be applied to a wide range of practice settings. Applying it can help clients to "reclaim a measure of personal power in their lives" (Weick et al., 1989, p. 354).

THINKING THINGS THROUGH

As a way to practice using the strengths perspective, assess the strengths that exist within yourself. Next, find ways to focus more on your strengths in your daily life. Can you offer some examples?

◼ AN OVERVIEW OF HELPING MODELS

Fundamental to professional helping is the ability to assist clients in coping with their difficulties and developing healthy problem solving skills. Along with establishing helping relationships, helpers often use specific interventions and techniques to help clients achieve their goals. Interventions, in this context, are "deliberate, planned actions undertaken by the client and the worker to resolve a problem" (Compton & Galaway, 1984, p. 11). Often, resolving problems involves change. The desire to change must come from within the client, but helpers may be called upon to help facilitate change. We consider this change process analogous to the problem-solving or helping process.

Understanding the basic premises of the problem-solving process is therefore essential to any helper attempting to work with people. Brill and Levine (2002, p. 126) consider the problem-solving process to be an adaptation of the scientific method, a "disciplined, orderly framework for worker's thinking and an overall pattern that can be learned and used to deal with problems. . . ." Problem-solving becomes a constant ongoing process used until termination.

We concur with Brill and Levine's contention that this process is basically a working structure for the helper, yet clients must become involved in each step as much as possible. In essence, "this way of working should be a growth experience for clients; only as they are involved in it will they grow."

Compton and Gallaway (1984) identify several prerequisites for this process to be effective:

- A working relationship must be established between client and helper.
- Helpers must adhere to the values of their profession.

- The helping process must be guided by appropriate knowledge and skill.
- Helpers must be aware that although the effectiveness of each stage depends on the successful completion of the preceding phase, the process will often not be linearly sequential, so flexibility and acceptance are necessary.

A myriad of problem-solving approaches are in use today. In fact, there are even specific approaches within the various disciplines of social work, human services, counseling, and counseling psychology. Although there are differences within each model, all share some similarities. For one, all models feature a beginning, a middle, and an ending. In most approaches, relationship building is the first step followed by problem assessment, goal setting, interventions, and finally evaluation and termination. Despite these similarities, there is some variation regarding whether the helper is considered primarily responsible for the change process or there is shared responsibility between helper and client. Here we will highlight the AGATE model, which is an integrated approach we have developed from our experience and review of the many models used today. In the AGATE model, we also focus on stages that are sequential and reflect a process from beginning to end. However, our emphasis during the initial phase is somewhat different. Our focus on client *concerns,* rather than problems, sets a more positive tone. Also, we highlight client strengths during assessment and throughout the process, and we advocate a mutual relationship wherein helper and client decide together on the direction of their work. We encourage respect for cultural diversity throughout. Some aspects of the TFA approach are incorporated, as well as an ecological viewpoint common to the biopsychosocial approach.

■ AN INTEGRATED APPROACH: THE AGATE MODEL

The AGATE model is a comprehensive approach that highlights client strengths and has broad applicability in most agency settings. The model is composed of five stages (see Figure 4.1):

Assessment of client's strengths and concerns
Goal setting: reorienting, shaping, and prioritizing
Action: planning and intervention
Termination: evaluation, review, and ending
Examination of process and self: follow-up

In the last stage of this process, helpers reflect on the previous stages in preparation for making any changes in their style that are necessary for their next client assessment.

The value of this model is that it accommodates change in the micro, mezzo, and macro levels of practice. Because roles and functions of human service professionals vary so widely, there is a need for a model that allows for many different types of interventions and involvement. The AGATE model grew out of the need for an approach that could have relevance for an array of helpers—social workers, case managers, counselors, psychologists, and general human service practitioners. As noted above, in developing the model, we considered the most valuable aspects of problem-solving models, the TFA approach, the biopsychosocial view, and the strengths perspective, as well as the need to be sensitive to cultural diversity.

Assessment of Strengths and Concerns

Developing a Helping Relationship
The foundation of the helping process is the development of a positive working relationship between client and helper, where rapport, trust, and effective communication are foremost. This relationship building begins at the start of treatment and remains important throughout the client-helper interaction. The three basic communication components of listening, responding, and expressing are all-important in promoting a positive

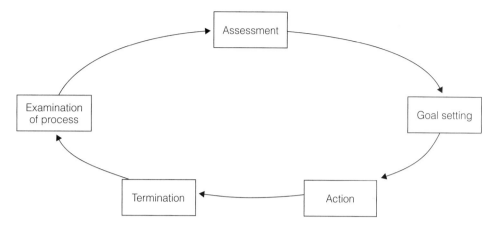

Figure 4.1
The AGATE Model

helping relationship. We also regard as indispensable such characteristics as accurate empathy, positive regard, genuineness, appropriate self-disclosure, and intentionality.

Sperry, Carlson, and Kjos (2003, p. 23) agree that "engagement is essential for therapeutic change to occur. . . . The client's level of engagement can be significantly increased by the therapist [helper], and by the nature and quality of the therapeutic relationship process."

Structuring the Interview

During this initial stage, the helper and the client should discuss some basic issues related to the helping relationship. Part of this is informed consent, which we will elaborate further when we deal with ethical and legal issues. Key issues are the limits of confidentiality and of the relationship, the purpose and length of the interview, the worker's qualifications and orientation to helping, the client's and the helper's expectations, the agency's fees for service and its rules, and any other concerns the client may have. Special attention must be paid to the setting where helping takes place and the atmosphere of the first few minutes of the interview because both help set the tone for the rest of the helping process. Remember that this first encounter can give the helper pertinent information regarding the client's motivation and capabilities, which can provide clues about prognosis. Similarly, the client can also size up the helper to judge whether there is an appropriate fit.

Neukrug (2002, p. 44) agrees that there are "a number of technical issues that are crucial to the development of an effective working relationship. . . . The helper should be concerned with assuring that the physical environment feels safe, offering a professional disclosure, and obtaining informed consent."

Determining Areas of Concern

Once these initial conditions are met, the helper's focus turns toward identifying the client's concerns. Hackney and Cormier (2005, p. 27) state, "Even as you and your client are in the process of establishing a relationship, a second process is under way. That process involves the collection and classification of information related to the client's life situation and reasons for seeking counseling." The worker asks what led the client to seek help, emphasizing the positive nature of seeking professional guidance. To ensure accuracy, the helper should tell the client how his or her concerns are being interpreted. If the client indicates that the helper has misunderstood, the helper asks for further elaboration and attempts to redefine the concerns until there is mutual agreement.

We refrain from using the term *problem* as much as possible because it tends to communicate that something is "wrong" with the client. We focus on assessing clients' concerns; stressing that it is good that they seek help and are willing to be active participants in the helping process; and emphasizing their current coping skills, past successes, and strengths. This does not mean paying no attention to their concerns. We also make a concentrated effort to define these and work toward goals that will alleviate them. In this regard, the AGATE model deviates slightly from the strengths perspective in that we do not focus exclusively on client capabilities. Our approach embodies a dual focus: Client concerns are identified and explored, and strengths are also highlighted.

Client Strengths

Human service workers must include client strengths in the assessment process. To put these strengths to good use, workers must first identify them. This not only fosters the development of a positive relationship but also enhances the client's confidence and self-esteem. Corey and Corey (2003, p. 120) stress that "to create an effective helping relationship, it is essential for the helper to assist clients in becoming aware of their assets and strengths rather than concentrating on their problems, deficits, and liabilities."

We also strive to acknowledge clients' diverse backgrounds and remain sensitive to possible obstacles related to cultural views regarding seeking professional help. Here are typical questions and statements to use in assessing strengths:

1. What do you most like about yourself?
2. Can you list the positive qualities you feel you possess?
3. What types of compliments do others give you?
4. Ask others in your life to give you feedback about your strengths.
5. Describe healthy coping skills you have used before.
6. If asked to describe yourself by focusing on the positives, what would you say about yourself?

Client's History

It is also vital to learn as much as possible about the history of clients' concerns and their life as a whole, both present and past. Assessment begins in the initial phase of treatment. In fact, "[m]uch of the assessment stage is inseparable from the relationship-building stage because helpers are observing their clients and collecting information from the moment that they first meet" (Young, 2001, p. 36). Gathering information about the client will help you form a sound impression of him or her and develop an appropriate plan of intervention or treatment. A working hypothesis about a client's difficulties and resources is based on the current data obtained. When the biopsychosocial approach is applied to this assessment process, the benefits are even greater because of the comprehensive picture that it affords. We believe that assessment must always include this person-in-environment view.

The selection of goals and interventions hinges primarily on the assessment. If you do a thorough assessment, it will help you refrain from reaching a premature hypothesis regarding the client's situation; you gain the opportunity to discover who he or she really is. Unless you obtain the necessary information about the presenting concern—its duration, precipitating factors, and past history—we believe that you will not truly understand the client or recognize the underlying dynamics. The result is likely to be an inaccurate or incomplete assessment, often leading to inappropriate goals and interventions, which will ultimately hinder positive changes in the client. On the other hand, a thorough assessment helps you understand the client, no matter what type of agency you work in or what role you play.

Assessments conducted at different types of agencies vary in nature. In some settings, helpers make an independent assessment; in others, they are part of a team working together with the client. Regardless, we encourage attempting to learn as much as you can about each client's history—interpersonal, family, social, educational, medical, drug or alcohol use, and psychiatric. Only certain information may be available or relevant. But in most cases, the better you know your clients, the better you can serve them. Gathering background data enables the helper to determine whether the services available are appropriate for a particular client and serves to ensure that the helper is not missing a serious condition (Young, 2001).

Sources of Information

Zastrow (1995) and Young (2001) discuss ways to achieve a complete and professional assessment. They concur that information used in making an assessment is derived from various sources, both formal and informal, including the client's verbal report, self-assessment forms, collateral sources, psychological tests, helper observation of nonverbal behavior, interactions with significant others, home visits, and helper intuition from direct interactions.

Use of Lazarus's BASIC I.D.

Lazarus (1989, 1992a, 1992b, 1997a, 1997b, 2000a) developed the multimodal approach that divides a person's complex personality into seven major areas of functioning as follows:

B = behavior (overt behaviors)
A = affective responses (emotions, moods, and strong feelings)
S = sensations (five basic senses of touch, taste, smell, sight, and hearing)
I = images (how individual pictures himself or herself)
C = cognitions (insights, philosophies, ideas, opinions, self-talk, and judgments that constitute one's fundamental values, attitudes, and beliefs)
I = interpersonal relationships (interactions with others)
D = drugs/biology (drugs, biological functions, nutrition, and exercise)

A comprehensive assessment of these seven modalities of human functioning and the interaction among them can provide important information during the assessment phase. Because the acronym BASIC I.D. is easy to remember, we encourage using it early on as a guide when beginning to evaluate a client.

Knowledge Needed in Making an Assessment

Knowledge of human behavior, development, and the social environment is necessary for conducting assessments. Also useful are knowledge about the biopsychosocial aspects of an individual's life and information about diversity issues.

Often, the purpose of the interview determines what factors are to be examined and hence what areas of knowledge the helper needs to master. The approach, theories, and techniques the helper applies may also affect what is examined when conducting an assessment. For example, a psychodynamic clinician focuses on how the past has affected the present, whereas a behaviorist may concentrate more on current behaviors.

Assessment Outlines and Forms

There are many different types of outlines and assessment forms. Your agency may specify guidelines that include certain forms, or you may be able to develop your own assessment guidelines.

Goal Setting: Reorienting, Shaping, and Prioritizing

When helpers and clients have achieved a mutual understanding of the issues, and when sufficient information about client strengths and history has been obtained, it is time to incorporate direction and change into the process. This involves goal identification, goal setting, careful planning, prioritizing, and decision making. Before this takes place, however, clients must be reoriented to the helping process. Young (2001, p. 37) concurs that "once a helper has gained an understanding of the client's problems and the important background issues, it is time to identify the helping goals."

Reorienting Clients to the Helping Process

Clients often possess unrealistic expectations for change. It is critical to educate them early on about the nature of helping, in which there is joint involvement but they are primarily responsible for change. Some clients' preconceived notions about helping place all the responsibility on the helper. To minimize this, we often tell our clients that we are there to help guide them, but they are the ones who must work toward change. Similarly, when clients credit us for their change and growth, we stress that they are primarily responsible for the success.

Reorienting them to basic ground rules of the helping process is important. At this time, clients must learn some of the ground rules: use of time, promptness, agency policies, confidentiality, professionalism, dual relationships, and meeting place. This often prevents future conflict, misunderstanding, and disillusionment.

Explaining General Benefits and Risks

Once there is agreement on the preceding issues, helpers can present clients with an overall explanation of the benefits of engaging in the helping process. Clients can then begin to develop their own positive coping skills, which will enhance their ability to navigate through life and foster a positive outlook characterized by hope and confidence.

Discussing risks is equally important because it provides clients with an honest appraisal of the possible drawbacks involved in change. It is helpful if they can understand concepts of the systems approach, in which change in one part of a system affects all other parts. They should also be prepared for the emotional turmoil they may undergo as they pursue change. To share the potential risks of helping is the ethical thing to do, and it is likely to increase the level of trust between client and helper.

Shaping Goals

A logical next step is for the helper to enlist client participation in shaping and prioritizing goals. Certain aspects of goal setting are critical for both client and helper to understand:

1. Goals must be shared, allowing collaboration to continue.
2. Goal setting helps establish a framework for movement toward attaining the goals.
3. Goals increase the likelihood of success in the helping process. Without goals, both client and helper may become lost, frustrated, or disillusioned.

Goal selection may be clear cut: "I need to learn communication skills that can help me relate better to those in my life." Sometimes, however, this process is not so simple. Certain clients may be unclear about their situation and unsure of what they want. The helper must then be more active in helping the client decide which goals are to be pursued.

Setting specific, clear, and attainable goals often requires ongoing collaborative work between helper and client and continual evaluation (Corey & Corey, 2003; Faiver, Eisengart, & Colonna, 2004; Hackney & Cormier, 2005, Neukrug, 2002; Young, 2001). Many clients state such broad goals as "I want to be happy." Although this expresses the client's desire, it is too broad and ambiguous a statement; it may ultimately hinder efforts toward a resolution of the client's concerns. Encourage such a client to be more realistic and specific. You could say, "List five things that would make you happy," or "What do you mean by being happy?" She may respond by stating, "Being happy means I would be more social, be more motivated to pursue college, and be more active in taking care of myself." From this response, specific treatment goals can be set, such as the following.

1. Client will engage in a social activity one time per week and will attempt to contact old friends.
2. Client will call the local college and make an appointment with a career counselor, who can help her begin the process of determining a career choice and setting up a curriculum to follow.

3. Client will take a nutrition class to incorporate healthy eating into her life and will begin to exercise 20 minutes three times a week.

As this client begins to follow through with these goals, it is likely that she will begin to feel better and more fulfilled. Educating the client about the common pattern of "one step forward, two steps back" will help prevent her from becoming angry at herself or feeling defeated when she regresses or makes slow progress.

Prioritizing

Once goals have been defined, the client and the helper prioritize them by considering which are important to pursue first and which are secondary. At this point, clients and helpers often have different views. The client may view a certain goal as the most important, and the helper may feel that other goals merit primary consideration. This can present an ethical dilemma, which the helper must carefully examine. If the client's well-being is at stake, the helper should work with him or her to redefine the concerns to allow a redetermination of the goals. In general, though, clients need to "own" their goals, so their views should take precedence. Brill (1990, p. 131) echoes our view: "Goal setting is most effective when it is a shared process, when the client has a major voice in deciding what needs to be achieved and how it is to be done. Motivation and independence are strengthened by this involvement."

This process of prioritization is also useful in helping clients to learn skills so that they can eventually engage in this process on their own. It also conveys the message that goals must be approached in an active manner.

Examining the Consequences of Goal Attainment

Often, helpers overlook the need to examine consequences of goal attainment. In actuality, this is quite important. Clients may proceed with set goals, only to find family, friends, or coworkers to be upset by their changes. Negative consequences may result.

SCENARIO

Susie's goal is to become more assertive in all her interpersonal relationships. But her spouse, children, extended family, friends, coworkers, and employer may view such a change in a negative light. In an attempt to learn to be assertive, she may act too aggressively, or others may be frustrated at no longer being able to relate to Susie as a passive person who struggles with saying no. When clients consider such consequences, they become more realistic and prepared for possible setbacks.

In addition to possible negative consequences, clients must consider positive ones. Sometimes, even positive consequences can create the stress, fear, and anxiety that often accompany change.

Armed with this information, clients can make conscious decisions regarding which goals they wish to pursue and which they need to redefine. Helpers should not fear that this process will alienate clients from following up on goals. Most clients seem to value the process and appreciate the helper's honesty and directness. In a sense, considering consequences can be a preventative measure.

Establishing a Contract

When clients and helpers have been successful at goal setting, a contract is established to cement their commitment to working together on goal attainment. Some helpers may actually list goals and have clients sign a formal contract. Others may feel that a verbal agreement suffices. In any case, the contract can instill hope, provide structure and direction, and enhance the client's understanding of the helping process. It also reinforces the fact that the helping relationship is a partnership.

During the goal-setting stage, there is a concentrated effort to shape, specify, and prioritize goals. As goals are solidified and prioritized, the action phase can begin.

Action: Planning and Intervention

Action implies actual implementation and planning to meet the goals identified. Various other names for this stage have been coined such as "encouraging client exploration and taking action" (Corey & Corey, 2003), "initiating interventions" (Hackney & Cormier, 2005), "intervention and action" (Young, 2001), or simply "work" (Neukrug, 2002).

McClam and Woodside (1994) outline three activities of this phase:

1. Reviewing perceptions of concerns because the client or his or her environment may have changed since initial assessment and planning
2. Overcoming resistance—the notion of change may frighten and confuse clients, so it is essential to develop a positive attitude toward change
3. Developing the implementation plan, according to which specific action is to be taken

A review of concerns is important to be sure the set goals remain applicable. Resistance may appear at this stage; the helper must maintain a positive attitude and be aware of potential obstacles. An action plan involves the development of effective procedures, although positive change may actually occur at any stage of helping.

Identification of Existing Resources and Support

Clients may not be fully aware of the resources they personally have or the support of others in their life. The helper may need to assist the client in exploring these areas. Engaging in such identification early on in the action phase can facilitate a collaborative effort and a better outcome.

Resources within oneself Examining which resources exist within ourselves entails using concepts of the strengths perspective. Clients are asked to look within to determine their strengths. They may become aware of how they have showed courage in the past to survive previous struggles. Once uncovered, their past successes in coping and resolving conflicts can be reinforced. Some clients may have been oblivious to their strengths and therefore find this process enlightening. Clients who struggle with low self-esteem may experience difficulty in acknowledging and accepting their positive qualities. In this case, helpers must be persistent but gentle in helping clients begin to accept their strengths.

Resources within one's environment In many cases, clients can enlist support from their environment. Hutchins and Cole (1992) have grouped such resources into three broad categories: people, data, and things. Clearly, not every individual will require all these resources nor will they all be available to each client. Nonetheless, with the helper's assistance in suggesting options and making appropriate referrals, clients are likely to gain support in their change process. McClam and Woodside (1994, p. 130) see this as "building a support system," where clients and helpers work together in exploring "who might give support, what kind of support the client needs, and when the client will need it."

Ultimately, helping clients build such support keeps them from feeling alone in their struggles. As a final note, helpers should be aware that not all support is positive. To better determine the nature of the client's support, helpers should study the dynamics at work between the clients and their support systems.

THINKING THINGS THROUGH

Map out your own support systems. Evaluate what resources exist within yourself and within your environment. Compare and contrast your lists. Which area do you believe you may need to improve in? How might you achieve this?

Procedures for Goal Implementation

To develop effective procedures, helpers must be willing to explore their full range of expertise and possibly to research how best to work with particular clients or concerns.

In this phase, helpers may question which intervention strategies are most appropriate, as well as which theories might be most applicable to individual cases. This may require the helper to examine assumptions about the client's capabilities, to ensure their success in using specially designed procedures. Sometimes, we may have such high expectations of clients that we omit certain necessary steps. In contrast, we may expect so little that we fail to challenge or stimulate clients into action.

Being adept at applying insights from a wide range of theoretical orientations can certainly enhance a helper's ability to tailor treatment to the needs of individual clients. Being open to using a variety of techniques is also beneficial. However, be careful not to attempt an intervention about which you are not knowledgeable. Also consider that clients must be comfortable, willing, and in agreement with your approach.

Decision-making steps At this juncture, decisions are made regarding viable alternatives for attaining the goals. Along with rating and selecting alternatives, determining client satisfaction is integral to this phase of treatment.

Planning for small and achievable steps With mutual commitment from helpers and clients, the task of fine-tuning the plan begins. Together, helpers and clients consider steps and procedures that will help bring about the set goals. An essential skill is the ability to break down larger steps into smaller, more manageable and achievable ones. (This resembles the shaping and prioritizing of goals mentioned earlier.) This fosters a sense of progress in the client and can also help prevent frustration and feelings of failure resulting from attempting lofty steps that are too large in scope. We often use the analogies of baby steps or a staircase to illustrate the value of proceeding slowly. Completion of one step leads to another.

SCENARIO

Visualize yourself climbing a ladder and attempting to skip several steps because you so much want to reach the top. This may lead to losing your balance and ending up on the ground below the first step. In a client's pursuit of set goals, the same can happen: Attempting to eliminate basic steps can undermine the process and even result in unwelcome setbacks.

Use of the TFA model in practice Application of the TFA model often occurs during the action stage. Understanding how thoughts, feelings, and actions relate to behavior is essential in effecting change. Hutchins and Cole (1992, p. 155) state, "This knowledge and understanding of the client's behavior is absolutely essential because it is on this foundation that effective procedures can be designed to help the client change behavior."

The TFA model can be applied in the selection of procedures and interventions by starting with techniques that directly relate to the client's primary mode of expression. If a client operates in feelings more than thoughts or actions, then techniques from affective or humanistic approaches may be more helpful. Similarly, if the client is more focused on thoughts, cognitive or rational approaches would be applied. Lastly, if clients are primarily action oriented, then behavioral approaches may be most effective.

As the relationship develops and the client appears ready, added procedures are introduced that can tap into the other areas of the TFA triad that the client does not typically use. For instance, a client whose primary frame of reference is feelings can be encouraged to engage in cognitive or behavioral approaches. An integration of thoughts, feelings, and actions can then take place, helping the client become more balanced: "[B]ehavior that is different from the client's current TFA triad is exactly what is needed to help the client shift from a comfortable (well practiced, [yet ineffective]) behavior to more effective behavior" (Hutchins & Cole, 1992, p. 159).

Change can be effected if clients think more positive thoughts about themselves, behave more assertively in their relationships with others, become more aware of what they feel, and learn to manage their emotions more effectively.

Indications of Potential Difficulties

Throughout the helping process, the helper must remain alert to possible difficulties. The anticipation of possible obstacles in the process takes on an even more critical role during the action phase, when procedures are actually implemented. Clients may seem to be content with the procedures used, but one must be alert to subtle cues that indicate possible struggles. By paying attention to specific expressions of how the client thinks, feels, and acts, the helper can detect potential problems. For instance, paying attention to how clients think involves noticing how they say things (enthusiastically or hesitantly). Clients' feelings can be suggested by their energy level and verbal and nonverbal behavior. How clients act (the way they talk, their body language, and their completion or disregard of homework assignments) can also give helpers clues to possible conflicts.

In responding to any difficulty, be flexible enough to consider alternative solutions that might prove to be more effective. For example, incomplete homework assignments may suggest several problems: difficulties with follow-through, resistance to change, or simply fear that the helper will recognize poor reading and writing skills. Consider other treatment approaches that do not involve structured written homework for such clients.

Continuing to build on client strengths remains critical at this stage. It fosters positive self-esteem and increases the client's confidence in following through with steps that will eventually lead to goal achievement. Conversely, if clients are not positively reinforced and the focus is on their deficits, they may not have sufficient faith in themselves to follow through.

Encouraging Active Participation

Striving to maintain clients' active participation is often a key to successful interventions. When clients are encouraged to be active in the helping process, they remain stimulated and challenged and also feel ownership for their change and growth. In contrast, neglecting to promote clients' participation can lead them to feel bored, disillusioned, and uninvested in the helping process. This is often the case when a client prematurely withdraws from treatment, feeling hopeless or apathetic toward the helping process.

Homework can keep clients motivated to work on their issues so that they remain active participants. Homework is valuable for several reasons:

1. Keeping the client actively involved between sessions
2. Providing clients with the ability to learn more about themselves and giving the helper detailed accounts of certain incidents
3. Allowing clients to see patterns in their lives
4. Clueing the helper to the benefits of homework or to its ineffectiveness
5. Enhancing clients' sense of responsibility—that they ultimately have the power to control their situation
6. Allowing clients to practice behavior changes in their natural setting

Types of homework may include keeping a journal or log, charting behavior, reading books on certain relevant topics, and practicing assertiveness or other social skills at home, at work, or in other settings. Finally, successful completion of homework might indicate that the helping process is working effectively.

THINKING THINGS THROUGH

What types of homework would be appropriate to use in your fieldwork with clients? Have you observed that others in your agency use homework with clients? Has it been effective?

As with other techniques, the use of homework may be helpful for some but inappropriate with others. Helpers should not automatically assign homework without considering the client's receptiveness, abilities, or circumstances.

As clients begin to achieve their goals, they must receive positive reinforcement to encourage continued progress. When goals have been met or nearly so, the helper may begin to broach the topic of termination.

Termination: Evaluation, Review, and Ending

Termination usually results from one of three factors:

1. The goals have been met.
2. The circumstances have changed. Clients have moved or separated; their health has improved; they have returned to school or obtained a job; their relationships have improved; the helper is no longer available; or a referral has been made to someone who may be better able to help.
3. The client no longer feels invested in the helping process and either withdraws prematurely or seeks closure.

Of course, the ideal condition is the first—clients meeting their goals. Here, we discuss termination from this perspective.

The termination process is most valuable when started early, not on the last visit. This allows clients to experience a range of feelings that may not be dealt with if the helper waits until the end to discuss everything involved in termination. Moreover, both clients and helpers can work through the process more effectively. Some of you may be placed in agencies where you may see a client only once or twice; the focus of the termination process is then different. You may not need to deal with termination issues or may do so much more briefly.

The Evaluation Process

Intrinsic in the AGATE model is *evaluation,* an ongoing, continuous process. We believe that evaluation begins at the start of treatment, continues throughout the process, and ends upon termination. Brill (1990, p. 135) states, "A continuous and ongoing review of what has occurred, an analysis of success and failure which attempts to understand the factors involved and the results, should point to continuation, termination, or redesign of the working plan." McClam and Woodside (1994) have classified evaluation as either formative or summative.

Formative evaluation This is the ongoing, continuous type of evaluation described above. This involves simultaneously evaluating the client's behavior and the success of the helping process. According to Young (2001, p. 38), "[h]elpers regularly ask their clients to evaluate by reflecting on the progress made toward the treatment goals." Flexibility is necessary because helping often leads to unexpected change, which may result in unforeseen difficulties. Helpers must take responsibility for initiating evaluation and continuing it throughout the helping process.

Actively involving clients in the evaluation process contributes to effective outcomes. Ask clients to assess how they are experiencing their progress; their point of view may differ from that of the helper. The evaluation process allows the helper to adjust the theoretical approach and the techniques applied. Thus, evaluation facilitates ongoing correction of the helping process.

Summative evaluation This occurs at the end of treatment and involves determining the outcomes of the process, assessing the client's satisfaction, and checking for success (McClam & Woodside, 1994). "Evaluation also occurs during the final sessions when the helper and the client review and celebrate the resolution of problems. When client and helper agree that the goals have been reached, either new goals are set or the relationship is terminated" (Young, 2001, p. 38).

The helping relationship itself can also be evaluated. In a sense, summative evaluation serves as a review of the helping process. Determining outcomes and checking for success entails examining the original goals to establish whether they have been met. This involves asking clients directly as well as soliciting feedback from relevant sources (family, friends, teachers, employers, physicians). Clients should be encouraged to list areas where there has been change and improvement. This is likely to lead to increased self-esteem, a continued commitment to apply skills they have learned, and a greater likelihood of returning for additional help in the future. A good match between helper and client facilitates change and improvement. Also, clients can be once again validated for their strengths, commended for their courage to participate in the process, and reinforced for their ongoing commitment to change.

Examining unanticipated outcomes is also important. Clients may need to talk about such outcomes, especially those that were undesired. If dissatisfied with the process, clients may discontinue treatment early on, feeling insecure about their abilities or questioning the value of helping. This could discourage them from seeking help in the future. Disappointment and unsuccessful outcomes can be caused by incomplete assessment, unclear or unrealistic goals, unrealistic expectations about the helping process, and mismatch between helper and client. Also, the client may not have been completely ready or able to engage in the helping process.

When success has been limited, helpers can sometimes discuss possible reasons and encourage continued involvement, referral, recognition that change takes time, or acceptance either that the goals were unrealistic or the helping relationship itself was not conducive to the change desired.

Helpers may feel uncomfortable in exploring the reasons for a client's dissatisfaction. Nonetheless, if the helper can encourage clients to share their honest perceptions about the helping relationship, this will affect the evaluation process positively. This feedback also gives the helper important information about skills or interventions that may require closer scrutiny. In our experience, to work with clients' anger or dissatisfaction, helpers must have a nondefensive attitude and a willingness to look objectively at themselves and the agencies where they practice.

Review of the Process

An overall review is essential in helping clients express their feelings about the process, recognize their areas of growth and strengths, and learn from both their successes and their unmet goals. The review can serve as an educational experience for both helpers and clients and can promote healthy closure. This often leads to clients' skills and strengths being reinforced. Clients can also be helped to engage in anticipatory planning—considering future situations in which their positive coping skills could be advantageous. We often ask clients to consider all the possible scenarios that could arise in the future and to be ready to apply the skills they have developed. Recognizing that life will present them with additional stressors, conflicts, and concerns but that they will be capable of coping, clients can face the future with hope.

Ending

It is important to assess the client's readiness for termination, to discuss unresolved issues, to reinforce client skills, and to encourage future involvement in helping relationships if needed.

Client's readiness for termination   Helpers must ask themselves if the client is ready to terminate. Asking clients questions such as the following may be of use (Maholick & Turner, 1979, pp. 588–589; Hutchins & Cole, 1992, p. 180):

"Have your initial concerns been reduced or eliminated?"
"Has stress or pressure been reduced?"
"Are improved coping skills present?"
"Has self-understanding increased?"

"Do you believe you are capable of living effectively without involvement in the helping process?"

Discussion of unresolved issues This step often begins when a summative evaluation takes place. Dealing with unresolved feelings is essential to allow client and helper to acknowledge concerns and attempt to come to some resolution. Perhaps the helper and client were not in synch, or there was a lack of clarity regarding goals. This process can give the client and the helper a more positive outlook; they may even decide to continue because a new direction has been set. Helpers and clients must not think they have failed or are inadequate. It is most beneficial to view the helping process as a learning experience.

Dealing with loss, grief, and sadness Allowing clients to express their feelings about the approaching closure of the helping relationship is extremely important. For some, closure may be welcomed because it indicates success and progress. Others may feel ambivalent about ending; they may be ready but still feel sadness over terminating the relationship. Still others may feel quite shaken by this step. It may trigger old feelings of loss, abandonment, and rejection. Given these possible reactions, allow clients to express their emotions, have their capabilities validated, and recognize that the door is open for future contact or involvement in other helping relationships.

If loss and abandonment issues are triggered, the helper can take the opportunity to help resolve those feelings. Helpers should also reaffirm that ending is a result of the client's progress, not an abandonment or a rejection. Helping clients deal with these issues can be an added bonus of the helping process; they may apply this lesson to future situations that involve loss and separation. Discussion of loss issues also allows helpers to express their feelings and grieve at the termination of the relationship.

Examination of the Process, Self, and Follow-Up

Self-Evaluation

After closure is complete, helpers can engage in more successful self-evaluation. They can gain insight about which methods were effective, which techniques were useful, what aspects of the relationship reinforced positive change, and which helping skills were the most beneficial. They can then be better prepared to work with similar clients, realizing what changes may need to be incorporated in their future work.

Helpers can begin to consolidate their skills and develop a repertoire of interventions that can be used wisely with different populations. It is an ethical responsibility for helpers to continually evaluate their roles. This self-examination is necessary if helpers are to continue to be competent, qualified, and well prepared to work with their clients sensitively and effectively.

Follow-Up

We have advocated leaving the door open for future contact or further help. The client may later reconnect with the helper or engage in another helping relationship, and the helper can follow up with the client. Such follow-up can also:

- Enhance clients' growth
- Convey that they are valued and that others care
- Help determine the success of the process
- Reinforce the need to continue to implement skills learned
- Alert helpers to the need to reestablish a relationship with clients who are not doing well
- Solidify the message that the door remains open for further interaction
- Communicate that clients are not alone and deserve to be helped

Follow-up may not always be possible because of the heavy caseloads many human service workers have. For various reasons, it may not be possible or appropriate to contact former clients. In any case, whenever feasible, we encourage you to incorporate follow-up into the helping process.

Summary of the AGATE Model

This model has much to offer because it is comprehensive and incorporates many perspectives on helping that are not included in other models. As with any model, however, it has its limitations:

- In work with certain clients, not all the stages are relevant, nor can they be followed in the sequential manner described.
- This model may not be appropriate to use in some agencies and in some forms of helping.
- There is overlap in some of the stages presented.

Use of this model has proven to be effective in our work with diverse client populations in many types of agency settings. In some cases, we have had to adapt parts of this model to fit the specific needs of our clients or the demands of a particular agency. We encourage you to consider applying this model in your practice.

THINKING THINGS THROUGH

Consider a case you have worked with or are currently involved in. Think how it would work if you tried to apply the AGATE model. Once you have done so, describe your reactions to using this approach. Was it helpful? If so, how? If not, why not? Do you have any particular recommendations?

Also consider how you might apply this model to other types of helping that do not involve a treatment approach.

■ DEVELOP YOUR OWN STYLE AND APPROACH

As a helper, you must work to develop the style and approach to helping that best suit you. Attempting to copy others can lead to a lack of genuineness and the use of ineffective interventions. Helpers are often role models; by being true to themselves, they show their clients the value of being unique. Working in a way that is congruent with who you are also allows you to feel more free and uninhibited.

However, this does not mean that helpers should not try new techniques or borrow any. If you are tempted to emulate your supervisor or a valued coworker, this can be constructive. But it is wise to recognize that you will gradually develop a style of your own. Helpers should not try to be someone they are not or feel insecure about who they are.

Three main areas will have some bearing on which style and approach you develop: self-awareness because your style and approach will be partially based on your own personality and interests; your work setting because your approach may depend in part on what kind of agency you work for; and your experience level, which can also dictate, to some extent, what helping approach you apply. Consider these aspects regularly in your work with clients. Be open to altering your style and approach to fit their particular needs. Also, recognize the need for additional training, supervision, or assistance in certain situations. If you remain open to learning and changes, you will have a better chance at continued growth, heightened self-awareness, and the realization that becoming a helper is an ongoing, evolutionary process.

THINKING THINGS THROUGH

Reflect on your developing style. Can you identify any specifics? How do you feel about your approach to helping? Identify your strengths, as well as areas you would like to work on further. Also identify which mentors, supervisors, coworkers, and theoretical orientations have had the greatest influence on the development of your present style.

◼ CONCLUDING EXERCISES

1. According to the strengths perspective, focusing on one's strengths is crucial. As role models for our clients, we must be open to identifying and appreciating our own strengths. Take a moment to identify the strengths you are most aware of. Ask your closest friends, partners, supervisors, and coworkers what they consider to be your greatest strengths. Make a list of all the things that they tell you about yourself. Devote one whole day to appreciating yourself and all the strengths that you possess. Focus on your qualities and be proud of your achievements.

2. Imagine that at least seven years have passed since you completed your education. You are now employed as a full-time professional in a helping capacity. (You decide the specifics.) Because of the vast experience you have gained, you feel more competent, knowledgeable, and skilled. Furthermore, you are well respected and valued at the agency you currently work in. You are honored by your supervisor's decision to recommend you as an agency-based supervisor for incoming human service students doing fieldwork. You are excited, but you find yourself experiencing normal anxious feelings regarding this new role. As students interview you to select their fieldwork sites, you find yourself unprepared for a question many seem to ask: "What is your style of helping?" What would you say? How would you define *helping?* What specifics would you share about the helping process? Lastly, what approach or methods would you say are the most effective and applicable to your work at this agency? Please elaborate.

CHAPTER FIVE

The Diversity of Human Services

▪ AIM OF THE CHAPTER

The slogan "celebrate diversity" emphasizes the recent trend toward appreciating our individual differences. It is no longer considered appropriate for people to hide, play down, or feel ashamed of their heritage, views, beliefs, preferences, and lifestyles. This is not to say that prejudicial attitudes are a thing of the past. Unfortunately, racism, sexism, heterosexism, ageism, classism, and ableism still persist; in some cases, they may be on the rise. We who work in human services must maintain a keen awareness of issues related to diversity. We must recognize which groups in our society are the most vulnerable and therefore subject to discrimination, negative stereotyping, exploitation, and oppression.

As you engage in fieldwork, you will be exposed to diversity. You must have the knowledge and skills to be effective and sensitive when working with diverse population groups. In this chapter, we first address issues of ethnic diversity. Next we focus on general issues of diversity—its elements and impact on human services. We then discuss three components of diversity sensitivity: self-awareness, knowledge, and skill. Lastly, we examine challenges, barriers, and benefits involved in working with diversity in human service agencies.

▪ VIEWS OF DIVERSITY

Diversity is a mosaic of people who bring a variety of backgrounds, styles, perspectives, values and beliefs as assets to the groups with which they inteact. Diversity extends far beyond the obvious dimensions of ethnicity and gender. People are similar and different on an infinite number of dimensions. By viewing the idea of "valuing diversity" as something that is equally relevant to all of us, it becomes inclusive to everyone [Kaiser, 2004, p. 2].

Diversity has been considered primarily an ethnic and cultural concept. Such terms as *ethnic-sensitive, cross-cultural, multicultural,* and *cultural pluralism* have become commonplace in the literature, emphasizing the importance of culture, race, and ethnicity in the helping process. Lum (2004) writes of "ethnic consciousness"—an increased awareness that "distinctive ethnic cultures" exist. Such a focus is a natural response to the steady increase of various ethnic and racial populations in the United States. It is a fact that "America is becoming increasingly diverse" (Neukrug, 2004). A U.S. Census Bureau of 2001 finds that approximately 30 percent of Americans are racial and ethnic minorities with an increase to 50 percent estimated midway through this century (U.S. Census Bureau, 2002a, 2001b, 2001c). Various reasons for these changing demographics include higher birthrates of culturally diverse populations, the fact that most immigrants no longer come from Western countries, and the fact that immigration rates are the highest in American history (Neukrug, 2004, p. 192). At some point, most human service workers will encounter clients who are from different cultural, ethnic, or racial backgrounds. Therefore, it is ethically and practically imperative to be sensitive, knowledgeable, and skilled in working with culturally diverse groups. Cormier and Hackney (2005, p. 9) concur that as diversity continues to grow in this century and "as the dimensions of client diversity expand, the competence of counselors [helpers] to deal with complex cultural issues must also grow." For some, the current emphasis on cultural diversity is welcomed, promising a more open, aware, and accepting climate for helping. For others, however, learning to engage in ethnic-sensitive practice may be more challenging. Our hope is that you, as beginning helpers, share our excitement about learning to work with diverse populations. If this is viewed as a positive learning experience, it becomes much easier to integrate newly acquired skills into daily practice.

Important Concepts

A thorough understanding of the multicultural perspective requires awareness of such concepts as ethnicity, ethnocentrism, culture, cultural tunnel vision, cultural pluralism, and cultural competence.

Ethnicity

This term has been defined in many different ways, yet typically includes concepts of heritage, ancestry, and tradition. Gilman and Koverola (1995) characterize ethnicity as traits that make someone different from ourselves. Brill and Levine (2002, p. 249) define ethnicity as "characteristics of people that distinguish them from others on the basis of religion, race, or national or cultural group." Basically, an ethnic group shares some common ground—their religious beliefs and practices, language, historical continuity, and common ancestry or place of origin (Devore & Schlesinger, 1995).

There is often confusion regarding the distinction between ethnic minorities and racial minorities. Neukrug (2004, p 195) states that "ethnicity, as opposed to race, is not based on genetic heritage but on long-term patterns of behaviors that have some historical significance and may include similar religious, ancestral, language, or cultural characteristics." Taylor (1994) differentiates the two as follows: An ethnic minority is distinct in its cultural criteria, whereas a racial minority differs from the majority in its physical characteristics. Not all ethnic groups are minorities; a group may be vast in number as well as economically and politically powerful. According to Taylor (1994, p. 2), "[a] group's restricted access to sources of economic and political power and marginal location in the social order define it as a minority."

Ethnocentrism

This is defined as "the tendency to view the norms and values of one's own culture as absolute and to use them as a standard against which to judge and measure all other cultures" (*Encyclopedia of Sociology*, 1974, p. 101). Lum (2004, p. 196) adds, "In ethnocentrism, the individual's ethnic groups forms his or her central point of reference, to the exclusion of other groups." Certain ethnic groups feel superior to others. Clearly, such an elitist attitude could result in discrimination against other ethnic groups. As human service workers, we must guard against developing or perpetuating ethnocentric beliefs and instead work toward dispelling myths and stereotypes about cultural, ethnic, or racial groups. In particular, we must be aware of any prejudiced views we may hold, because they may affect our work with certain clients.

Culture

This term refers to "a learned worldview or paradigm by a population or group and transmitted socially that influences values, beliefs, customs, and behaviors, and is reflected in the language, dress, food, materials, and social institutions of a group" (Burchum, 2002, p. 7). Simply put, "[c]ulture represents the common values, norms of behavior, symbols, language, and common life patterns that people learn and share with one another" (Neukrug, 2004, p. 194).

Cultural Tunnel Vision

This term characterizes members of the majority culture who have limited cultural experience and see it as their purpose to teach others their view of the world. This is related to ethnocentrism because it implies that the majority's ways are superior. You can imagine the potential damage a helper could do when operating with cultural tunnel vision. Clients might easily feel inferior, misunderstood, or even rejected. In working with people from other cultures, we must try to understand their culture and see the situation from their frame of reference. If we ask clients to help us understand their world, they often feel honored, respected, and genuinely enthusiastic about educating us.

Cultural Pluralism

A culturally pluralistic society is made up of diverse ethnic groups, each preserving its own traditions and culture but also loyal to the nation as a whole (Brill & Levine, 2002). A truly pluralistic society is the goal behind present efforts to encourage ethnic sensitivity and a multicultural perspective. We may have a long way to go, but at least we are on the road to working together. In this effort it is important to remain realistic. Lum addresses this issue in his statement that "cultural pluralism initially envisioned a utopian society in which ethnic groups would maintain their distinctiveness and coexist in harmony and respect," yet a "realistic view of cultural pluralism recognizes the right of multicultural groups to coexist as recognized entities within a larger society that tolerates and encourages multiple cultures" (2004, p. 99).

Cultural Competence

Becoming culturally competent is essential for anyone hoping to become an effective helping professional. According to NASW's National Committee on Racial and Ethnic Diversity (2001, p. 11): "Cultural competence refers to the process by which individuals and systems respond respectfully and effectively to people of all cultures, languages, classes, races, ethnic backgrounds, religions and other diversity factors in a manner that recognizes, affirms, and values the work of individuals, families, and communities and protects and preserves the dignity of each." Similarly, in the APA's code of ethics, psychologists are required to obtain training, experience, consultation, or supervision necessary to ensure competence in their services to clients of different ages, genders, gender identity, race, ethnicity, culture, national origin, religious orientation, disability, language, or socioeconomic status. The ACA echoes the importance of counselors only engaging in services for which they have competence, including cultural competence. Lastly, NOHSE's ethical standards highlight that "human service professionals are knowledgeable about the cultures and communities within which they practice. They are aware of multiculturalism in society and its impact on the community as well as individuals within the community. They respect individuals and groups, their cultures and beliefs" (NOHSE, 2004, p. 2).

In all of these professional ethical codes, the issue of nondiscrimination is well stated in that all helping professionals "provide services without discrimination or preference based on age, ethnicity, culture, race, disability, gender, religion, sexual orientation, or socioeconomic status" (NOHSE, 2004, p. 2).

Clearly, engaging in ethnic-sensitive practice and maintaining a multicultural perspective are worth striving for as you begin your fieldwork experience. Being sensitive to your clients' ethnicity requires that you focus on both the individual and the ethnic group in question. Such a dual focus will allow you to show your concern for the individual client, while still addressing the consequences of racism, poverty, and discrimination for the group to which the individual belongs. Work can take place on several levels: micro (with the individual), mezzo (interpersonally, with the group), and macro (in the community and policy arenas), as discussed in Chapter 4.

In the last decade, considerable literature has been published about training and skill development for ethnic-sensitive practice. Numerous writers have made note-worthy contributions in which they emphasize the importance of acceptance as described by Cournoyer (2005, p. 58): "Cultural competence requires respect for and acceptance of others. Acceptance of others involves the process of self-understanding, cross-cultural understanding, valuing, and joining with people regardless of the degree of similarity or difference from oneself." We begin with the work of Paul Pedersen, who addresses common culturally biased assumptions that we must be aware of and challenge if we are to be ethnic sensitive.

Culturally Biased Assumptions

Pedersen (1988, 1991, 2000) elaborates on the many culture-specific assumptions we all are apt to make. He underscores the potential negative consequences of allowing these assumptions to go unexplored. Examine the following list of assumptions and consider how you might challenge your current views:

- "There is a single standard for what is normal." Is there? Are all cultures' standards the same?
- "Individuals are the basic building blocks of society." What about the family or the community? Aren't they important too? In some cultures, isn't the focus on the welfare of the family or community, not the individual?
- "A problem can always be defined by clear-cut academic boundaries." Do problems always fit nicely in one category—psychology, sociology, medicine, religion, and so on? How about taking a multidimensional perspective?
- "Others will understand our abstractions in the way we do." Really? What about language and syntax differences, slang, and untranslatable words?
- "Focus on independence as universally desirable and healthy." Is this always true? Can you name cultures where interdependence is valued just as much, if not more?
- "Formal counseling is the only viable way to help clients." Is this always neces-sarily true? What about the value of support systems?
- "Linear thinking, based on cause and effect, is how all people come to understand the world around them." Is this so for all people? What cultures might not use this type of logic?
- "Clients must change to fit the system." What about helping change the system to fit the client? Why would this not work? Or would it?
- "The emphasis should be on immediate events, not on a person's history." Isn't history sometimes the reason for a current crisis? Why would ethnic history not be relevant?
- "We already know all of our assumptions." Are you sure? Isn't this a cop-out? Perhaps you are "culturally encapsulated"—unwilling to challenge your assumptions?

THINKING THINGS THROUGH

Make an open assessment of yourself. How close to home did you find the preceding assumptions? How might you go about challenging them? As an exercise, pair off with a classmate and challenge each assumption in a logical and rational manner.

Characteristics of Culturally Skilled Helpers

Sue and Sue (2003) have identified attributes of culturally skilled counselors. Even human service workers who are not counselors can use Sue and Sue's views as a frame-work for assessing their strengths and limitations. Sue and Sue stress three categories: awareness of the helper's assumptions, values, and biases; understanding the world-view of culturally different clients; and developing appropriate intervention strategies and techniques. Later in this chapter, we will use this framework to consider more general issues of diversity.

A Broader Definition of Diversity

As mentioned, the term *diversity* is often associated with culture, ethnicity, or race. However, these aspects do not encompass the entirety of diversity. Human service providers must be aware of other facets of human diversity that are integral to an individual's development and current status. Lum's definition of *human diversity* reflects our view (2004, p. 6):

> *Human diversity is* an inclusive term that encompasses groups distinguished by race, ethnicity, culture, class, gender, sexual orientation, religion, physical or mental ability, age, and national origin. In its broadest sense, human diversity covers the various people who comprise a society. Diversity includes all of us, because each individual is unique, one of a kind, and has distinguishing characteristics, yet there are similarities between persons that connect us in many unusual ways.

Pedersen (1991, 2000) also defines culture more broadly, incorporating numerous variables:

- Demographic variables such as age, sex, and place of residence
- Status variables—social, educational, and economic standing
- Affiliations, formal and informal
- Ethnographic variables–nationality, ethnicity, language, and religion

Differences between the terms *diversity* and *multiculturalism* are addressed in the literature: "Diversity describes the experiences of clients who are different across dimensions such as age, gender, race, religion, ethnicity, sexual orientation, health status, social class and so on" (Cormier & Hackney, 2005, p. 9). Corey and Corey (2003, p. 193) add that "diversity refers to other individual differences and characteristics by which persons may self-define." Multiculturalism "puts the focus on ethnicity, race, and culture." It also involves an "awareness and understanding of the principles of power and privilege" (Cormier & Hackney, 2005, p. 9). According to Robinson (1997, p. 6), multiculturalism is "willingly sharing power with those who have less power" and using "unearned privilege to empower others."

When taken together, these variables define the individual in a more comprehensive way, allowing the helper to get a more complete picture when considering appropriate interventions. Certainly, inherent in the process of becoming an effective and diversity-sensitive helper is the ability to understand these different variables relating to diversity. Baird (2002, p. 779) writes, "Understanding diversity is not a 'luxury,' it is an essential."

Elements of Diversity

Many aspects of diversity must be considered if we are to work successfully with people of backgrounds different from ours. Dixon and Taylor (1996) distinguish between primary and secondary characteristics, as outlined in the following lists.

Primary Characteristics

Often visible
May differ for specific groups
Not choice-driven, as these are the given traits of particular groups

Examples
Race or ethnicity
Gender
Age
Physical and mental abilities
Sexual orientation
Size

Secondary Characteristics

> Often invisible
> May change by individual choice or due to certain circumstances
> *Examples*
> Political ideology
> Geography
> Marital/family status
> Socioeconomic status
> Religion
> Occupation and skills
> Experiences
> Education

This approach to diversity can help us recognize that although we are all born with certain traits, we also have the capacity to make choices. This can instill hope and a sense of direction because our destiny is not predetermined. However, the intricacies of determining who can make such choices and how they are translated into action are not as clearly defined. There may be a fine line between primary and secondary characteristics, with one affecting the other. Although some characteristics are considered secondary, they may still have an impact on a person's ability to overcome obstacles.

SCENARIO

An individual who is bright and highly capable of attending college may come from an immigrant family that is poor and unable to provide the means to pursue higher education. He may want to attend a university to better himself, but he may feel that his duty is to remain at home, where he can work and help provide for his family. This does not necessarily mean that he does not want to succeed in college. Rather, due to his circumstances, he is not able to follow through.

SCENARIO

A young, bright, affluent gay male wishes to pursue an education in theology, with the ultimate goal of entering the ministry. However, because of heterosexism, he is not able to practice his lifestyle openly. Ultimately, he may be barred from becoming a minister or priest. Although he has the ability, the means, and the desire, circumstances may prevent him from reaching his goal.

Clearly, diversity can be a complex matter, but we must start with the basic awareness that diversity exists. Beginning with this awareness and a respect for differences, we can learn to understand how the many facets of individuals' lives can influence them, directly or indirectly.

THINKING THINGS THROUGH

Provide examples, from personal experience or from your fieldwork, of how primary and secondary characteristics affect an individual, positively or adversely.

Diversification

Atkinson, Morten, and Sue (1993) call current population trends "the diversification of the United States." As discussed earlier, several forces are primarily responsible for the population changes that are diversifying our nation. First, the current immigration wave (including documented and undocumented immigrants and refugees) is the largest in U.S. history. The U.S. Census Bureau (2001b) reported that 30 percent of the nation's population growth resulted from the surge of immigration during the 1980s. . . . Immigration patterns have affected certain states more than others; various immigrant groups tend to settle in certain areas. Noteworthy is the fact that "most immigrants no longer come from Western countries" and "are now Asian (39 percent) or Hispanic (43 percent) as compared with past immigrants who were mostly white European, and they have greater tendency to assert their cultural heritage than be swallowed up by the Western-based American culture" (Neukrug, 2004, p. 192).

A second significant population trend relates to different birthrates among racial and ethnic groups. Many of the immigrant populations, as well as other ethnic groups already in residence, have higher birthrates. This is certain to result in a dramatic increase of non-white populations, which accentuates the importance of understanding diversity.

In addition, many other factors have contributed to the diversity of today's United States, including the civil rights movement in the 1960s, the changing role of men and women, increased emphasis on the family and child welfare, heightened awareness of the "graying of America" (the aging population), open activism by gay and lesbian people, changes in the religious composition of the country, and a trend toward returning to our roots.

Brill (1990, p. 226) capture the plight of the most vulnerable groups in our society:

> To be poor in an affluent society; to be old and slow in a world that values youth and speed; to be a child when family structures are weakening and changing and when proliferating no-fault divorces are mainly concerned with the rights of adults; to be a member of a minority group that is considered inferior; to be so handicapped as to be unable to use available resources—to be any of these is to be isolated, stereotyped, and dehumanized.

Sadly, many people are at risk for some form of exploitation due to their powerless position or ascribed status. Such individuals are subject to further pain and suffering, which may or may not be inflicted intentionally. Because they feel vulnerable, they are unlikely to reach out for support and assistance—or, in some cases, even to feel worthy of being helped. We who are committed to helping must be especially sensitive to who these individuals are and must be willing to reach out to them rather than waiting for them to reach out to us.

Hoffman and Sallee (1994) have identified five major populations that are at risk:

- *Women:* poor women and their children, sexism, social problems particular to women, gender relationships
- *People of Color:* institutional racism, individual racism, and subtle forms of discrimination
- *Gay Men and Lesbian Women:* AIDS, homophobia, prejudice, and discrimination
- *The Elderly:* ageism, policies adverse to the interests of the elderly, inadequate care of the elderly
- *Physically and Mentally Challenged People:* discrimination, policies that harm people who are physically disabled, mentally retarded, learning disabled, or mentally or emotionally impaired

This list in not exhaustive by any means, but it does give a sense of who in our society is most at risk for some form of abuse and may therefore need our services more than others do.

We regard diversity as a comprehensive concept, encompassing the same areas cited earlier (age, gender, race, culture, and ethnicity, socioeconomic status, sexual orientation, disability) as well as other areas such as religious preferences and geographic location. Human service providers must be aware of all areas of diversity, not only the most apparent ones. It is also important to consider diversity within the profession because this can have both positive and negative implications.

Diversity within the Profession

Human service workers are also likely to work with professionals from different disciplines, which enhances the services provided to clients. Each discipline brings unique theoretical viewpoints and treatment perspectives. Working together in a multidisciplinary manner can be enriching for client and helper alike. Unfortunately, helping professionals can also fail to appreciate the differences among the disciplines and thereby sabotage the process of working together. The same resistance toward diversity among clients may apply to human service workers themselves. For instance, "degreeism" and "classism" are forms of discrimination often seen in agency settings.

SCENARIO

In many large organizations, there is a status hierarchy based on what degree the helpers hold. In many cases, those with a higher degree not only have more authority and power but are also felt to be more competent.

Is this always the case? How can this perception influence service delivery?

SCENARIO

There is often an unspoken rule that only those of the same status should interact socially. Clerical staff may not mingle with clinical staff, or clinical staff may engage in separate functions from medical staff.

Have you noticed this in your agency? If so, explain how. Describe how you have felt as a beginning worker or intern.

The implications of these prejudiced behaviors can be far reaching; they may ultimately influence the effectiveness of service provision. If teamwork is undermined, services may be fragmented and clients left feeling rejected. How can we hope to work effectively to help others when we are unable to build cohesion, trust, and respect among our colleagues? What sources of resistance might exist in your agency with regard to the different disciplines represented?

THINKING THINGS THROUGH

List all the professional disciplines represented at your agency. Follow by discussing how they relate to one another. Is there any discrimination on the basis of credentials? How about other categories of diversity (e.g., lesbian therapist, African-American doctor, female administrator)? If there are conflicts, how do they affect services to clients? What type of resolution would you recommend?

Impact on Human Services

Historically, certain time periods have seen great changes in demographics, roles, views, and lifestyles. Such shifts have a profound impact on individual lives and society. In U.S. history, the last several decades of the 20th century are destined to be written about in terms of the nation's vast diversity. During the 1920s, when immigration was then at its peak, Jane Addams of Hull House and other settlement house workers were challenged by the flood of immigrants. They learned to work with the immigrants as well as with natives affected by their arrival. Colburn and Pozzeta (1979) recount how the settlement house workers learned to respect the diversity of the immigrants and to view their cultural heritage as unique and valuable. They endeavored to understand the customs and traditions of each group. They then acted as advocates for the immigrant groups, developing policies to protect them, helping them survive the adjustment to a different society, and giving them opportunities to prosper in their new country. Human service workers today have a similar role because of the increasingly diverse U.S. population.

A process similar to what immigrants underwent in the 1920s can be seen with regard to various other groups. African Americans fought for civil rights in the 1950s and 1960s. Women have progressed through the suffragette movement, the struggle for equal rights, and women's liberation. Men have been joined by women in the workforce. The oldest segment of our population has increased many times over, requiring increased attention to the needs and rights of the aged. Children's rights have evolved through the years from the advent of child labor laws to child-abuse prevention. People with disabilities have fought and won several battles over access issues, reimbursement for certain physical and mental conditions, and equal rights in the workforce. Homosexuality has become more open, although the atmosphere is not yet totally accepting. Many gay and lesbian people feel less inhibited about their sexual orientation. Together, all these changes have had an enormous impact on human services, with important implications for anyone working in the field. With regard to race, culture, and ethnicity, we must be culturally aware, sensitive, and understanding to meet the needs of culturally diverse populations. Services need to be structured to optimize use by ethnic minorities and other vulnerable groups. Local, state, and national policies and laws should address all forms of discrimination and the evils of exploitation and oppression. Brill and Levine (2000, p. 67) highlight the importance of understanding diversity:

> Human diversity is a significant factor in working with people. It is not only a determinant of individual and social functioning; it also affects every aspect of practice in human service. Only as workers are sensitive to differences among people, knowledgeable about their causes and effects, and skillful in recognizing and working with them will practice be effective.

In the helping professions, working with diverse population groups is a given. According to Baird (2002), learning to work with differences is both a challenge and an opportunity for helpers. Successfully helping diverse clients requires not only the desire and willingness to help but also sensitivity to diversity itself. Sue and Sue (2003) emphasize awareness, knowledge, and skills as keys to becoming competent in working with people from different cultures. Many other authors also highlight the importance of being aware of one's attitudes and beliefs toward culturally diverse populations, possessing a knowledge base that supports one's work with diverse clients, and having the necessary skills to apply to clients with diverse backgrounds (Pedersen, 1990, 2000; Sue, Bernier, Durran, Feinberg, Pederson, Smith, & Nuttall, 1982; Sue, Arredondo, & McDavis, 1992).

■ DIVERSITY: THE NEED FOR SELF-AWARENESS

The value of self-awareness for human service professionals in general was discussed in Chapter 1. Here we will address self-awareness as it pertains to diversity. Many authors have emphasized the process of developing self-awareness as a vital element in learning to work with clients of different backgrounds. For instance, Wintrob and Harvey (1991, p. 11) state, "An understanding of the [helper's] feelings about his or her own social class, racial, ethnic, and religious characteristics and points of convergence with or divergence from a patient is critical to therapeutic effectiveness." Similarly, Jenkins (1993) considers diversity sensitivity to be nurtured through self-knowledge. Lum (2004, p. 5) characterizes self-knowledge as "an awareness of one's own perceptions of difference on the basis of race, ethnicity, gender, and other factors; positive identification with one's racial or ethnocultural group helps to develop an appreciation of diversity in others."

From these statements, it seems clear that the first step toward developing a diversity-sensitive attitude is to look at oneself. Becoming cognizant of who we are and where we have come from makes it possible to recognize that our roots are indeed in an ethnic group. Lum (2004, p. 3) points out that the United States is composed of ethnic groups: "[G]eneration after generation of families from other cultures have assimilated into American society." We may be proud of our ethnic and cultural heritage and continue to honor certain customs, values, and beliefs. On the other hand, some people have either dismissed any information about their roots or simply do not know about their heritage. It is important that we learn to appreciate our own past if we hope to appreciate that of others.

Corey (2005, p. 24) stresses that effective counselors or helpers have moved beyond being culturally unaware "to ensuring that their personal biases, values, or problems will not interfere with their ability to work with clients who are culturally different from them. They believe cultural self-awareness and sensitivity to one's own cultural heritage are essential for any form of helping." Similarly, Neukrug (2002, p. 142) highlights that "culturally skilled human service professionals have an awareness of their own cultural backgrounds, biases, stereotypes, and values, and such helpers should have the ability to respect differences." As human service professionals, we need to recognize that we may possess certain values that are different from our clients and accept these differences. We must appreciate our own diversity, understand how it relates to our view of others, and learn to accept and appreciate this diversity in others. Awareness also involves recognition of our limited knowledge at times of other cultural traditions.

THINKING THINGS THROUGH

Discuss your own roots and cultural heritage. How do you believe they have affected your development and outlook on the world?

We should also take time to explore our family history from a generational vantage point. Most likely, there will be stories about an "odd" uncle, an eccentric aunt, a disabled cousin, or a grandparent who was discriminated against. This can help us to see that diversity also characterizes our own family background. Some suggestions: Trace family lines via a family tree; do a genogram; or simply ask older relatives to tell stories about their past. You may learn that a relative was subjected to exploitation and oppression. Personalizing such aspects of diversity can help us understand and empathize with others.

THINKING THINGS THROUGH

Discuss a family story about a relative who was once disadvantaged in some way. If you are unaware of any, provide an example from a client's experience.

When exploring family issues, we may become aware of biased beliefs and attitudes that are rooted deeply in our family history. You may learn that your family was a victim of discrimination or a participant in it. You cannot change history. But as Sue and Sue (2003) point out, you can be aware of your own values and biases and how they might affect your clients. Such acknowledgment allows you to work toward increased sensitivity and acceptance. Some helpers may find that they are unable to remain objective when working with certain client groups. In such a case, the best course of action may be to refer the clients in question to someone of their own group or at least to another helper who is unbiased.

Reflecting on our own circumstances can help us understand how certain of our primary or secondary characteristics may have an impact on who we are today. Changing a few aspects could alter the entire picture. For instance, if you belong to an ethnic minority and picture yourself as part of the dominant group instead, what differences would there be? What if you were born the opposite gender? What implications might this have for your development? What if your birth family had been affluent? Or a single-parent, lower-income family? What difference would this have made? Recognize that changing a single facet of your life could have wide-ranging effects, for better or worse. In this light, being sensitive to people who are different does not seem so difficult. Why, then, does there seem to be so much reluctance to understand and accept our differences?

Resistance to Diversity

People's resistance to becoming diversity sensitive can be viewed in various ways, as outlined in the paragraphs that follow.

Avoidant Behavior

This consists of minimizing, ignoring, or denying the existence of a problem related to diversity (Rowe, Bennet & Atkinson, 1994).

THINKING THINGS THROUGH

"The fact that you are an older woman has nothing to do with your not getting the job." "Your ethnicity has not affected your ability to succeed in this country." "Your disability has not influenced our decision to promote someone else."

These statements all indicate a lack of acknowledgment of the actual facts. Why? What leads to such avoidant behavior?

"Color Blindness"

This approach seeks to treat others with equal regard as people. Certainly this is commendable, but this approach may fail to acknowledge important issues of culture, race, gender, or sexual orientation that are essential to understanding clients and helping them effectively. Helpers must be sensitive to differences and address them openly in working with clients (Proctor & Davis, 1994, p. 316).

SCENARIO

Some agencies make no effort to translate forms into Spanish or Vietnamese even though many potential clients are monolingual Hispanic and Vietnamese people. Supposedly, all clients are treated equally, but are they really? How is this an example of "color blindness"?

SCENARIO

Some helpers believe that there should be no differences in how they treat their clients of different backgrounds. They do not consider altering their helping approach in any way. How might such helpers be considered "color-blind"?

The "One for All" Approach

The assumption here is that the same helping methods, techniques, and approaches should work for everyone. This seems unlikely to work in the complex world of the helping professions, but many in the field make this faulty assumption. Such an approach reflects a narrow viewpoint and a lack of sensitivity to others' differences.

SCENARIO

Workers who learn new techniques, such as Gestalt therapy or family sculpturing, may attempt to apply these with all their clients, regardless of appropriateness or the client's willingness.

In what way does this represent the "one for all" approach? How do you think some minority groups might react to such treatment?

The Egocentric Style

"My way is better than yours" is the attitude—placing oneself above all others, taking a narcissistic perspective. This alienates clients because it is a clear sign of disrespect for them. We must guard against unexamined ethnocentric assumptions of all kinds.

SCENARIO

Some workers may feel that their approach to problem solving is the only acceptable way. They may expect their clients to respond and behave in set ways; this can make clients feel inferior or inadequate. How might you feel if seeking help from someone who acted this way?

SCENARIO

A helper of a certain culture may feel that people from other cultural backgrounds are not as capable of succeeding as those from his or her own culture. Is this discriminatory?

The "Learned Helplessness" Viewpoint

Unfortunately, many believe that people's problems are perpetuated by their tendency to wallow in self-pity and see themselves as victims of circumstances. This viewpoint fails to acknowledge the historical discrimination, injustice, and socioeconomic inequalities that have plagued minority cultures (Sue & Sue, 2003). This lack of sensitivity and awareness can be quite harmful to the clients. Helping professionals must instead try to examine the underlying reasons for behavior that some call "learned helplessness."

SCENARIO

Some helpers tend to blame the victim in cases of spousal battery or rape. For example, the battered woman may be asked such questions as, "Why do you allow him to beat you? Why do you stay in that abusive relationship? What is wrong with you that you don't leave? You must want to be abused at some level—otherwise why do you keep going back?" Or a rape victim may be asked, "Why didn't you stop it? What were you wearing"? If you were a client seeking help due to battery or rape, how would you feel when questioned in this manner?

Blaming Society

On the other extreme is the view that society is to blame for the individual's problems. Environmental stressors, institutional discrimination, and other social forces are considered the causes of all the client's difficulties. In truth, neither the individual nor society is 100 percent responsible for how a person's life develops. As Longres (1990) points out, there is an interaction between individual change and social change. Human service professionals are encouraged to focus on strategies that link the two.

SCENARIO

Helpers who work with adolescents may blame society, focusing on TV, drugs, alcohol, divorce, increased freedom, working mothers, single-parent families, and lax discipline as the reasons for teen pregnancies and juvenile delinquency. Although these may be factors contributing to those problems, other underlying factors must be examined. What additional issues do you believe should be explored?

THINKING THINGS THROUGH

Provide examples of resistance from your own experiences, either personally or in your fieldwork setting. How did you feel on encountering these situations? How did you react? Discuss ways you intervened or could have intervened to increase sensitivity and acceptance.

Awareness of Belief Systems

A person's belief system may also contribute to insensitivity toward diversity. Directly or indirectly, our beliefs are communicated to our clients and influence the helping process. Awareness of your beliefs is necessary to your development as an effective helper. The more you know about yourself and your own background, the more easily you can relate to diverse clients and be responsive to their needs. Similarly, the more you understand possible reasons for your own resistance, the sooner you can begin to challenge your own biases. Gaining added knowledge about diverse populations is a key to dispelling myths and stereotypes. Corey and Corey (2003) encourage helpers to become aware of their own cultural assumptions to see how they might interfere with effective helping in multicultural situations. They highlight the following cultural assumptions that may require further scrutiny:

- Assumptions about self-disclosure: Although highly valued in Northern America, in some cultures self-disclosure may be discouraged and even viewed as a taboo. Keeping problems in the family may take precedence and therefore must be respected by the helper.
- Assumptions about family values: Family values may differ from culture to culture. For instance, in Chinese-American families the values of family bond, unity and respect are superior to the values of self-determination and independence held by many North Americans (Jung, 2000).
- Assumptions about nonverbal behavior: The fact that clients may disclose themselves in many nonverbal ways requires helpers to be aware of the particular cultural views on such behaviors as eye contact, physical distance, facial expressions, and use of silence. Because there may be many variations among cultures, as we have discussed earlier, cultural expressions are subject to misinterpretation. "Helpers need to acquire sensitivity to cultural differences to reduce the probability of miscommunication, misdiagnosis, and misinterpretation of nonverbal behaviors (Corey & Corey, 2003, p. 202).
- Assumptions about trusting relationships: Often helpers expect that clients will readily develop a trusting relationship with them. Many clients from different cultures find it difficult to talk about themselves openly or to share private matters freely. Also, they may take longer to trust and develop meaningful relationships with helpers. Being sensitive to this is important in earning a client's eventual trust.
- Assumption about self-actualization: Although many of us value the notion of self-actualization and encourage clients to do the same, many value working toward collective goals (Eastern orientation), and the betterment of the tribe rather than their own gain (Native Americans) (Anderson & Ellis, 1988). Understanding the client's views on this issue is critical before attempting to set goals or proceed with interventions.
- Assumptions about directness and assertiveness: Being direct and assertive may be a highly valued concept for Westerners, yet there are cultures (Asians, Hispanics) where being direct and assertive is considered rude and unprofessional. This is especially true for women who come to America from other countries where being quiet and unassertive is respected; while they struggle with becoming more Americanized, one must be careful in considering how their family and surrounding community might view their new behavior. In some cases, women who succeed in becoming more assertive face conflict and pressure from their family

and friends. We agree with Corey and Corey (2003) that "it is crucial that both you and your clients consider the consequences of making changes that challenge their cultural values" (p. 203).

THINKING THINGS THROUGH

Consider which of these assumptions you may hold and ways in which they may be detrimental to your work with clients from different cultures. Explore ways that you might work toward becoming more culturally sensitive with regard to these assumptions.

DEVELOPING A SOLID KNOWLEDGE BASE

Developing a sound knowledge of diversity is crucial to fully understanding your clients. According to Cournoyer (2005, p. 127), "[K]nowledge involves appreciating and understanding clients' cultures. Such understanding is based on knowledge gained from the professional and scientific literature, resource persons from other cultures, and findings from research studies about communications between your culture and the clients'." Sue and Sue (2003) state that you must be "knowledgeable and informed about the particular group you will be working with." For instance, if you will be working with children, you must know about child development to determine what is appropriate and what deviates from the norm. Similarly, understanding the aging process and related issues is a prerequisite to effective work with seniors. In addition to understanding issues specific to race, gender, culture, and language with respect to older adults, one must "also understand the diversity specific to this population's life stage: increased disability, increased health problems, increased trauma due to loss of significant others, lowered socioeconomic status due to retirement and living on a fixed income, and decreased self-importance stemming from living in and with a society that places a more positive emphasis on youth" (Capuzzi & Gross, 2003, p. 19). Further examples follow.

SCENARIO: Working with Adolescents

You are a new intern at a teen shelter, where you will be working on the girls' unit. You are excited but anxious because this is your first experience working with adolescents. During your group work with the teens, you are struck by the mood swings many exhibit. Also, you worry about their ongoing identity struggles. You begin to confront them about their moodiness and identity issues. Your supervisor is helpful in teaching you about the role of hormones in emotional stability and the difficulties many teens face regarding identity development. She is able to help you see that the group members' behaviors are appropriate and need to be acknowledged as such.

How would you feel if you were an intern in this situation?

SCENARIO: Working with Men

You are a female counseling intern, treating a male client who is struggling with intimacy issues. From the beginning of treatment, you have noted how intellectual he is, and you determine that this is contributing to his problems. You encourage him to express his underlying pain and sadness and become frustrated with his reluctance to deal with these emotions. He seems most comfortable talking about his anger and resentments. Your frustrations prompt you to do some additional reading about men's issues, where you learn that many men do not have experience with expressing such emotions as sadness and hurt. Anger, however, is a feeling men have been encouraged to express more freely.

How might you incorporate your new knowledge regarding men's issues to help your client? What if you had continued to expect him to talk about all his feelings? How do you think he might have felt and reacted?

SCENARIO: Working with the Homeless

You are placed at a homeless shelter, where your role is to help clients "get back on their feet." You work on job skills training and help them learn to find and use community services. Your work with a particular young mother disturbs you. You do not feel that she is motivated to change her situation. You push her, only to feel even more exasperated. The shelter manager educates you about the many types of homeless clients the shelter serves. She explains that some truly do want to return to working and living on their own, but others have a mental illness or substance abuse problem that hinders their ability to function independently. Others prefer to remain homeless for various reasons and thus are not motivated to change their situation. You realize that your client may not want to change her circumstances—or is afraid to change because this is all she knows.

How can this information be helpful to you in your continued efforts to work with this client?

SCENARIO: Working with Quadriplegic Patients

You are hired to work in a rehabilitation facility with people who have spinal cord injury. You are struck by the denial of many of the new patients you interview. Despite the serious nature of their injuries, they seem to believe that they will walk in a short time. On various occasions, you have felt compelled to confront this denial because you worry that it will do them more harm. With your supervisor, you talk about this strong level of denial, and he assures you that this is part of the normal process of grieving a loss. He helps you understand that these patients cannot yet deal with this traumatic event, so they deny it as a way of coping during the transition from being able bodied to being bedridden or wheelchair bound. He encourages you to allow them time to come to terms with their loss and to be supportive during this period. He assures you that they will eventually break through the denial. The individuals who remain in denial for an extensive time are those you need to be most concerned about.

What might have happened if you had continued to push a client to face his quadriplegic status?

SCENARIO: Working with a Lesbian Client

Imagine that a lesbian client attends a session at a family service agency where you are an intern. She wants to talk about her decision to "come out," but you focus on exploring underlying reasons for her sexual preference. You suggest that perhaps this is due to rebellion against parents, or maybe a result of her being molested by a male, or perhaps just a fad. As the session progresses, you notice that she becomes reluctant to share much. You discuss this session with your supervisor and learn that due to your lack of knowledge about gay or lesbian issues, you may have alienated this client. Your supervisor explains that "coming out" is different from questioning one's sexual identity. She encourages you to take a look at your own possible homophobia or heterosexism. You appreciate your supervisor's feedback and learn more about gay and lesbian issues. You contact the client and apologize for not being sensitive to her needs or aware of her issues. She agrees to return to see you to work through feelings touched off by the last session.

How would you feel if you were the helper and your client had returned to work through her issues?

SCENARIO: Working with Minorities

You are a Caucasian intern, raised in a middle-class, primarily white suburb of Milwaukee. You are placed in the inner-city school, where you are exposed for the first time to minorities. You work with students who are having difficulties in school or excessive absences. You are allowed to make home visits when contact by phone is not possible. One of your first such visits is to the Gonzales home, in a barrio on the outskirts of town, to find out more about Anna, an 8-year-old. As you approach the address, you begin to feel anxiety and trepidation. Graffiti and boarded up or dilapidated homes are everywhere. At Anna's home, many children are playing outside in the dirt. Although they are carefree and apparently content, you question whether this environment is adequate. You also note how shabbily they are all dressed. As you approach the door, an older child, about 12, asks who you are. You tell him that you are here as a representative of the school and wish to speak to Mrs. Gonzales. At that moment a matronly, Hispanic woman approaches the door. Her hands are covered with flour, and the smell of tortillas is in the air. She seems hesitant to speak, but motions you in to sit down. She offers you a beverage, but you decline because you are worried about germs.

The home is sparsely decorated, but it does seem fairly clean. Nonetheless, you feel uncomfortable. When you ask about Anna, she takes you into one of the bedrooms in the small duplex. Anna is sleeping at one end of the bed. The older boy says that Anna has been ill these last few weeks and has been sleeping a great deal. The local health clinic diagnosed her as anemic. She was told to rest and to take iron pills it provided. You notice that this room is cluttered and ask who else sleeps here. The boy states that Anna shares this room with her five younger siblings. You are struck by the fact that there is only one bed. You quickly conclude that this family is "enmeshed" and needs help with boundaries. You also wonder whether Anna has been undernourished due to improper cooking. You write your assessments on the referral form and leave abruptly.

The next day, your supervisor calls you in to discuss this assessment. She asks that you expand on your comment about "an enmeshed family with no boundaries." You explain your observations. She gently chides you for lack of sensitivity to diversity issues and proceeds to explain that this family is not enmeshed, but just too poor to afford a larger home or more beds. She provides insight into cultural and socioeconomic issues you had not been aware of before. You are visibly upset. She comforts you, stating, "That is why you are here: to learn!"

How might you feel in this situation? How could this have been handled differently?

SCENARIO: Working with Unfamiliar Religions

You are working at a local women's clinic. You have been assigned the case of a young foreign woman. She feels that certain spiritual rituals are necessary to overcome her current medical problems. You question the appropriateness of these customs and consult with a coworker after your visit with the client. This coworker has worked in the clinic for some time and is quite knowledgeable, but she is not sure of the significance behind these rituals. She encourages you to find out more through research. Later that week, you meet a friend and one of her coworkers for lunch. Your friend's coworker tells of having just engaged in similar rituals. You inquire further and learn that these are common in Eastern religion and philosophy and are appropriate within that context. When you return to work, you share your findings with the coworker who encouraged you to engage in further research.

What might the client have felt in this situation?

SCENARIO: Working in Rural Areas

You have lived most of your life in large cities. Your internship requires that you work at a satellite clinic located in a rural area. You take on your new assignment with excitement and some anxiety. Your first day is overwhelming. You arrive in your new Toyota, dressed in a very professional manner as usual. You are met by stares, which you experience as dirty looks. The secretaries are friendly but dress casually and do not seem as professional as those at the main community center. Your new supervisor's attire is also entirely casual. Throughout the day, you are struck by differences in the dress code, norms, and relations between staff and clients. Later, when you discuss your observations with a fellow intern, he says that he too felt uneasy the first few times he went to outlying centers. He has learned that this is common in rural settings and encourages you to reconsider your dress on your next trip. Although uncomfortable at first, you follow his suggestion and find that you actually feel more at ease and are accepted more freely by staff and clients alike.

Would you have acted any differently? If so, how?

THINKING THINGS THROUGH

Can you identify with any of the scenarios presented above? Use this list to provide your own examples of how lack of knowledge about a specific area of diversity can hinder the helping process.

Learn as much as you can about the aspects of diversity that are applicable to your clients; then gain as much detailed information as possible about the specific traits and circumstances of each client. Taken together, this knowledge can provide you with a more complete understanding of your clients. Many clients may be members of several minority groups at once, for example:

- A Latina female who is lesbian
- A disabled African-American adolescent male
- A poor Caucasian male
- A Native-American family that has recently moved to the city from the reservation

THINKING THINGS THROUGH

Using the above list, consider some issues that might affect these clients, based on their status in various groups. What helping approaches would you recommend?

Understanding Worldviews

Besides a sound knowledge base regarding the needs of certain client groups, one must be able to understand their worldview. Sue and Sue (2003) believe that helpers must have a good understanding of the sociopolitical system's operation in the United States with respect to its treatment of minorities and be aware of institutional barriers that prevent minorities from using human services.

Meeting these criteria will require you to understand the forces of oppression and the role of racism, sexism, heterosexism, ageism, discrimination, and stereotyping in the development of ethnic or group identity.

Oppression

Oppression creates a hierarchical system "whereby acts are designated to place others in lower ranks of society" (Lum, 2004, p. 199). According to Turner, Singleton, and Museik (1994), oppression is a systematic state wherein one segment of the population prevents another from attaining access to scarce and valued resources. In particular, oppression occurs when a person is deprived of some human right or dignity and is (or feels) powerless to do anything about it (Goldberg, 1978). Historically, oppression has been manifested in numerous ways. Extreme forms of oppression include genocide of European Jews and enslavement of Africans. More subtle and insidious are the political, economic, and social forms of oppression of undocumented immigrants, welfare recipients, and minority groups. Atkinson, Morten, and Sue (1993) cite specific examples: under-representation of certain groups in the census, paying undocumented workers wages that are illegally low, racial and ethnic slurs in everyday communication, and physical attacks by racists. Oppression can easily have a negative influence on the helping process. Discrimination, poverty, stereotyping, exploitation, and feelings of powerlessness make many members of oppressed groups hesitant to trust others different from themselves or to seek help from the outside.

THINKING THINGS THROUGH

Have you worked with any clients who have experienced oppression? Please explain. How might you be sensitive to this issue in your work?

Development of Group Identity

Erickson (1950) considers identity to be a process at the core of the individual and at the heart of his or her communal culture. (The term *communal culture* is more or less equivalent to *ethnic group.*) When an ethnic group is considered inferior or is discriminated against by others, the group identity is certain to be adversely affected. McGoldrick (1982, p. 5) writes:

> If people are secure in their identity, then they can act with greater freedom, flexibility, and openness to others of different cultural backgrounds. However, if people receive negative or distorted images of their ethnic background or learn values from the larger society that conflict with those of their family, they often develop a sense of inferiority and self-hate that can lead to aggressive behavior and discrimination toward other ethnic groups.

Newman and Newman (2003, p. 324), in their discussion of adolescence, state:

> In the United States, a history of negative imagery, violence, discrimination, and invisibility has been linked to African Americans, Native Americans, Asian Americans and Hispanics. Young people in each of these groups encounter conflicting values as they consider the larger society and their own ethnic identity. They must struggle with the negative or ambivalent feelings that are linked with their own ethnic group because of cultural stereotypes that have been conveyed to them through the media and the schools, and because of the absence of role models from their own groups who are in positions of leadership and authority.

Personality formation, career or vocational choices, and psychological problems are all tied to identity issues. One's ethnic group becomes a significant reference group whose values, outlook, and goals are taken into account as one makes important life choices (Newman & Newman, 2003, p. 324). Phinney (1996) stresses that because ethnic identity varies across ethnic groups and among individuals within groups, ethnic identity is therefore more aptly viewed as a psychosocial rather than a demographic variable (see Newman & Newman, 2003, p. 324). Understanding issues tied to ethnic identity and ethnic group formation can help you be more receptive to working with diversity and develop effective approaches to helping. However, caution must be taken in not generalizing based on group membership; we cannot assume that just because clients come from a specific cultural background or belong to a specific ethnic group, they necessarily have specific characteristics. "In other words, human service professionals have knowledge of the group from which clients come, yet do not make assumptions about clients' way of being" (Neukrug, 2003, p. 142).

SCENARIO

You work in a college career center with minority students. Because you are the only bilingual member of the staff, you work primarily with the Spanish-speaking students. One day, you meet with a high school senior who is considering attending college. She is somewhat hesitant at first to share her concerns. She doubts that she can get accepted and succeed because of her poor English. Her family and peers say it is ridiculous to think she could ever make it in college. Another fear relates to her undocumented immigrant status. Apparently, she would have to return to her country, apply for a student visa, and register as a foreign student, paying much higher fees than a resident. Furthermore, she is ineligible for loans. She begins to cry upon leaving, saying, "Thank you for your time. I guess I should have listened to my family when they told me I wouldn't be able to go to college." You feel helpless yet offer her support. Months later, you happen to see her at the grocery store, where she tells you of her recent marriage and pregnancy. She seems content enough, but a statement about being destined to stay home and raise kids haunts you.

How were her self-concept and confidence level affected by her ethnic development? What other forces are at work here? How might you have felt if working with her?

THINKING THINGS THROUGH

From your agency experience, personal life, or academic learning, provide an example of the effects of ethnic development on an individual. Were they negative, positive, or both? Elaborate.

In addition to understanding ethnic group identity issues, Corey (2005, p. 25) adds that culturally skilled practitioners "possess knowledge about the historical background, traditions, and values of the client populations with whom they work. They know about minority family structures, hierarchies, values, and beliefs. Furthermore, they are knowledgeable about community characteristics and resources." Learning about these issues takes both time and diligence, yet the more the helper knows about the client's history, culture, family, and community, the more effective he or she is likely to be. McGill (1992, p. 340) suggested the idea of the cultural story as a nonthreatening way to find out relevant background about a client: "A cultural story refers to an ethnic or cultural group's origin, migration, and identity. Within the family, it is used to tell where one's ancestors came from, what kind of people they were and current members are, what issues are important to the family, what good and bad things have happened over time, and what lessons have been learned from their experiences. At the ethnic level, a cultural story tells the group's collective story of how to cope with life and how to respond to pain and trouble. It teaches people how to thrive in a multicultural society and what children should be taught so that they can sustain their ethnic and cultural story."

THINKING THINGS THROUGH

Can you think of a cultural story from your family or one you have heard? Please elaborate.

Cultural Assessments

Sperry, Carlson, and Kjos (2003) stress that cultural factors and dynamics reflect the individual's cultural identity. Similar to the cultural story is a cultural assessment. Accordingly, in conducting a complete cultural assessment, the following factors are important:

- Client's ethnic identity
- Client's age
- Client's race
- Client's gender and sexual orientation
- Client's religion
- Client's migration and country of origin
- Client's socioeconomic status
- Client's acculturation in terms of assimilation (strong affiliation with dominant culture), tradition (rejection of the dominant culture), and biculturalism (assuming the cultural norms of both cultures without denying either)
- Client's language
- Client's dietary influences,
- Client's education
  (Sperry, Carlson, and Kjos, 2003, p. 61.)

Taking time to gather this information is essential in better understanding one's clients from a cultural vantage point. The results will likely add dramatically to the knowledge base of the client and his or her culture. Additionally, observation of "client's speech, client's clothing, grooming, posture, build and gait, facial expressions, other bodily movements, and general appearance" (Young, 2001, pp. 155–156) is helpful in completing a picture of your client.

Assessing One's Knowledge Base

Many of you may choose to work with a population you are familiar and comfortable with, so you may already have a sound knowledge base to draw from. However, you may decide to take the risk of working in unfamiliar territory, with population groups you know little about. For you, it is especially important to learn all you can about your clients. Willingness to learn, together with a concerned and caring approach, can help instill trust in your clients and promote a positive working relationship.

Even when you are familiar with a certain area of diversity, there is still much to learn. It is important not to generalize or stereotype. Recognizing your limitations, you can explore ways to learn and understand more about your clients. Asking clients directly for information about their group can often be useful. It has the added advantage of conveying to them that their views are important and valued.

THINKING THINGS THROUGH

Identify one area where adding to your knowledge base could help you in your fieldwork placement or in your career as a human service worker. List potential avenues for gaining this knowledge.

Ethnography

Ethnography is a method used in anthropology. Day (2004, p. 409) adds, "Ethnography is the description of culture, an ethnographic perspective seeks to comprehend the viewpoint of people inside the culture." Ethnographic researchers seek to learn about other cultures, trying to understand their worldview (Locke, 1992). Thornton and Garrett (1995) elaborate an ethnographic research method for studying cultural groups. This method develops skills that help students understand the knowledge, values, and skills needed for practice with people of different cultures, as well as giving students a cross-cultural experience. Ethnographic researchers "seek to learn from those they are studying and to understand the perspectives, customs, behaviors, and meaning within the context of their culture" (Thornton & Garrett, 1995, p. 68).

Ethnographic researchers try to avoid passing judgment, responding to preconceived stereotypes, or interpreting behaviors based on their personal experience. Such an open-minded approach can give social science researchers invaluable information about other cultures. Ethnographic interviewing and informal observation of clients in their environment can yield important knowledge about a person's or group's cultural values, family structure, child-rearing practices, religious beliefs, experiences with discrimination and oppression, and level of acculturation. As human service workers, you may not be able to engage in specialized ethnographic interviews to observe all your clients in their own setting. However, you can incorporate some ethnographic concepts into your work. By making an effort to understand your clients' point of view, you can convey the message that they are the experts and you the learner. Moreover, by initially being inquisitive about their culture, you indicate that you are truly interested in understanding their situation and respect their worldview. Also, you can learn from careful observation of your clients while they are at the agency—for instance, observing their style of dress, their interactions with others, who accompanies them, and what mode of communication they primarily use. You may also want to visit their community to learn more about the environment in which they live because this may teach you about other aspects of their culture and lifestyle. The main emphasis in ethnography is being open, flexible, and willing to understand people in *their* world, not yours.

Learning as an Ongoing Process

You can also learn about diversity through literature and textbooks, research, specialized courses, workshops, and interviews. Because change is ever present in working with people, learning must be an ongoing process. Remaining open and flexible are therefore essential to your success in working with human diversity. Awareness and a solid knowledge base are both necessary, but they are not sufficient if you lack the ability to apply them.

Skill Application

Skill application entails incorporating awareness and knowledge into the services provided; this includes "those interpersonal abilities based on awareness and knowledge that enable you to communicate effectively and provide helpful services to people from other cultures" (Cournoyer, 2005, p. 127). Sue and Sue (1990, p. 170) refer to this third component of becoming diversity sensitive as "developing appropriate intervention strategies and techniques." The effectiveness of working with diverse client groups "is most likely enhanced when the [helper] uses [helping] modalities and defines goals consistent with the life experiences/cultural values of the client. . . . Because groups and individuals differ from one another, the blind application of techniques to all situations is ludicrous . . . differential approaches" are needed. This requires knowledge of some basic facts, which are discussed in the paragraphs that follow (adapted from Sue & Sue, 2003).

Economically and educationally disadvantaged clients may not be oriented to engaging in advanced treatment modalities, such as journaling, logging, charting behavior, reading self-help books, or using cognitive restructuring techniques. When efforts are made to work with clients at their level, they will be more comfortable and respond better.

SCENARIO

You are working with a depressed middle-age woman from Eastern Europe. She is fairly articulate in English and works as a housekeeper at a local hotel. Her primary problem is related to her low self-esteem, which seems rooted in negative messages she received as a child. Previously, you have had success working with these issues by using such cognitive interventions as rational emotive therapy and journaling. You assign her homework based on these techniques. The next few times you see her, she says she has forgotten her homework. On several occasions, she cancels her appointment with you. Finally, you confront her about her apparent resistance to completing the homework. She becomes tearful and tells you that she cannot read nor write. You are surprised by this revelation. In your eagerness to help this client, you were not careful to consider her limitations and needs. How might you proceed at this point? What other techniques might you have used from the outset, had you been more conscious of this client's issues?

In some cases, *encouraging self-disclosure* may be futile or even detrimental to the helping process because it may be incompatible with the client's cultural values. Many cultures and families believe that talking about personal issues is not appropriate. Some Asian Americans, Latinos, and Native Americans consider this to be exposing the family's private affairs.

SCENARIO

You are working with a female Asian college student in a study skills group. She appears shy, reserved, and overly compulsive about her studies. You feel that she is pressured from home to be the best; you ask her to tell you more about her family circumstances. She is very reluctant and says only, "Everything is fine at home." Discussing the group with a coleader, you become aware that asking this client to self-disclose may have been contrary to her cultural values. You hope your approach has not frightened her away from the group.

What do you think? What would have been a more culturally appropriate approach? Even when self-disclosure is accepted, many groups fear the repercussions of sharing personal information.

SCENARIO

With the recent increase of border patrol activities in California, undocumented immigrants from many countries (not only Mexico) are fearful of deportation. You notice a decrease in attendance at the neighborhood community center where you conduct support groups for Spanish-speaking women. Those women who do attend seem more reluctant to talk about themselves or their families. When you ask them about their reactions to recent news stories regarding deportation, they hesitantly share their fears.

How might you respond to these fears? Are they rational?

The *process-oriented, time-consuming, and ambiguous nature of the helping process* may be contrary to the values of your clients. Many clients value being highly involved with their families, which could be misinterpreted as dependency. It is important to be aware of your own values as well as those of your clients.

SCENARIO

An Egyptian woman applies for a volunteer position in the agency you are managing. You become her supervisor and gradually get to know her quite well. At 25, she still lives at home with her parents and siblings and seems reluctant to move out on her own. Also, she is very hesitant about dating anyone outside of her culture. You tend to view her behavior as weak and immature.

Is this an appropriate assessment? What cultural factors may be at work?

Some clients prefer a more *active or directive approach,* different from the more inactive or nondirective approach that many helpers use. Such clients may seem demanding and dependent because of their desire for answers and specifics on how to cope. It is important not to jump to conclusions by misjudging this wish for more directive intervention.

SCENARIO

Many cultures do not emphasize awareness and insight building. Clients may instead want specific directives on how to cope or concrete feedback. Providing them with the information they need might then be an appropriate goal.

Competencies

Translating this knowledge into action requires ongoing self-evaluation and a willingness to seek client feedback about your effectiveness. You should also be open to using a variety of response modalities (Sue & Sue, 2003) in working with diverse clients. Effective helping professionals are able to "apply generic interviewing and helping skills with culturally diverse populations while being aware of and able to apply specialized interventions that might be effective with specific populations" (Neukrug, 2002, p. 142). Soloman (1976, p. 26) advocates interventions that "enlist the mutual involvement of the client and the worker"; he also acknowledges the client's experience with oppression. Similarly, Lum (2004, p. 254) advocates engaging in "joint intervention strategies," in which the following themes are considered:

- Oppression versus liberation
- Powerlessness versus empowerment
- Exploitation versus parity
- Acculturation versus maintenance of culture
- Stereotyping versus unique personhood

To incorporate knowledge of cross-cultural issues into practice, helpers must not only "start where the client is at" but also help clients move beyond that starting point (Goldstein, 1993). This entails developing goals that are based on the client's perception of needs, which may differ from that of the helper. Similarly, "strategies for change must also fit within the client's culture" (Thornton & Garrett, 1995, p. 69) and be mutually agreeable to the client and the helper. It is helpful here to apply the ethnographic principle discussed earlier: Clients are viewed as the experts on their lives and cultures, and workers are the experts on the problem-solving and helping process. In this framework, clients and helpers work together to meet client needs in ways that are both effective and culturally acceptable (Thornton & Garrett, 1995). Corey (2005) concurs that multiculturally sensitive practitioners are those who use methods and strategies and define goals consistent with the experience and cultural values of their clients. Furthermore, he contends that practitioners must continually modify and adapt their interventions to accommodate cultural differences.

According to Sue (2001), becoming culturally competent is multidimensional in that one must become competent in areas of attitudes, knowledge, and skills, as well as the specific race and cultural attributes of various minority groups. Keeping in mind that not all cultures focus on individual change, Sue (2001) also underscores the importance of considering culture in targeting change where the system becomes the focus. Sue (2001, p. 816) asserts that "cultural competence for one group is not necessarily the same for another group." Cormier and Hackney (2005), Daw (1997), and Sue (2001) all address the importance of cultural competence. Here we offer some of their recommendations:

1. Learn about your own cultural background and affiliations and how these might have any impact on your work with clients.
2. Be willing to learn about others different from you by becoming immersed in other cultures. Pursue opportunities to interact with others from different cultures and maintain an open attitude about what they have to say.
3. Remain honest and realistic about your own range of experiences. Be aware of issues tied to power, privilege, and poverty and how these might affect different ethnic groups. Ask yourself if any of the positions you hold contribute to oppression, power, or privilege.
4. As a helper recognize that it is your responsibility, not your clients', to become educated about the various dimensions of culture. Seek out information to help you become more knowledgeable and always attempt to demonstrate a sincere interest in your clients' cultures.

Whatever helping approach is used, the helper must be skillful at communicating in a way that the client can relate and respond to.

Effective Communication

It is essential to understand communication patterns—both nonverbal and verbal (Corey, 2005; Cournoyer, 2005; Lum, 2004; Neukrug, 2002). Human service workers need to develop and maintain clear communication with clients, colleagues, and the community at large. Brill and Levine (2002, pp. 88–91) have identified salient aspects of communication that are useful in working with diverse clientele. We present them briefly in the paragraphs that follow.

Attitudes and beliefs of both client and helper help in understanding the messages being sent and received Beliefs affect all communication and can so distort sending and receiving that breakdown occurs. Earlier, we examined the impact of belief systems in general, but such beliefs can hinder or enhance the communication process in particular.

SCENARIO

A Caucasian helper who is working with an Asian woman views her as resistant for not wanting to confront her husband about his excessive time at work, which leaves her with the children at home. What belief system does the helper hold that may differ from the client's? Could this viewpoint damage the helping relationship and process?

Recognizing similarities and differences between people This can help you determine whether there is a common understanding and acceptance of which modes of communication are to be used. Brill and Levine (2002, p. 88) state, "Differences in age, sex, cultural identification, social position, ethnic background all affect our ability to communicate with each other. Awareness of and sensitivity to the meaning of these differences is the first incentive for open communication channels."

SCENARIO

In our work, we often misjudge clients' comprehension. Words such as *ambivalent* and *insight* may be commonplace for us, but some clients do not really understand their meaning. Many clients do not ask for clarification, so helpers must be sensitive regarding the appropriateness of their vocabulary.

Have you ever experienced a situation where you used words or phrases that clients did not understand? How did you know? How did you respond?

Understanding a client's capacity to use both verbal and nonverbal communication This may require that you learn about any impairments the client may have, such as reading, hearing, or speech problems. Awareness of the client's level of education and ability to verbalize will help you tailor treatment approaches to meet individual needs.

SCENARIO

If you are working with a client who is unable to read, try using techniques that do not involve reading or writing. More action-oriented approaches may be more appropriate.

Using various means of communication Besides spoken language, this includes tone of voice, facial expressions, silence, gestures and movements, physical appearance, body sounds, and touch.

SCENARIO

Touch is a very important way to communicate with the elderly because tactile stimulation is one of the most lasting senses in most people. However, one must take care to use touch appropriately.

Willingness to seek feedback This is essential in preventing misinterpretations and feelings of being misunderstood.

SCENARIO

Both Lupe and Randy have experienced cases where clients misinterpreted something that was said. By active listening and observation of nonverbal communication, the confusion became evident. The helpers could then check for comprehension and clarify their intended message.

Ensure that gestures and symbolism are understood Slang may not be transferable from one culture to another, even if everyone speaks the same language. For instance, certain Spanish words can have different meanings in different Latin cultures. Being sure to use the correct word is important. Some words may be insulting or have a totally different meaning when used with other Latin cultures.

Communicate at the client's pace Providing too much complex information too quickly or being overly simplistic can erode communication. Clients may feel overwhelmed, resentful, and angry. They may leave without having understood what was said or may withdraw from the helping process altogether as a result of their discomfort.

SCENARIO

When children are referred for an evaluation for possible attention deficit hyperactivity disorder (ADHD), the helper may sometimes quickly give parents all the information about this condition. This tempo may not allow them to express their feelings about their child's condition. Can you think of a situation from your agency where you provided too much information too quickly? Or where you were too simplistic in your approach? How do you think this affected the client?

Interference and distractions It is sometimes impossible to isolate clients from distractions, but at least be aware that their comprehension may be affected.

SCENARIO

A client brings her infant daughter and toddler son into a session. Both the client and the helper are unable to give their undivided attention to the session.

THINKING THINGS THROUGH

Conduct a self-assessment regarding your communication skills. Which are strongest? Which are weakest? How might you improve your overall communication abilities?

Alternative Approaches to Helping

Human service professionals working with diversity should consider alternative treatment options or out-of-office strategies. Examples include engaging in outreach, acting as a consultant or an ombudsman, and acting as a facilitator and advocate for the client. These functions may differ from what you are accustomed to or comfortable with, but they might be conducive to helping clients from different backgrounds. In later chapters, we will elaborate further on the role of outreach, advocacy, and consultation in human service work.

To become more adept at understanding clients from different cultures, we suggest that helpers become more involved in activities with minority individuals at community events, celebrations, and neighborhood groups. Also, a willingness to seek out educational, consultative, and training experiences "to enhance their ability to work with culturally diverse client populations" (Corey, 2005, p. 25) is recommended. Finally, we encourage being open to consult regularly with other "multiculturally sensitive professionals regarding issues of culture to determine whether or where a referral may be necessary" (Corey, 2005, p. 25).

SCENARIO

Taking on an outreach role, Randy visited people with mental illness who had not shown up for their psychiatric appointments after discharge from the hospital. The clients could then be treated in a nontraditional way, which seemed to be instrumental in preventing rehospitalization.

SCENARIO

Working with elderly and disabled people, Lupe found that it was difficult for many of them to attend sessions at the guidance clinic. She obtained permission from the administration to conduct home visits to give these clients counseling and case management services they might otherwise not have received.

Can you think of an alternative treatment approach you have used? Were you comfortable with it? How do you feel that it benefited the client?

Engaging in alternative treatments may mean persuading the agency to allow you the flexibility to use different approaches. Some organizations may be rigid and closed in their thinking. By remaining persistent, assertive, and committed to working with diverse clients, you may sometimes gain the support of your agency.

THINKING THINGS THROUGH

Describe a case where you worked with a minority client who benefited from alternative methods of treatment. Elaborate and discuss the barriers to helping that you faced. If you have no actual experience in this area, use the knowledge presented thus far to give some examples.

To be a competent human service worker, be aware of your helping style, recognize your limitations, and anticipate their impact on your clients. Each of you will develop an individual style, based on your unique personality and experiences, both positive and negative, in working with clients. Your style will evolve as you gain more experience and diversify. Remain alert to cues that indicate problems. This can help you explore alternatives and understand the underlying reasons for the problems. Perhaps you are unaware of a biased attitude that gets in the way of the helping process. Or maybe your approach is not appropriate to the particular client group, although it has worked with others with a similar background. In some cases, referral is the most appropriate plan of action. Learn to accept that you cannot be successful with every client or group. This will minimize feelings of inadequacy and enable you to work on strengthening your skills.

Practical Suggestions

Neukrug (2004, pp. 206–208) and Egan (2002) offer the following suggestions for working with individuals from different ethnic and racial groups:

1. Encourage clients to speak their own languages if at all possible.
2. Learn about the culture of your clients.
3. Assess the cultural identity of your clients.
4. Check the accuracy of the nonverbal communication of your clients.
5. Use alternative modes of communication.
6. Encourage clients to bring in culturally and personally relevant items.
7. Vary the helping environment; advocate an environment that supports professional tolerance.
8. Don't jump to conclusions; resist cultural stereotyping or generalizations.
9. Know yourself; be aware of your own cultural roots and your own biases.
10. Know appropriate skills; seek opportunities for additional training in the working knowledge and skills related to becoming diversity sensitive.

Other aspects of diversity that are important to consider include helping clients with diverse religious backgrounds, working specifically with women or men, helping gay men and lesbians, working with those afflicted with HIV, helping the homeless and the poor, assisting older persons, treating the mentally ill, and helping those who suffer from disabilities. For a more in-depth view of these specific areas of diversity, please refer to Neukrug's (2004) *Theory, Practice, and Trends in Human Services*, Chapter 7, where he offers practical suggestions that we have found to be very helpful.

▪ DIVERSITY IN AGENCIES

Two important practice issues must be taken into account in working with diversity: agency barriers to effective service delivery and the risk of helpers becoming encapsulated.

Barriers

For many minority groups, the road to help is not an easy one. It can be further complicated by agency barriers that hinder effective service delivery. Sue and Sue (1990) write that helpers must "be aware of institutional barriers that prevent minorities from using [human] services." Similarly, Adams and McDonald (1968) address the "cooling out" of poor clients in some clinics: Poor people are said to be "unworkable" because of their differences from clinic staff. This "cooling out" process enforces a disengagement, and the poor are not allowed to conceive of themselves as adequate (Adams & McDonald, 1968, p. 140). This process continues to occur today in some agency settings. Besides poverty, this process could be applied to any other kind of diversity, given the continued discrimination on the basis of age, gender, race, socioeconomic status, and other factors in our society.

The relevance of this process can be better understood when seen from an agency's perspective. The following list summarizes Adams and MacDonald's (1968) discussion of tactics used by some agencies, consciously or unconsciously, to "cool out" certain client populations. Consider how they may apply to either your agency setting or others you know.

Agency Tactics for "Cooling Out" Certain Clients

Ambiguous Standards: There is no consistency or uniformity, creating confusion for clients, who may feel ambivalent about returning for further assistance.

Status Differential: Helpers may use a large vocabulary and pose as having a higher status than their clients. Clients feel inferior and reluctant to return for help.

Inappropriate Use of Inquiry: Helpers may ask clients to expose all when they are not ready, willing, or capable. The system is insensitive to the client's needs and capacities.

Inflexible Approach to Helping: This involves demanding that clients comply with the worker's or the agency's expectations.

Being Tentative About Services: The worker or agency may be tentative about providing services, unsure about how to help, and reluctant to engage fully with the client. This behavior may create a sense of futility in the client.

Gradual Disengagement: Helpers may gradually disengage from clients in an effort to discourage them from continuing treatment.

Sidetracking: Clients may be prematurely referred elsewhere for a variety of reasons: not feeling comfortable with the client, not knowing how to help him or her, not believing the client is capable of changing.

Denial of Services: Clients are denied services because of the worker's or agency's discomfort with or negative views of the client's group.

▏ THINKING THINGS THROUGH

Can you identify with any of the tactics listed? Which ones apply to your agency? Which of these behaviors have you witnessed in yourself? What has been the impact? What efforts might you make to change this?

It is important to recognize that many of these behaviors are not deliberate on the worker's part. They may be ingrained in informal agency policies. Most helpers do not intentionally discourage certain population groups from seeking assistance. Also, note that referral or denial of services is sometimes the most appropriate course of action. Nonetheless, take a critical look at whether the "cooling out" process may be taking place at your agency. Because this process is often subtle, it may be quite difficult to detect. If you remain sensitive to any of these behaviors in yourself or others at your agency, you can work on corrective action.

Working in the Community

Also consider aspects of work in the community. The human service worker can serve as a community organizer, mobilizing diverse ethnic groups in a community to achieve reciprocal empowerment that encourages working together toward social justice. For this to occur, the helper must understand and respect diversity and use it in forging unity.

To be effective within the community, human service professionals must be open to learning more about collaborative, empowerment-based practice. This may require nontraditional participation as activists, advocates, and mobilizers of community potential.

The Risk of Encapsulation

New workers faced with the difficulties of becoming professionals may become static in their work. They may feel so threatened and frightened by the difficult tasks they are asked to perform that they quit learning and growing—thus becoming encapsulated (Woodside & McClam, 1994, p. 274). Cultural encapsulation involves "the tendency to judge other people's behavior from our own narrow perspective (Young, 2001, p. 154). Hackney and Cormier (2005, p. 9) further stipulate that "cultural encapsulation involves defining reality according to one set of cultural assumptions and stereotypes, becoming insensitive to cultural variations among individuals and assuming that one's personal view is the only real or legitimate one, embracing unreasoned assumptions that one accepts without proof." It goes without saying that becoming encapsulated severely impairs the helping process.

The term *encapsulated counselor* was proposed by Wrenn in 1962, which suggests that this problem is not a new one to the field of human services. Corey, Corey, and Callanan (2003, p. 113) list the following characteristics of an encapsulated counselor:

- Defining reality according to one set of cultural assumptions
- Showing insensitivity to cultural variations among individuals
- Accepting unreasoned assumptions without proof and without regard to rationality
- Failing to evaluate other viewpoints and making little attempt to accommodate the behavior of others

According to Sue and Sue (2003, p. 9?), "[e]ncapsulation results in helper's roles that are rigid and impart an implicit belief that the concepts of 'healthy' and 'normal' are universal." Pedersen (2000) adds that the encapsulated helper, by thinking in such a narrow manner, is trapped, likely to resist adaptation, and unlikely to consider alternative approaches to working with different populations.

Does encapsulation occur only with counselors? Can only a majority-culture helper become encapsulated? The reality is that no one in the human services field is automatically immune to encapsulation. Various authors (Ho, 1991; Pedersen, 1991; Ridley, 1989) indicate that helpers with minority backgrounds are just as likely to become so narrowly focused in their work that they become encapsulated. Terms such as *unintentional racist* and *reverse discrimination* have been used to describe this process whereby people from minority cultures begin to discriminate against those who are different from them.

Examples of Encapsulation

- Working with a Latino family where machismo is strong, a majority-culture helper may try to change this cultural value.
- You believe that once a person turns 18, he or she should be either working and living independently of his or her parents or in college. In many cultures, moving out is not the norm. Because of the high cost of living, many young adults are unable to make it alone; they cannot move out, or they may have to return home.
- A foreign student says, "All Americans are the same. They only think of themselves, not their families or their community." This is a case of reverse discrimination.
- You attempt to lecture a gay client to change his sexual orientation because you do not believe it is morally correct.
- You have deep religious beliefs and attempt to sway clients to believe as you do, regardless of whether or not they are interested.

Regardless of our cultural background, we are all potential victims of encapsulation. Self-awareness and cultural reevaluation can help us avoid becoming encapsulated. Woodside and McClam (1994, pp. 274–275) state:

> One way to address the problem of encapsulation is through a commitment to lifelong learning. Part of learning is making changes in behaviors and attitudes: change and growth can be very exciting parts of the learning process, though difficult at times. Part of the change process is unlearning, which can counteract the encapsulated worker's tendencies to cling to truths and to accept the values of the traditional culture. To avoid or counteract encapsulation, workers must constantly try to develop professionally by revising and updating knowledge, skills and values of the profession. Clients may need to work on unlearning old, negative, maladaptive coping before they are able to learn more adaptive coping skills. Likewise, we ourselves are constantly challenged to unlearn old ways of working in agency settings in response to change.

THINKING THINGS THROUGH

Can you think of any examples of encapsulation in yourself or in others? Explain why these might affect the helping relationship negatively. Offer recommendations on how to cope with this problem.

We conclude by advocating that "culture responsiveness begins with RESPECT." A model for respecting culture initially developed by the Boston University Diversity Taskforce has been edited and adapted by Butterworth (Kaiser, 2004) as follows:

R   recognize and respond to differences
E   explore assumptions, beliefs, and social structure
S   seek options and solutions
P   pursue the "power" differential
E   express empathy and educate
C   consider concerns and fears of patient and family members
T   talk to foster a therapeutic alliance

We find this model useful in our work with clients, colleagues, other organizations, and the community at large.

■ CONCLUDING EXERCISES

1. Describe diversity in your fieldwork agency—your coworkers and the clients your agency serves. What differences are you aware of? How do you see the professionals in your agency accepting and dealing with diversity?

2. Describe diversity in your personal life—your immediate family, friends, or relatives. How do you see these people, and yourself, accepting and dealing with the difference?

3. Imagine that you are traveling abroad with a group of coworkers to a conference on human relations. Your agency is paying your way in hopes that you can conduct in-service training on pertinent issues. As a foreigner in this country, you feel anxious because of your lack of familiarity with the language, customs, and culture. On the second day of the conference, your supervisor calls from the United States with an urgent request. One of your fellow travelers had been assigned to prepare for a 30-minute round-table discussion on diversity issues at the conference. This person has suddenly fallen ill with apparent food poisoning. He asks that you please take over. You agree, yet soon after feel completely overwhelmed due to your lack of preparation and resources. And the coworker has misplaced his notes, which you might have been able to draw from. You have only one day to prepare. A coworker encourages you to use your knowledge and experience.

   How would you proceed? Describe what introduction you would give at this round-table discussion. What exercises or activities might you prepare to involve the participants and cover the topic effectively and thoroughly? Remember, your audience is international and representative of various disciplines in the human relations field. Many are not even familiar with the concept of diversity itself.

CHAPTER SIX

Ethical and Legal Issues

AIM OF THE CHAPTER

The topic of ethics in the human service field has great importance because helping others involves a vast array of challenges. What's more, each profession (counseling, social work, psychology, and others) has developed its own standards, guidelines, and codes. This underscores the need to understand what the issues are and how they might affect your work with clients. As you enter your fieldwork, you must be fully oriented to the ethical and legal issues encountered in the human service field.

In this chapter, our discussion will be twofold. First, we will discuss some common ethical and legal issues you are likely to face in the field, and then we will present some special ethical issues you may also encounter. Our discussion of common ethical and legal issues will orient you to those issues covered in most professional codes of ethics. We will address informed consent; various aspects of confidentiality (and exceptions to it); abuse issues of children, dependent adults, and the elderly; danger to self (suicide); intent to harm others (homicide); and legal proceedings.

Our discussion of special ethical issues will focus on emotional abuse, domestic violence, and AIDS. In considering ethnic-sensitive practice, we present some important aspects to consider in working with diverse populations. We also explore dual relationships and their impact on work with clients.

You may not be confronted with every ethical and legal issue discussed here. Nevertheless, it is valuable to have an overview as you begin your fieldwork experience. Knowing the fundamentals can help you work more effectively with any ethical issues that may arise. As you begin your fieldwork, you must be familiar with the tenets of sound ethical decision making.

ETHICAL DECISION MAKING

The values and ethics of the human service profession must govern in every aspect of practice, each decision, and every action you undertake as a professional. Cournoyer (2005, p. 82) writes:

> This dimension supersedes all others. Ethical responsibilities take precedence over theoretical knowledge, research findings, practice wisdom, agency policies, and, of course, your own personal values, preferences, and beliefs. Ethical decision making is a central component of professionalism. . . . Because you affect, for better or worse, the lives of the people you serve, you bear a substantial burden of personal and professional responsibility.

Ethical decision making is fundamental. Without it, you may not be effective in your work, and you may not "legitimately claim professional status. To be effective and professionally responsible, you need means to identify, address, analyze, and resolve the ethical issues inherent in virtually every action you take as a social worker" (Cournoyer,

2005, pp. 82–83). Human service workers must understand the dimensions involved to confront the complex ethical issues that often occur.

Dimensions of Ethical Decision Making

According to Cournoyer (1996, 2005) the following essential dimensions must be considered when making ethical decisions:

> Understanding your legal obligations as a professional
> Understanding the fundamental values and ethics of your profession
> Identifying ethical and legal implications
> Knowing how to determine which obligations take precedence in cases of conflicting obligations

When the ethical and legal issues are clear, conforming to stipulated standards is a straightforward matter. However, in real life, the applicable principles are often in conflict, creating confusion and uncertainty about which direction to take. "Deciding which obligation takes precedence is the most complex and challenging aspect of ethical decision making" (Cournoyer, 2005, p. 83). Corey and Corey (2003, p. 217) concur that ethical practice dictates much more than simply "knowing and following a professional code of ethics." Similarly, Neukrug (2002, p. 161) states, "Although the use of ethical guidelines is crucial when dealing with ethical dilemmas, good ethical decision making should be based on more than the sole use of these codes."

Lowenberg and Dolgoff's "ethical principles screen" (1992, p. 60) is a useful tool to consider in determining which ethical obligations conflict:

> Principle of the protection of life
> Principle of equality and inequality
> Principle of autonomy and freedom
> Principle of least harm
> Principle of quality of life
> Principle of privacy and confidentiality
> Principle of truthfulness and full disclosure

In applying these principles, one must consider the totality of the case at hand. Although you may fear that you are violating some of the client's basic rights, this may be justifiable to uphold higher principles such as protection of life.

Principle Ethics and Virtue Ethics

The distinction between *principle ethics* and *virtue ethics* is discussed by numerous authors (Cottone, 2001; Corey, Corey, & Callanan, 2003; Forrester-Miller & Davis, 1995; Hill, Glaser, & Harden, 1995; Jordan & Meara, 1990; Loewenberg & Dologoff, 1992; Meara, Schmidt, & Day, 1996). Corey, Corey, and Callanan (2003, p. 13) describe principle ethics as "a set of obligations and a method that focuses on moral issues with goals of (a) solving a particular dilemma or set of dilemmas and (b) establishing a framework to guide future ethical thinking and behaviors." In short, they are "rational, objective, universal, and impartial principles in the analysis of ethical dilemmas." Typical questions asked include: "What shall I do?" and "Is this situation unethical?" The focus is on acts and choices. In virtue ethics, the focus is on "the character traits of the [helper] and nonobligatory ideals to which professionals aspire rather than on solving specific ethical dilemmas" (Corey, Corey, & Callanan (2003, p. 13). The emphasis is on the character of the helper. He or she consciously aspires to behave ethically, asking, "Am I doing what is best for my client?" Woodside and McClam (1994) consider moral responsibility to be crucial in ethical decision making. They contend that "a commitment to rational thinking and a knowledge of moral principles are necessary components of moral reasoning" (p. 267).

Recognizing the role of principles, virtues, and morals will give you a foundation on which to build. Additional assessment skills will equip you to make more ethical decisions when confronted with dilemmas you may not be sure how to handle.

Ethical Dilemmas

Even if you are aware of ethical guidelines and principles, you will likely encounter situations that such codes do not specifically address. Such ethical dilemmas are often difficult to tackle. Unlike clear-cut issues, ethical dilemmas are confusing. Corey, Corey, and Callanan (2003) note that rarely will one find clear-cut answers when dealing with ethical dilemmas, as these dilemmas are too complex and ambiguous to solve simply. Perhaps the greatest challenge most of us will face is the translation of ethical principles into direct practice. Gibson and Pope (1993, p. 330) echo this, stating, "It is the translation of a code's principles into practical directions for conduct that is the greatest challenge for most of us."

Many authors (Corey, Corey, & Callanan, 2003; Faiver, Eisengart, & Colonna, 2004; Herlihy & Corey, 1994; Russel-Chapen & Ivey, 2004) suggest various constructs that may guide you when making decisions about ethical dilemmas. These include:

1. Autonomy that encourages helpers to assist clients in developing a sense of empowerment, independence, self-determination, and an acceptance of different values
2. Nonmaleficence, which refers to doing no harm, even inadvertently
3. Beneficence that encourages promoting mental health and wellness
4. Justice, which relates to a commitment to fairness, quality of service, allocation of time and resources, establishment of fees, and access to services
5. Fidelity, which implies that helpers honor commitments to clients, students, and supervisees

We encourage considering each of these general ethical principles when encountering ethical dilemmas. Faiver, Eisengart, and Colonna (2004) recommend the use of various resources in resolving ethical dilemmas. Seek guidance from:

> Supervisors and colleagues
> The ethics committee of your profession
> State licensure or certification boards
> Professional journals and texts that address ethical issues

Ethical Decision-Making Models

Numerous authors have written extensively about ethics in the helping professions and offered models for ethical decision making. Most models emphasize the importance of being both thorough and honest. Taking one's role seriously is essential. Corey, Corey, and Callanan (2003) and Corey, Corey, and Haynes (2003, pp. 20–21) list ethical decision-making steps that take into account many of the various models:

Ethical Decision-Making Steps

> Identify the problem or dilemma
> Identify the potential issues involved
> Review the relevant ethical guidelines
> Obtain consultation
> Consider possible/probable courses of action
> Enumerate the consequences of various decisions
> Decide on the best course of action

Another ethical decision-making model proposed by the U.S. Congress (2000, p. 10) that is worth mentioning is the ETHIC Model of Decision Making, which considers the following steps or processes:

E – Examine relevant personal, societal, agency, client, and professional values.

T – Think about what ethical standard of [the professional] code of ethics applies, as well as relevant laws and case decisions.

H – Hypothesize about possible consequences of different decisions.

I – Identify who will benefit and who will be harmed in view of social work's commitment to the most vulnerable.

C – Consult with supervisor and colleagues about the most ethical choice.

SCENARIO

You work in a teen shelter on the evening shift. Your supervisor is very close to one of your co-workers. The longer you work the night shift, the more concerned you become about possible indiscretions. You notice that your coworker often comes in late, leaves early, and spends endless time on the phone. As a result, you often work alone and are beginning to feel exploited. You want to speak up but are unsure of how to broach this issue with your supervisor, who is often the one to drop this coworker off, pick her up, or call her on the phone.

In what way is this an ethical dilemma? How would you proceed?

SCENARIO

You are working with an adolescent girl whose parents are both law enforcement officers. You find yourself caught in the middle. The parents regularly ask about their daughter's behavior and are especially concerned about her relationship with a boyfriend. They inform you that he is being investigated in a murder case. They have prohibited her from seeing him, which she readily agreed to: "I'm over him anyway." In sessions with you, however, she tells you about her escapades with the boyfriend; she routinely sneaks out to meet him.

How would you cope? Would you feel torn between the interests of your client and her parents? What are the ethical issues involved? How would you proceed?

SCENARIO

After a peer review meeting, one of your colleagues approaches you and asks to consult with you about a case. She does not give names. Her concern is for a client who is being battered by her spouse but is afraid to leave, as he has threatened to kill her if she does. He collects guns and has ample ammunition available.

What would you suggest to your coworker? How do you think she should handle this? How would you deal with it?

SCENARIO

You are leading parenting classes at a child guidance center. At an intake interview on a referral, you obtain basic information from the client. You learn that she has a son in the same Boy Scout troop as yours. Your husband usually takes your son to meetings, but you are often involved in field trips and other activities. You bond quickly with this client. You will be the one facilitating her parenting classes.

How do you feel about continuing to treat this woman, either individually or in your parenting class? Should you refer her elsewhere? Why or why not? What dilemmas does this case present?

THINKING THINGS THROUGH

Choose an ethical dilemma you have personally experienced, one you are aware of, or one from the preceding scenarios, and apply the steps identified. What is the result? Did it prove helpful to follow the guidelines?

How to Handle Unethical Behavior

In fieldwork, beginning helpers may encounter some type of unethical behavior. To avoid illegal or unethical behavior, helpers must adhere to ethical guidelines and to criminal and civil statutes (Ogloff, 1995). Such guidelines and legal stipulations by themselves are insufficient, but the human service worker must nonetheless be well informed about them. He or she must also hold morally sound principles from which to make competent judgments and decisions.

Unethical Behavior in Oneself

If you find yourself bordering on acting unethically or illegally, stop and take a close look at yourself and your possible motives. It is important to engage in ongoing self-inquiry to assess your underlying motivations. Seek consultation with others (e.g., supervisors, legal counsel, or ethics board) as a way of seeing the situation objectively and considering appropriate courses of action. Helpers who fail to explore their own issues are often those who show a negative pattern of unethical behavior. This can lead to malpractice suits—and, most unfortunately, harm to the client.

In our practice in agency settings, sometimes certain staff tend to focus on the behavior of fellow colleagues rather than to focus on themselves. According to Corey, Corey, and Callanan (2003, p. 221), "[i]t is easier to see the shortcomings of others and to judge their behavior than to develop an attitude of honest self-examination. You can control your own professional behavior far more easily than you can that of your colleagues, so the proper focus is to look honestly at what you are doing." We advocate taking time to evaluate your own practice patterns because this is where you possess the most likelihood to make changes. When we spend time needlessly focusing on others, we are delaying the opportunity to make improvements in ourselves. This type of behavior may also create undue stress in the agency and serve to hamper efforts for cohesion building and teamwork.

SCENARIO

You are working at the university children's center, where students and faculty leave their children while they are in class or working. You work with the 4-to-5-year-old group. Usually, you have good rapport with the parents when they drop off or pick up their children. One day, you notice bruises on the back and thighs of a little boy. He is the son of a professor with whom you have had classes in the past. You are reluctant to ask the child any questions regarding the bruises, yet you eventually do inquire about them. He tells you that his father sometimes gets upset and hits him with a belt. Upon hearing this, you become increasingly anxious and wish you had not asked any questions at all. This appears to be a case of reportable child abuse. However, you worry about the impact of making such a report. You decide to ignore the situation for fear of creating conflict within the university.

Do you think this was handled properly? Would you handle it differently?

THINKING THINGS THROUGH

If you were to find yourself in a similar situation to the above scenario, how might you respond, considering the stressful circumstances as well as the legal and ethical implications? Role-play the interaction between you and your supervisor when you share your concerns about following through with a child-abuse report.

Unethical Behavior in Others

In contrast to simply not agreeing with a colleague who differs in his or her practice patterns, if you are aware that a colleague or administrator is engaging in unethical behavior, the best course is to broach the issue directly with the individual and encourage immediately discontinuing the behavior. Sometimes, when there are "serious accusations such as observing someone in the agency misappropriating client funds, fondling a client, or snorting drugs" the helper has the responsibility to take these to their supervisor (Royse, Dhopper, & Rompf, 1999, p. 123). Dependent on the nature of the situation, you may be obligated to report your colleague's misconduct to the professional ethics board (Ogloff, 1995). The National Organization for Human Service Education (2000) summarizes its recommendations in the following standard:

> Human service professionals respond appropriately to unethical behavior of colleagues. Usually this means initially talking directly with the colleague and, if no resolution is forthcoming, reporting the colleague's behavior to supervisory or administrative staff and/or to the professional organizations(s) to which the colleague belongs.
>
> As with most ethical dilemmas, it is rarely a pleasant experience. Anxiety, confusion, frustration, and anger may all be felt when one encounters unethical behavior in a colleague. There must be ongoing self-awareness, supervision, consultation, and commitment to the values of the human service profession.

SCENARIO

You work in a county mental health center where much of the treatment is multidisciplinary and team oriented. It is all-important to have a good working relationship with coworkers. On a weekend trip with your family, you see a coworker dining with one of the clients. You are at first shocked, and then begin to feel anxious and confused. You are unsure what type of relationship they have. It is unclear whether it is just social or more intimate. Nonetheless, it is obviously a dual relationship. You decide not to acknowledge them at that time, although they both saw you as well. Upon returning to work after the weekend, you find yourself feeling unusually tense and unsure of how to proceed.

How do you feel about this situation? Do you feel unethical behavior is involved?

THINKING THINGS THROUGH

In the situation above, would you confront your coworker directly, make an indirect comment in passing, or proceed as if nothing had occurred? Consider a role-play between yourself and a supervisor or trusted colleague.

INFORMED CONSENT

Whether in a clinical or nonclinical capacity, your work with people requires you to understand informed consent because it is a very important client right. "Informed consent involves the client's right to know the purpose and nature of all aspects of client involvement with the helper and within the helping relationship" (Neukrug, 2002, p. 162).

Respect for the client is inherent in the notion of informed consent. Day (2004, p. 49) concurs: "Respecting and valuing the client includes letting him or her in on what will happen in [treatment]." The ethical standards of the National Organization for Human Service Education further address this issue of respect in its statement, "Human service professionals are to respect the integrity and welfare of their clients and treat them with respect, acceptance and dignity." Social workers, likewise, are to "respect the inherent dignity and worth of the person" (National Association of Social Workers [NASW], 2000, p.3). The American Counseling Association (ACA, 2002, p.1) Code of Ethics also supports that "the primary responsibility of counselors is to respect the dignity and to promote the welfare of clients".

Aspects of Informed Consent

In effect, according to the NASW Code of Ethics (1999) informed consent requires human service professionals to "use clear and understandable language to inform clients of the purpose of the services, risks related to the services, limits to services because of the requirements of a third-party payer, relevant costs, reasonable alternatives, clients' right to refuse or withdraw consent, and the time frame covered by the consent." Similarly, in the Ethical Standards of Human Service Professionals, Statement 1 highlights the importance of all human service professionals gaining consent by negotiating "with clients the purpose, goals, and nature of the helping relationship prior to its onset, as well as informing clients of the limitations of the proposed relationship" The ACA Code of Ethics describes informed consent similarly as "disclosure to clients." For psychologists, in addition to informing clients about therapy, informed consent is required for assessment and testing and for research purposes. Here we address the specific aspects of informed consent in more depth.

Informing the Client about Your Role, Qualifications, and Experience
Clients who seek help deserve quality treatment and should be told about the helper's training and qualifications. Interns must also inform clients of their status.

SCENARIO

A client enters a shelter for battered women where you are an intern. During the intake interview, you neglect to inform her of your limited experience and intern status. She seeks help for her depression and thoughts of suicide. You have never dealt with a suicidal individual and abruptly refer her to your supervisor. She seems offended and later tells your supervisor that she felt you didn't like her. Talking with your supervisor, you realize that informing the client of your qualifications might have avoided her becoming upset. You realize that your own insecurities about your intern role interfered.

How could you have handled this situation differently? Can you think of other circumstances in which you would hesitate to tell a client about your role, qualifications, and experience?

Informing Clients about Services
This includes educating clients regarding assessment procedures, intake policies, techniques to be used, testing to be performed, and referrals to be made.

SCENARIO

A client, distraught about extensive damage to her home due to flooding, seeks assistance at your facility. You hand her various forms to complete. She becomes angry when she learns that processing of her claims will take two to six weeks.

Having informed her of this time factor at the outset might have helped clarify the process. Would you have thought to inform the client?

Giving Information about Goals, Limitations, and Potential Risks of Treatment
This involves ensuring that the client understands the implications of treatment and diagnosis, reports to be made, and referrals to be given.

SCENARIO

A client needs help with completing medical and Social Security forms, which is not within your job description. When the client becomes aware that you are not able to help him, he feels frustrated and disillusioned with the system.

How might you sensitively tell the client about your role and its limitations? Would a referral be appropriate? Would you consider going ahead and helping this client with the forms, although that's not your job?

It is important to recognize what cultural implications may be involved when considering a treatment plan and objectives. Working with clients from multiethnic backgrounds requires you to be sensitive to underlying cultural issues. Sometimes, in our eagerness to help others, we may encourage clients to pursue certain goals without informing them of the consequences of attaining them.

SCENARIO

A helper encourages a Latina client to become more independent and assertive in her interactions with family, friends, and her employer. At her next visit, the client is very upset. She has created much interpersonal conflict in her attempts to be more direct. If you had understood her culture better, you might have helped to prepare her for possible outcomes in this situation. Orienting this client to the potential positive and negative consequences would give her the choice of deciding whether this was, in fact, the direction she wished to take.

Can you think of a situation where you encouraged change without taking into account the possible ramifications of this change for the client's life?

Clear Information about Confidentiality

This entails spelling out all the instances when information may be shared with others—members of the treatment team and supervisors.

SCENARIO

A client believes that whatever he tells you will not be shared with anyone else. When he learns that you shared an important aspect of the treatment with your supervisor, he is angry. How might you avoid such a situation? Would you consider explaining confidentiality at the very beginning?

Sharing Exceptions to Confidentiality

Although this may make the client hesitant to share potentially reportable information, it remains an important part of informed consent. It has been our experience that clients will usually share such information even if they know it may be reported because they basically want help.

SCENARIO

A longtime client reveals that she recently lost control and physically abused her young son, leaving visible marks. She feels tremendous guilt and seeks your emotional support and guidance. You notify her that you must report this incident to child protective services. She becomes irate because you had never before disclosed this exception to confidentiality.

What could have been done to prevent this situation? Do you remember to inform clients about confidentiality at the beginning of your contact? Do you feel this is enough? Should clients be reminded periodically during their treatment?

The Requirement of Written Consent to Release Records or Information

Clients often become upset when they ask you to release or obtain information about their situation, but you cannot until they sign a release form. Wilson (1978, p. 57) writes, "This form exists in many versions in as many different settings. Very few agencies and practitioners use it appropriately and abuses of consumer privacy occur almost daily." Wilson maintains that general forms are all but useless and do not constitute true consent. To meet the criteria for informed consent, it is best to develop a document tailored to the client's unique needs.

SCENARIO

A man brings his daughter to the children's center where you are working. Explaining the family circumstances, he tells you that the mother and he are divorced because the mother was using drugs and physically abusing the girl. Later, the mother phones, inquiring about details of the current situation with the child and the father.

How would you handle this telephone call? Should you talk to her?

Client's Right to Obtain Case Records

Helpers may be unaware of this right and may unknowingly discourage clients from viewing their records.

SCENARIO

A client who is emotionally disturbed (paranoid) insists on seeing your evaluation, progress notes, and treatment plan. You are very uncomfortable with this demand; you feel that the client may misinterpret certain information that is documented and become even more irrational and upset.

How do you handle this? What might you do?

Allowing Clients to Participate in the Treatment Plan

This encourages a cooperative atmosphere, enhances the helper-client relationship, and facilitates evaluation of the treatment strategies used.

SCENARIO

You are working with a teen client in a chemical dependency unit. You are new to the unit and are unfamiliar with the procedures. You coddle the client. Later, her progress is evaluated and found to be minimal. She criticizes you for not making clear that she was to be an active participant in her treatment. She had assumed you would be the one to make her well again.

How would you handle this situation?

Client's Right to Refuse any Recommended Services

Often clients are not aware that it is their choice to continue in treatment. Some may do so although they do not really want to. Others may believe they are required to continue treatment. Conversely, clients may refuse treatment without being apprised of the possible consequences of withdrawing.

SCENARIO

You are working in an agency that monitors the visitations of parents whose children have been removed from the home due to child abuse. The parents must be informed that they have to comply with court stipulations—for example, attending parenting classes, undergoing counseling, attending drug and alcohol diversion programs, and making monitored visits with the child. By letting them know that not following through with the requirements may jeopardize the reunification process, you help the parents fully understand the choices they have.

How would you deal with a parent who wanted to drop out early?

Benefits of Informed Consent

Besides being an ethical and legal requirement, providing informed consent is viewed as an integral part of the therapeutic process, regardless of your professional role or theoretical framework.

Establishment of Trust

Informing clients about the services to be rendered often goes a long way in establishing trust. It demonstrates respect for their ability to make informed choices and encourages them to entrust you with their confidence. Accordingly, Corey (2005) says that by educating your clients about their rights and responsibilities, you are both empowering them and building a trusting relationship with them. He also believes that by "providing clients with information they need to make informed choices tends to promote the active cooperation of clients in their [treatment]" (2005, p. 40).

Discovery of Possible Problem Areas

While informing clients about their care, you may be alerted to possible conflicts or areas of difficulty. You can begin early exploration of such areas and provide education regarding possible treatments.

Agency Responses to Informed Consent

Agencies require different levels of disclosure, depending on the types of services they provide. Some agencies may be rigid about these policies, whereas others may be indifferent or even unaware of the disclosure requirements.

In our experience, some agencies have been resistant to the idea of informing clients about the treatment process and the limitations of confidentiality. Other agencies have developed formal procedures in this regard, including standardized forms for clients to sign as well as informative brochures and pamphlets. Of course, there may also be unspoken norms about how informed consent is to be handled.

One way to learn about your agency's view of informed consent is to ask professionals within the agency how much they disclose to their clients. You might be surprised to discover that different levels of disclosure exist within a single agency.

If you feel confused by this complex issue, recognize that this is normal and make sure to seek out consultation and supervision. Be aware that even if you are not in a counseling position, client's rights must still be protected.

▓ CONFIDENTIALITY

According to ethical guidelines of the various helping professions, the human service provider is "generally obligated to treat the client's communication in a confidential manner; that is, the [helper] agrees not to share information given by the client with other persons" (Cormier & Hackney, 2005, p. 62). Of course, there are exceptions, which we will elaborate in a moment. "It is the helper's assurance that information the client divulges will remain between the two of them" (Woodside & McClam, 1994). The right to privacy is one of the most discussed ethical concepts in human services. It is a significant legal issue; specific penal code sections apply in many jurisdictions. Ethically, confidentiality is considered the cornerstone of the helping relationship and is included in the ethical standards and codes of all professional associations (Leslie, 1989). Professional relationships with clients are built on trust; emphasizing the right to privacy helps build a strong foundation for such trust. Corey and Corey (2003, p. 231) emphasize this issue of trust: "The helping relationship is built on a foundation of trust. If clients do not trust their helper, they will be unlikely to engage in significant self-disclosure and self-exploration. Trust is largely measured by the degree to which clients feel assured that what they share will be kept confidential."

Hepworth, Rooney, and Larsen (2002, p. 71) also address trust: "From a practical standpoint, confidentiality is a sine qua non of the helping process, for without assurance of confidentiality, it is unlikely that clients would risk disclosing [certain] private aspects of their lives." Clearly without a trusting environment as a foundation, much of the helping process could not proceed.

The timing of when to discuss confidentiality issues is typically early in the formation of the relationship; it is part of the informed consent process addressed earlier. Cormier and Hackney (2005) believe that the early timing is important for two reasons: (1) Clients are given enough "up-front" information regarding the various aspect of confidentiality that allows them to make an informed decision regarding their consent and participation in the helping process, and (2) it prevents situations where helpers make necessary disclosures without the client being aware of the legalities involved. Failure to inform clients about the exceptions to certain disclosures is unethical and contributes to a client's sense of betrayal that may further erode the helping relationship.

In some agencies, helpers do not discuss confidentiality with their clients. This may not be a problem, but in some instances it could become a major obstacle to a positive working relationship. In certain settings, the issue of privacy requirements has become important since the recently enacted federal government rule called the Health Insurance Portability and Accountability Act (HIPAA), which affects helpers and confidentiality. "The act involves regulations about electronic as well as traditional paper health-care related charts. Part of the HIPAA regulations are designed to protect client privacy in that they delineate the steps that providers have to take to secure client information" (Cormier & Hackney, 2005, p. 63). Because HIPAA regulations are complex, please consider consulting with a supervisor, a colleague, a trusted attorney, or the specific law. The "Help on HIPAA" section of the American Counseling Association's website at www.counseling.org/publications;hippa.htm. is one of various resources we recommend using if there are questions or confusion regarding this law.

Release of Information
"No one other than the client should be given any information—written, verbal or otherwise—about the client without explicit written and signed permission from the client" (Baird, 2002, p. 32). Release-of-information forms are commonly used in agency settings. If no such forms exist at your placement, we suggest that you consult with your supervisor.

Safeguarding Records
This can be problematic in large agencies, where files, case notes, and all other client records must be easily accessible to various staff members. Sometimes a breach of confidentiality can result. To ensure confidentiality, Baird (2002) suggests the following commonsense precautions:

Keep all case notes, records, and other written or recorded information in a locked file cabinet. In many agencies, a checkout system is used to request charts. Either way, make an effort to keep records in a safe place.

Take care not to leave client files and any other client information anywhere they may be seen by others, even if inadvertently.

Client records may be labeled "confidential records."

Any client information that is computerized should be safeguarded as well. Consider ways of allowing access only to those working directly with the client. Many agencies are in the process of developing special guidelines to maintain confidentiality with regard to computerized records.

As students, you may choose or be asked to keep journals or class notebooks that contain case information. To ensure that confidentiality is not breached in the event that you lose or misplace these personal materials, it is wise to disguise the identity of the client or agency.

Sharing Information with Colleagues

Working in agencies requires collaborative work with others. Some coworkers may not work directly with your clients, so sharing of information is necessary. Also, if you are an intern or a new employee under close supervision, client information is also shared with your supervisor. The rule for sharing information with outside sources is relatively clear: It should not be released without the client's written permission. However, this rule is not so clear cut within an agency. Baird (2002) offers a general framework of six questions to ask yourself in making such decisions about sharing client information:

1. *Who is* the person you will be sharing this information with?
2. *Why* does he or she want the information? Why are you considering giving it?
3. *What is* the information to be shared? Is it specific information or general?
4. *Where* will you be when the information is shared? In a private office, in the lounge, in the hall, at a restaurant?
5. *When* are you planning to discuss the information? Is the timing right, or could there be another more appropriate time?
6. *How* will you share the information? In writing, on the phone, in person?

SCENARIO

You are a social work intern at a clinic that provides both counseling and case management services. One day you are chatting with coworkers in the front office, which adjoins the waiting area. A clinician speaks of her reluctance to see her next client. She proceeds to talk openly about this client in front of everyone present. You become uncomfortable and notice that people in the waiting room might be listening. Then the client in question angrily approaches the receptionist to cancel her appointment; she has overheard what was being said about her. The clinician is dismayed as she suddenly becomes aware of her confidentiality violation.

How would you have handled this situation? What do you feel could have been done differently to avoid this result?

SCENARIO

You and your coworkers on the crisis hotline frequently go out to lunch together as a stress reliever. Often this is the only time you have together, given everyone's hectic schedule and odd hours. Sometimes people consult or just vent about their difficult cases. One day a colleague tells of a call from a client distraught because her husband is leaving her to live with a man. The colleague divulges that the husband has been cross-dressing. The group discusses this unusual situation, not taking care to disguise the name. In the next booth are two businessmen. You later find out that they are the client's spouse and his lover.

What went wrong? Was this a breach of confidentiality? What could have been done differently?

SCENARIO

You are a newly appointed caseworker at a women's clinic, where you work primarily with teen pregnancy. The staff is cohesive, and the secretaries and receptionists are all very helpful and involved. As you interact more with them, however, you notice that they are often overly inquisitive about the private lives of clients. On one occasion, after a 17-year-old scheduled her second abortion in a year, a secretary begins to ask you specific questions about the client. She suggests that the girl is misusing the agency's services and has clearly not learned her lesson regarding birth control. You chat a little with her, then retreat to your office. You feel uncomfortable about having engaged in this conversation.

Was confidentiality breached? Would it be appropriate to share information with the secretary at all? How would you have proceeded?

SCENARIO

You have been asked to present a talk to the staffs of local preschools on child development and associated problems that may require professional attention. As you share various stories involving clients, you use initials but do not disguise other identifying data such as gender, age, ethnicity, residence, and school. A participant states that, from your case description, she believes one of your clients is her next-door neighbor. You are astonished and try to change the subject, dismayed at this carelessness on your part.

How could this have been handled differently?

THINKING THINGS THROUGH

Can you think of any occasion when you unknowingly violated confidentiality by sharing informa-tion about a client? Can you think of any situations where this might be appropriate? Under what conditions?

Breach of Confidentiality

The concept of confidentiality is not always obvious and clear. "Although most [helpers] agree on the essential value of confidentiality, they realize that it cannot be an absolute" (Corey, 2005, p. 41). For instance, a worker may unknowingly break the rule of privacy by releasing certain information about a client without formal authorization. It might be an innocent or minor disclosure, such as telling a family member who calls that the client has already left the agency. Often, a husband calls to ask whether his wife showed up for her appointment. Technically, just giving a yes or a no answer, and thus revealing that the wife is indeed a client, constitutes a breach of confidentiality. The sharing of sig-nificant information with an unauthorized party might lead to serious repercussions that could ultimately affect the client's well-being. For example, a batterer might be told by an agency staff member that his spouse has sought refuge at a certain women's shelter. If he were to act violently, he might endanger his wife as well as others there.

Requests for information often originate from the following (Wilson, 1978, pp. 68–82):

Staff in the same agency
The general public
Researchers
Attorneys
Computer data banks
Other agencies and members of other disciplines
Professional standards review organizations
Family and friends of the client
Third-party payers
The news media
Police

Even if the client has given verbal permission for disclosure of information, he or she must still sign a release form.

As you begin your career, it is important to remain aware of confidentiality standards and related issues. Another important facet involves the exceptions to confidentiality.

Exceptions to Confidentiality

Although the law states that confidentiality must be maintained, it also specifies times when privacy between helper and client must be broken to protect all concerned. Such legally mandated disclosures are commonly referred to as *exceptions to confiden-tiality*. These may differ from state to state, and even from one county to the next. The most common exceptions relate to physical or sexual abuse of children, the elderly, and dependent adults; cases of potential harm to self and others; gravely disabled individuals; and court orders and subpoenas requesting client information, client waivers of confidentiality, and supervision and consultation. Additionally, disclosures may be permitted in cases of emotional abuse and spousal abuse and in situations where the reporting of AIDS is deemed necessary.

Each situation must be judged on an individual basis. State stipulations differ greatly with regard to mandated and permissible disclosures. Furthermore, individual agencies are likely to have their own procedures in the event that such a disclosure is necessary. Become informed about the policies and procedures of your particular agency, county, and state. Similarly, you must be familiar with the ethical code of your profession. For psychologists, social workers, and marriage and family counselors, the requirements are fairly clear. For other human service workers, there may be no clear guidelines; talk with your supervisor or consult the professional ethics board of your discipline. Woodside and McClam (1994, p. 260) write, "Agency guidelines, the profession's code of ethics, and laws of the state in which one practices are important considerations in the delivery of services."

The key elements of exceptions to confidentiality must be understood, as well as the underlying intricacies. First of all, one must have a clear sense of what *privileged communication* means.

Privileged Communication

Privileged communication is a legal concept that entitles certain professionals to withhold testimony in court proceedings. According to Arthur and Swanson (1993, p. 7), privileged communication is "a client's legal right that confidences originating in a therapeutic relationship will be safeguarded during certain court proceedings." This privilege is designed to protect the client from potentially harmful disclosure of personal information by people involved with him or her in a professional capacity (Corey, Corey, & Callanan, 2003). The law is for the protection of the client rather than the helper. Privileged communication affects helpers in many ways. For example, trust can more easily develop when the client knows that the helper will not be disclosing information to others without specific authorization. If the client does want information released to a specific person, the helper has no right to withhold that information. As professional helpers, we are responsible for understanding the ramifications of privileged communication.

The professionals who are most commonly allowed this privilege are attorneys, physicians, clergyman, psychiatrists, psychologists, social workers, chartered accountants, marriage counselors, and registered nurses (Beneke, 1998; Corey, Corey, & Callanan, 2003; Glosoff, Herlihy, & Spence, 2000; Wilson, 1978). This list is by no means exhaustive or definitive because there is great variance and frequent change in state statutes and relevant case law. Wilson (1978, p. 100) writes, "While state statutes may define the right to privileged communication generally for various professional groups, the courts are constantly interpreting the concept as it applies to specific case situations." There really is no consensus regarding the privilege. Corey, Corey, and Callanan (2003) share that privileged communication between a human service professional and client is not an absolute matter. Ongoing evaluation and consultation is therefore necessary. State statutes specify exceptional cases in which communication is no longer treated as privileged; as a human service professional, you will be legally bound to comply with such exceptions. Glosoff, Herlihy, and Spence (2000) offer the following guidelines for practitioners such as psychotherapists affected by privileged communication:

- Make use of informed consent statements regarding the parameters of confidentiality.
- Inform clients of the situation and involve them in the process.
- Take reasonable steps to protect client confidentiality.
- If information must be disclosed, provide only minimal disclosure.

- With permission, contact the client's attorney.
- Document all action taken and what information was disclosed.
- Make it a practice to consult with colleagues regarding clinical judgements and with attorneys regarding legal obligations.

Technology and Issues of Privacy

Issues related to confidentiality are even more challenging today given many technological advances. As a result, it becomes necessary to find ways to safeguard client confidentiality and privacy and to ensure that all mediums used are appropriate and quality controlled. According to Corey and Corey (2003, pp. 234–235), "[t]here are a host of ways to violate a client's privacy through the inappropriate use of various forms of technology. The use of the telephone, answering machines, voice mail, pagers, faxes, cellular phones, and e-mail can pose a number of potential ethical problems regarding the protection of privacy of clients." Donna DeAngelis (National Association of Social Workers, 2004), executive of the Association of Social Work Boards, urges giving serious thought to the following technology-related scenarios and issues:

- Misdirected e-mail accidentally read by third parties
- Family members eavesdropping on telephone calls
- Family members (or network administrators) gaining access to clinical e-mail
- Snoopers (or neighbors) intercepting mobile (cell) or cordless phone conversations
- Hackers uploading e-mail or "listening in" on PC-based videoconferences
- Misrepresentation by e-clients
- E-mail communication generating extensive clinical records

DeAngelis (National Association of Social Workers, 2004) goes on to say that "for every technology there are pitfalls involving matters of confidentiality and privacy. Without appropriate safeguards, unwanted third parties can gain illegal access to confidential records." Based on the potential hazards of modern technology, we recommend that caution be used when relying on the many technological advances that have become commonplace. Because client privacy can be easily and unintentionally breached, it becomes necessary to include a discussion of these potential problems regarding technology and privacy during the informed consent process. Electronic record keeping and online counseling are especially at risk of misuse if precautions are not taken. Practitioners must therefore take special care in using computers in these areas and remain abreast of all the relevant ethical and legal guidelines that apply. Here we will review each of these issues and other issues in further depth:

Telephone and answering machines

Often when returning a call to a client or calling to cancel an appointment, there is a need to identify oneself. How can this be done in a way where confidentiality is maintained? Remley and Herlihy (2001, pp. 124–125) recommend the following as guidelines when using the telephone or leaving a message for clients:

- Refrain from acknowledging that clients are receiving services and do not give out information regarding clients to unknown callers.
- Ensure that you are actually communicating with the intended person when you make or receive calls where confidential information may be discussed.
- Abstain from making any comments you would not want your client to hear or that you would not want to repeat in a legal proceeding.
- Do not allow unauthorized persons to hear answering machine messages in your office while they are being left or retrieved.
- Do not leave personal messages on an answering machine unless this is pre-arranged with your client because another family member may retrieve the message left for your client.

Cellar phones If you are talking on a cellular phone with a client, assume that he or she is not in a private place. Your conversation may also be intercepted by an unauthorized person. Avoid reference to information that might identify a client or otherwise compromise confidentiality.

Pagers When using a pager, exercise caution when retrieving messages. In sending a pager message to a client, be careful to ensure your client's privacy by exercising the same caution you would if you were sending answering machine or voice-mail messages.

Fax machines It's become commonplace to use fax machines to readily transmit information to a client or about a client. It is important to first check the number dialed before transmission to avoid an error. Also, ensure that the person for whom the fax is intended is present to personally retrieve the fax at the time of the transmission by calling before sending the fax, especially if the fax machine at the receiving site is in a common area.

Electronic mail (e-mail) Use of electronic mail to communicate with clients has become very common, yet it remains an area where issues or privacy and confidentiality are problematic. Because e-mails may be accessed by those other than their intended recipients, privacy is not guaranteed. Corey and Corey (2003, p. 235) state, "Since there is no reasonable expectation of confidentiality for e-mail, clients need to have input regarding how they want communication to be handled so that their privacy is protected." Ensure that clients have given their consent to transmit client information. Prior to sending client information, secure client consent, which states the client is aware of the inherent security risks in e-mail transmission. If possible, encode identifying information such as substituting the client's name and any other information that could specifically identity the client by using either numerical or alphabetical code and sending the decryption program separately to the reception site by another means. Deleting client identifying information prior to sending may also be necessary. In this case, the decryption program should not be stored in an electronic media, but securely and separate from the client files. It may be necessary to install an encryption program. Additionally, inquiring as to the security of the e-mail reception site is important. Specifically, it is prudent to ensure that the physical site of the receiving computer and the personnel who have access to the computer will respect both the confidential nature of the e-mail and be responsible for its secure management. Lastly, password protection at the directory level is often indicated as is ensuring secure diskette management at the sending site.

Online counseling Also referred to as eTherapy, web counseling, or "distance counseling," online counseling, despite its benefits (we will discuss these further in Chapter 7), is also fraught with potential legal and ethical problems. Potential ethical considerations include validating the identity of both the therapist and the client, supplying proof of the therapist's professional credentials, securing permission for counseling minors, ensuring client confidentiality, providing for emergency counseling services, verifying that liability insurance covers the client's geographical location, determining the appropriateness of online counseling for the client population, and developing a protocol for online technical failures, legal jurisdiction, and technical competence of the therapist (Franz, McCartie, Gold, & Skyward, 1999; http://www.gingerich.net/etherapy.htm; National Association of Social Workers, 2004). The NASW Code of Ethics has identified confidentiality, informed consent, professional competence, and client records as issues related to online counseling.

Franz, McCartie, Gold, and Skyward (1999, p. 7) cite the following precautions:

- Obtain information and follow professional, ethical, and legal standards of practice; also, be conscious of cultural considerations that may apply to the client's geographical location.
- Provide clients with addresses of all appropriate professional and registering associations related to the field of practice.

- Obtain written consent from the legal guardians of minors prior to engaging in web counseling.
- Implement an identification system between client and therapist such as the use of a code word or graphic.
- Supply a reference to an appropriate professional who can provide emergency counseling within the client's geographical location.
- Implement an intake and assessment tool for use with website clients.
- Identify those issues that may not be appropriate for website counseling (e.g., suicidality, psychiatric disorders).
- Develop and discuss preplanned communication strategies with the client in the event of online technical problems.

Use of chat rooms Because chat line transmission provides immediate communication, it can lead to certain issues that may be problematic to professional standards of practice. In short, chat lines are typically not confidential, unless, of course, you are using private and secure "chat rooms." Given this, it is best to not use Internet relay as a forum to discuss clients unless confidentiality can be assured; written permission has been secured by the client, and information that may expose the client's identity is excluded from all dialogue. Also, because chat lines are often informal, therapists must be careful to follow the usual parameters of conduct outlined in the Code of Ethics that are applicable.

Computer files, computers, and networks Confidentiality may be breached when storing client information on hard drives of a stand-alone computer or of a computer that is part of a network. Certain precautions need to be taken such as having a password to protect files. Codes are also recommended to identify clients so directories do not reveal client identities. Password protected log-on should be required when client's names or other confidential information of a clinical nature appears in directories. Stand-alone computers should require password-protected access. Additionally, a reasonable measure of protecting files may require creating and maintaining password-protected back-up diskettes stored in a secure location, such as a locked filing cabinet. In this case, ensure that the transfer of diskettes from one location to another is secure (e.g., avoid mailing diskettes).

Laptop computer security Both practitioners and clients often use laptops. However, the security of the client information stored in these portable computers may be compromised because of their vulnerability to loss, theft, and damage. Precautions we recommend include using a nonidentifying bag or carrying case for transporting, installing a code entry password and an encryption program, backing up data frequently on a diskette, and using a locking cable for use outside the home or office.

Technical failure and computer viruses Because of the risk of technical failure and computer viruses, the internal security of files must be safeguarded by installing an antiviral program on any computer that will be used for inputting client data, saving all files from the hard drive to a back-up diskette, and making back-up copies of diskettes if diskettes are the main source of data storage.

Maintenance of skills Finally, it is incumbent upon individual therapists to seek out adequate training and develop necessary experience to use the computer effectively. This entails ensuring a good working knowledge of the computer, including its software programs, securing support services necessary to protect file security in the event of a technical problem, upgrading computer training and skills in keeping with advances in technologies, and staying apprised of the use of technologies in the context of the code of ethics and the technology guidelines of their respective profession.

The code of ethics of the various helping professions as it relates to technology and privacy warrants further discussion. For instance, in the ACA Code of Ethics (1999), the following applies to computer technology:

a. Use of Computers: When computer applications are used in counseling services, counselors ensure that (1) the client is intellectually, emotionally, and physically capable of using the computer application; (2) the computer application is appropriate for the needs of the client; (3) the client understands the purpose and operation of the computer applications; and (4) a follow-up of client use of a computer application is provided to correct possible misconceptions, discover inappropriate use, and assess subsequent needs.

b. Explanation of Limitations: Counselors ensure that clients are provided information as a part of the counseling relationship that adequately explains the limitations of (i) computer technology in the counseling process in general and (ii) the difficulties of ensuring complete client confidentiality of information transmitted through electronic communications over the Internet through on-line counseling.

c. Access to Computer Application: Counselors provide for equal access to computer applications in counseling services.

In terms of confidentiality of records, counselors must maintain appropriate confidentiality in creating, storing, accessing, transferring, and disposing of counseling records. The ACA further discusses confidentiality in the areas of informational notices, client waiver, records of electronic communications, and electronic transfer of client information. Under "Establishing the on-line counseling relationship," there are codes on the appropriateness of online counseling, counseling plans, continuing coverage, boundaries of competence, and minor or incompetent clients.

The NASW Code of Ethics discusses the issues of online social work in its standards that deal with informed consent, professional competence, and client records. Specifically, the code requires social workers to ensure confidentiality and privacy of information transmitted through electronic or computer technology.

In sum, because of the heightened risk involved, helping professionals need to be cautious and aware when they engage in communication with or about clients using these media. Often, even if helpers had every intention of protecting their clients' privacy, confidentiality may be breached unintentionally. We recommend discussing these potential problems regarding technology with clients early on during the informed consent process. In many cases, such an open dialogue can result in helpers and clients taking preventive measures to avoid any unnecessary and illegal breach of confidentiality. We conclude with feedback from ethics expert Frederic Reamer, who offers the following considerations for practitioners using computer technology:

1. Provide clients with full disclosure of the possible benefits and risks.
2. Use comprehensive assessment tools along with a statement of how important it is for the client to be forthright in completing these skills.
3. Provide clients with specifics on confidentiality and disclosure of safeguards— "Here are the steps I've taken, along with the attendant risks."
4. Secure emergency contact from the client and develop a clear emergency plan in case online communication is interrupted.
5. Consult the code of ethics of the appropriate profession to ensure compliance with issues pertaining to confidentiality, privacy, informed consent, conflict of interest, misrepresentation of credentials, and more.
6. Consult state licensing provisions to check for applicable statutory regulations.
7. If necessary, consult a malpractice/risk management attorney to learn more about existing standards of care, malpractice issues, and more.

Lastly, we recommend inquiring with relevant associations of online professionals and health-care organizations that have developed codes of conduct for online services. The "International Society for Mental Health Online" and the "Health on the Net Foundation" are two resources to consider.

Supervision and Consultation

We have seen that the right of confidentiality is not absolute. There are many instances in agency settings when case situations are discussed either with supervisors or presented in case consultations and peer review. Ideally, disclosures of this nature should be restricted to instances when they enhance service to clients and when clients have understood the purpose and consented. Hepworth, Rooney, and Larsen (2002, p. 72) add that "the client has a right to be informed that such disclosures may occur, and practitioners have a responsibility to conceal the identity of the client to the fullest extent possible and also to reveal no more personal information than is absolutely necessary." Additional instances may exist when client records may be reviewed and others may have access to files and case information. For instance, administrators, clerical staff, consultants, legal counsel, and outside persons may review records for purposes of quality assurance, peer review, or accreditation. Once again, it is essential that "the access to information should be for the purposes of better serving the client, and these individuals should sign binding agreements not to misuse confidential information" (Hepworth, Rooney, & Larsen, 2002, p. 71). Remley and Herlihy (2001) stress the ongoing importance of consultation whenever experiencing doubt regarding one's obligations about confidentiality and privileged consent.

Child Abuse

During your fieldwork experience, or later when working in human services, you will probably be exposed to some kind of child abuse—in direct or indirect work with abused children or adult survivors, or in the policy arena. This is an emotionally difficult area for human service workers and also a very complicated matter. The overview that follows can help you learn how to detect less obvious cases of child abuse, intervene appropriately, and be prepared for the potential impact of such cases on you personally.

Working in this area can be stressful and emotionally draining. Even for an experienced professional used to dealing with child abuse, it is often overwhelming and anxiety producing. Often such cases are crisis situations that require immediate action and careful intervention. A clear understanding of your ethical and legal responsibilities can decrease your anxiety and help you respond more effectively.

State laws delineate child maltreatment and reporting responsibilities and procedures. The Department of Social Services in your state has manuals outlining these policies and procedures.

What is child abuse? The National Child Abuse Prevention and Treatment Act (PL 93-247) enacted by Congress in 1974 defines child abuse and neglect as follows:

> Physical or mental injury, sexual abuse or exploitation, negligent treatment, or maltreatment of a child under the age of eighteen or the age specified by the child protection law of the state in question, by a person who is responsible for the child's welfare, under circumstances which indicate that the child's health or welfare is harmed or threatened thereby.

Although all 50 states now have statutes that making reporting suspected or known child abuse mandatory (Hepworth, Rooney, & Larsen, 2002), the specific laws differ from state to state. The Penal Code of California (2001) defines child abuse as "a physical injury which is inflicted by other than accidental means on a child by another person." It also includes "emotional abuse, sexual abuse, neglect or abuse in out-of-home care." Child abuse *does not* include "mutual abuse between minors, reasonable and necessary force used by a peace officer" under specified circumstances, or spanking that is "reasonable and age appropriate and does not expose the child to risk of serious injury."

It is often difficult to differentiate acceptable parental discipline from abuse. Kamerman and Khan (1976, p. 46) identify three basic characteristics of abuse:

> The behavior violates the norm or standard parental conduct—for instance, of the parent's obligation to the child. This includes the child's right to have his or her basic needs met, not violated.
> The infliction is deliberate—that is, nonaccidental injury.
> The abuse or neglect is severe enough to warrant intervention of some type—medical, social, legal, or a combination.

Thus, child abuse is "any act of omission or commission that endangers or impairs a child's physical or emotional health and development" (Kamerman & Khan, 1976, p. 46).

Types of abuse include:

Physical abuse (mild to severe corporal punishment/physical assault)
Sexual abuse (incest/intrafamilial sexual abuse, child molestation, child rape, and exploitation/child pornography)
Neglect (general neglect/inadequate supervision, severe neglect)
Emotional abuse (emotional deprivation, emotional assault)

The California Department of Justice (1976) states that it is "the infliction of injury, rather than the degree, that is the determinant for intervention." Parents or caretakers may begin by unwittingly inflicting minor injuries, which over time may lead to more serious harm. Detecting small injuries early on can make preventive action possible. In some states, simply witnessing domestic violence in the home is reportable child abuse.

Culture and child abuse The role of culture is now being considered an important factor when assessing for child abuse and neglect. Some people argue that child abuse should be viewed in a cultural context. Because "what constitutes 'abuse' in American culture may not be viewed as abuse in another culture . . . and the language used to describe abuse differs from culture to culture" (Corey, Corey, & Callanan, 2003, p. 229), it is ethically responsible for human service providers to be alert to cultural influences as well as the specific laws of the state one is practicing in.

In addition to differences within ethnic minority groups, subcultural differences related to profession and social class may also exist. Clearly, assessing for child abuse and neglect can become quite complicated. Hepworth, Rooney, and Larsen (2002, p. 74) ask, "What course of action should practitioners take when community or subcultural definitions of child abuse diverge from generally accepted standards, and when loyalties may result in punishment to informants and victims rather than perpetrators?" They state, "If professionals are to be both genuinely helpful and realistically effective, there is often a need to make compromises related to treatment ideals. Thoughtful evaluation is required to allow for treatment strategies that can fit and be useful within a given subcultural context, and still maintain ethical and legal principles" (p. 74). In essence, it is important to "look beyond legal guidelines and discipline-specific prescription" (Long, 1986, p. 136).

SCENARIO: Child Abuse and Culture

Hang is an 8-year-old from Laos. He recently arrived in this country with his parents and younger siblings. Hang has adapted quite well and seems to enjoy learning English. One day Hang enters class with several visible red marks on his forehead and temples. When the teacher asks him what happened, he explains in broken English that his father did this. The teacher becomes concerned and makes a child abuse report. As an intern in the Healthy Start Program at this school, you are asked to follow up. When you conduct a home visit, you meet with both parents and Hang. You observe a very loving, close-knit family. Hang's father shares his upset with the child abuse report and informs you that a Child Protective Services worker has already been out. He explains that he did not intentionally hurt Hang; he was trying to relieve a headache Hang had been complaining of by "coining,"—or "cupping," an Eastern folk remedy of rubbing oil or ointment into the skin, usually neck, temple, or other body parts, using a coin or spoon. You feel badly for the misunderstanding and help Hang's parents better understand child abuse laws and the role of the mandated reporter. They seem appreciative of your time and efforts.

SCENARIO: Professional Subculture

Ann is an intern placed at a university children's center. She is assigned to children ages 3 to 4. One day when Ann begins her shift, she notices that Peggy, typically a very active and articulate 4-year-old, is quieter than usual. She also seems more anxious and fearful. When Ann asks Peggy why she's so sad, she proceeds to talk about "being afraid of Daddy cause he hurts my pee-pee."

Ann becomes alarmed and consults with her supervisor about the need to make a child abuse report given the suspicion of sexual abuse by Peggy's father. Ann's supervisor is very surprised since Peggy's father is a well-known and respected professor at the university. Given the delicate nature of this case, Ann's supervisor states that she will handle it. Ann is relieved at first. However, she becomes upset when she returns several days later to work and finds that no child report was ever made. When she approaches her supervisor, she is told that the situation was taken care of and not to worry. Ann feels confused and decides to bring this up in her internship class.

THINKING THINGS THROUGH

Can you think of a situation when reporting child abuse is complicated by the culture or professional subculture? How would you proceed if you were an intern in either of the preceding scenarios?

Characteristics of abusive families Recognizing the common characteristics of abusive families is also valuable in understanding this complex problem. Some traits of such families are listed below. (Note that the list is not exclusive or definitive; abuse can occur in families that do not exhibit these traits.)

- The needs for nurturance and dependence are unfulfilled. There is most likely a history of deprivation or abuse in the adult's own childhood. Such families often have negative attitudes and poor self-images.
- Because of poor role models from their own childhoods, the parents often lack the ability to engage in nurturing, healthy, and appropriate child-rearing practices. They may be confused about the difference between discipline and punishment or be unaware that child abuse is not the norm.
- Abusive families often fear relationships. Their history has undermined their ability to trust. They may isolate themselves and create obstacles to engagement.
- As a result of their isolation and negativity, such families often lack support systems. They may never have learned how to ask for and receive help and therefore find it difficult to reach out.
- Parents often have severe marital problems. They may choose mates much like themselves—victims of abuse and low self-esteem. Sadly, the child often becomes the means of communication in such situations.
- Crises seem to characterize these families. They often have few healthy coping skills to manage their lives. They may not have self-control and often suffer from the results of their impulsive behavior.
- If the parents are absorbed in their own struggles to survive, physically and emotionally, they may have little to give the child. Parents or caretakers in such families are often unable to care for and protect the child. They may see the child as "special" or expect him or her to fulfill adult needs. These unrealistic expectations often lead to role reversal. The child may become an extension of the parents, or the parents may find themselves living through the child.

Child abuse risk factors Identification of risk factors may be valuable in assessment and in working with the family and local authorities. In determining whether the risk in a case is high, moderate, or low, the following factors can serve as guidelines (Filip, Schene, & McDaniel, 1991):

Severity and/or frequency of abuse
Severity and/or frequency of neglect
Location of injury
Child's age: physical and mental ability
Perpetrator's access to child (opportunity)
Child's behavior
Child/caretaker interaction

Child's interaction with siblings, peers, other adults
Caretaker's capacity for child care

Legal issues The legal aspects that all human service workers need to understand are summarized by the following questions: What is the law? How does one interpret and abide by it? How does one cope with the emotional toll of working with child abuse?

The law Child-abuse laws vary from state to state. In general, professional helpers, such as physicians, nurses, teachers, social workers, and counselors, are mandated in all 50 states to report suspected abuse of children under 18 years of age (Kalichman, 1999). When you enter an agency setting, it is important to "familiarize yourself with your state laws pertaining to reporting abuse and neglect, collect the telephone numbers of the agencies that must be notified" (Faiver, Eisengart, & Colonna, 2004, p. 126).

Routinely in most states, professionals who work with families and children under age 18 are required to report known and suspected incidents of child abuse, including physical injuries inflicted by other than accidental means, sexual abuse, neglect, failure to thrive, and abuse in out-of-home care. The reporting of emotional abuse is discretionary except for willfully cruel and unjustifiable infliction of mental suffering. Typically, all mandated reporters are to telephone immediately to law enforcement, the Department of Children's Services, or Child Protective Services (CPS). A written child-abuse report, on a designated reporting form, is to follow within 36 hours of first suspecting or encountering the abuse. The original plus copies for police or sheriff and district attorney are sent to CPS, and the reporting party keeps one copy. In cases where violent harm is likely, the police are to be notified at the same time as the call to CPS is made.

It is all-important that you be well-informed of the child-abuse reporting requirements of the state in which you practice, as well as the policies of your agency. Keeping up with any changes in the law is also essential.

Legal implications In many states, failure of a mandated party to report a known incident of child abuse is a misdemeanor punishable by up to six months in the county jail and/or a $1,000 fine. When reporting as required by law, mandated reporters are immune from civil and criminal liability, even if the report proves to be unfounded. All other reporters are free of liability unless it can be proven that the report was false and the reporter knew it to be false.

The act of reporting a suspected or unsubstantiated incident of child abuse does not necessarily mean that civil or criminal proceedings will be initiated against the suspected abuser. Depending on the circumstances, the authorities' subsequent inaction may be quite upsetting—when the reporter believes action should be taken. On the other hand, it can come as a relief—when the abuse is questionable or the suspected abuser did not mean any harm and is remorseful and seeking help to prevent a reoccurrence, for example. In any case, once a report is made, the human service worker no longer has any control. He or she may or may not actually be involved in the formal investigation. It is in the hands of Child Protective Services and/or law enforcement, who will follow up as they see fit.

Making a child-abuse report Deciding when to make a child-abuse report is one of the most difficult responsibilities a human service worker can face. Determining whether the instance is reportable can be a complex matter; the cultural standards of a community may be involved. Once you have learned the general reportable situations, you may feel confident that a report has to be made. Sometimes, however, the situation is not so clear cut. It becomes a matter of judgment whether to make a report or not.

For a beginning worker, the supervisor may need to assist in dealing with the intricacies of child welfare. The four scenarios that follow illustrate various aspects of child-abuse reporting and pose questions to help you think how you might respond.

SCENARIO: The Unclear Case

A client at a teen shelter returns from a weekend visit visibly upset. She resists engaging in group activities. You suspect that something happened. The client makes insinuations but when asked about being abused, she refuses to give a clear answer.

How should you proceed?

SCENARIO: The Helper Fears Betraying the Client

You have been treating a client intermittently for several years at a community counseling agency. The client is a very caring parent, dedicated to improving his parenting skills and resolving his childhood issues. One day, he calls you, upset by a recent incident in which he lost control and hit his 10-year-old son, leaving a mark on his upper thigh. Legally, you are obliged to report this. Yet you feel badly, because the client clearly feels remorse and is willing to work on controlling his impulses. He would prefer that you not involve Child Protective Services.

How do you proceed? What issues are you facing? How do you feel?

SCENARIO: The Client Abused Someone Years Ago

You come into contact with an elderly client who is new to the senior center. During your informal chats, he begins to discuss areas of his life that he feels are unresolved. He tells you of an incident in which he fondled his daughters in the bathtub when they were little. He never again did this, but he still feels guilty. Through the years, his children have grown and moved away, and their contact has gradually decreased. Although he has several grandchildren, he has no direct contact with them. The senior center is his only contact outside the home.

Should you report what happened many years ago? What if there is no current risk? What do you think is the best way to proceed?

SCENARIO: Client Was Victim of Child Abuse

You are likely to work with many clients who have been victims of some form of child abuse. For some, the abuse was mild and may not even be viewed as such: "My parents were strict." For others, however, the abuse was severe and has left lasting psychological scars. This may be seen in difficulty with trust and intimacy, low self-esteem, poor impulse control, and other traits. A common scenario is for adult abuse survivors to recall memories of being abused as children, once they become parents themselves. Sometimes, the memories that surface are so frightening that it seems as if the abuse were reoccurring.

How might you handle a client who is sharing such memories and fears? How might you be supportive? When might you have to report the abuse?

THINKING THINGS THROUGH

Review the four scenarios. Based on the knowledge you have obtained thus far, decide how you might handle each scenario.

In cases of child abuse, work with the family is often deemed necessary. Efforts are made to help the family learn more adaptive ways of coping. A social worker may be assigned to work with the family and monitor the case, allowing the child to remain in the home. Often the court orders parents to seek counseling and take parenting classes. Parents with substance abuse problems may be asked to attend a rehabilitation program. In severe cases, children may be removed from their parents and placed in foster care. They become wards of the court until CPS and the court determine that it is safe for them to be returned to their parents. In the most extreme situations, a child is never returned to his or her parents. Such children may be placed in group homes permanently or adopted.

Emotional aspects of child abuse reporting Once you have decided that a formal child abuse report is indeed required, you may feel overwhelmed by the realization that you will have to be the actual reporter. Feelings of anxiety, fear, and sadness may arise. "Reporting abuse and neglect is a difficult task, because the counselor is likely to be distressed at hearing of the situation, and clients are often upset by the need to report it" (Faiver, Eisengart, & Colonna, 2004, p. 126). Human service professionals must then confront and overcome their own internal barriers to reporting, such as the following:

Denial by the worker of the abuse or its severity
Rationalizing by accepting unrealistic explanations for how an injury occurred
Betrayal felt when contemplating having to report a child-abuse case
Family breakup feared if report is made, although this is, in fact, relatively rare

When the abuse is severe, you may experience shock, disbelief, and anger and reporting the abuse may be traumatic. Matters are further complicated if you are directly involved in working with the abused child and the family. How can you fulfill your ethical and legal responsibilities as well as cope with such emotional turmoil? The following suggestions may be helpful.

Helpful Interventions

- Be open with the parents/caretakers and the child about the requirement that you report abuse. If possible, present them with this information verbally and in writing at the beginning of your contact with them. Making these statements usually does not scare the client away.
- Make contracts with clients, where goals of your work together and behavioral descriptions of expected outcomes are spelled out.
- Set limits on clients' self-destructive or self-defeating behaviors. They may feel protected if your limit setting is done in a gentle way that conveys the message that you care for them.
- Use authority appropriately to gain the information needed to make the report. The helper can be firm yet supportive in dealing with the clients. Recognize the clients' feelings of helplessness and anger. Knowing that the helper will be available through the process is often a relief and can be quite comforting and reassuring.
- Be open to facing denial in abusive parents; this is the most common reaction. Rather than forcing them to admit to the abuse, work past this by focusing on the individual's strengths and the underlying family dynamics.
- When involved in making a child abuse report, attempt to stay with the client emotionally if possible. Make an effort to be supportive. This may not always be possible, however. If you are not able to follow up immediately, connect the child and family with appropriate referrals and resources.
- Coordination is essential to preventing chaotic treatment and unnecessary disruption of a family that is already in crisis. Work with others in a coordinated team effort to optimize the treatment. This will make your job a little easier as well.
- Be open to consultation. Reach out for support, objective feedback, and guidance. This can ultimately make a difference in both the outcome of the case and how you feel about your professional competence. Recognize that your feelings and fears are normal; allow yourself to vent when appropriate.
- Seek continuing education in the area of child welfare. It is essential to be aware of changes in laws and treatment. The risk for burnout is high in this area.

SCENARIO

Johnny is a 12-year-old boy whose mother is Caucasian and father is African American. The parents separated last year because of mounting turmoil and pressures in their relationship. According to the father, Johnny's mother gradually became depressed after several miscarriages and years of taking Johnny for chemotherapy, fighting cancer that was detected when he was an infant. Although Johnny's cancer is in remission, his mother never seemed to recover completely. She became paranoid and isolated, exhibiting psychotic delusions and hallucinations. The father left because he could no longer cope. He would visit Johnny every weekend to make sure he was properly cared for. One Tuesday, he got a call. There had been a child-abuse report made on Johnny. He had been arriving at school filthy and was also visibly bruised. He resisted telling how he received the bruises. Then Johnny told about his mother "confusing him for a burglar" and beating him with a broomstick. This was later attributed to her delusional state and visual hallucinations. When Child Protective Services went to her home to investigate, they found a chaotic home environment. The mother was sitting in front the TV, talking to characters in a soap opera. She was incoherent and became hostile and aggressive when asked about Johnny's care. At this point, it was decided that Johnny would have to be removed from this environment, because of the mother's inability to care for him adequately and her recent physical abuse of him. Johnny was placed with his father, who is seeking full custody. Both of them have attempted to visit the mother because Johnny never had the opportunity to say good-bye. However, no one knows her whereabouts. Johnny is depressed. He misses his mother and is struggling to adjust to the structure and rules in his father's home.

THINKING THINGS THROUGH

What factors apply in this case? What types of abuse need to be reported? What about the fact that Johnny's mother did not intentionally abuse her son? How would you proceed? How might you deal with the additional issues of loss and separation?

Elder Abuse

The elderly (age 65 and older) make up another vulnerable segment of our society. As with children, they are often victimized and subjected to abuse, within and outside the family. Human service professionals must be aware of this serious problem. We have a moral responsibility to act as advocates for people who are susceptible to victimization. Legally, we have a requirement to report elder abuse, just as with child abuse. The main difference is that reports of elder abuse are directed to Adult Protective Services for further investigation.

Scope of the problem The dramatic rise in cases of elder abuse has been proclaimed a "horrifying domestic disgrace" by the U.S. Senate Subcommittee on Aging (1985). As the number of older adults increases, cases of elder abuse are also likely to mount. Today, abuse of the elderly is a national problem of increasing proportions. Approximately 2 million elderly per year are reported to have been victims of abuse (American Association of Retired Persons, 2003). Moreover, most statistics underestimate the prevalence of the problem. In 1986, it was estimated that only one of six cases are reported (Hudson, 1986). It is believed that this inaccuracy is related to insufficient data, due in part to differing state definitions of the term *elder abuse* (American Association of Retired Persons, 2003).

O'Malley (1979, p. 2) defines elder abuse as "the willful infliction of physical pain, injury or debilitating mental anguish, unreasonable confinement or willful deprivation by a caretaker of services which are necessary to maintain mental and physical health."

Types of elder abuse The term *elder abuse* is commonly thought to signify primarily acts of physical violence against older persons, but it also encompasses many other forms of dangerous behavior. Morales and Sheafor (1995, p. 270) list "physical assault, verbal harassment, malnutrition, theft or financial mismanagement, unreasonable confinement, over sedation, sexual abuse, threats, withholding of medication or required aids, neglect, humiliation, and violation of legal rights." Common forms of elder abuse and neglect are categorized as follows by the American Association of Retired Persons (2003):

Physical Abuse: the intentional use of physical force causing pain or bodily harm

Psychological Abuse: the intentional infliction of mental anguish by threat, intimidation, humiliation, or other abusive conduct

Financial Exploitation: the conversion or unauthorized use of an elderly person's money, property, or other resources

Caregiver Neglect: a caregiver's intentional or unintentional failure to fulfill a caregiving obligation needed to maintain an elderly person's well-being

Self-Neglect: an older person's failure to provide himself or herself with the necessities of life, such as food, clothing, shelter, adequate medication, and reasonable financial management

One form of elder abuse is parent abuse. According to Kock and Kock (1980, p. 14), the term refers to "elderly parents who are abused by their children with whom they live or on whom they depend." In research in Cleveland, they found four main types of parent abuse: physical, psychological, material, and violation of rights.

The literature (Sommers & Shields, 1987; Steinmetz, 1978; Roberts, 2000) highlights the following aspects of elder abuse (paralleling those commonly noted in child abuse):

The victim's dependent position

The presumption that the victims are protected by loving and caring people

Emotional, physical, and financial stress on the caretaker, especially when the elderly person is also disabled

Implications of elder abuse Failing to address such abuse has substantial negative consequences in both human and monetary terms:

Decline in health—requiring more costly medical care, hospitalization, or early nursing home placement

Exacerbated psychological problems, which can lead to addictive behaviors and ultimately suicide

A negative message about our national attitude toward the elderly, which reinforces existing tendencies toward abuse and neglect of elderly people

Perpetuation of the stigma associated with aging

Helpful Interventions

Knowledge of state laws on elder abuse, crisis intervention techniques, and professional case management skills

Intervention with the family (in both assessment and follow-up)

Advocacy of the interests of elderly people

Preretirement planning, including education—this can help reduce the risk of elder abuse and prepare seniors for the issues they will face

Examination of our own values, attitudes, and beliefs about growing old; recognizing that we too are aging can make us more empathetic

SCENARIO

An 80-year-old Caucasian female (Mrs. S.), recently diagnosed with dementia, was brought to an adult day-care center by her 63-year-old daughter Jan, who had heard about the center from her physician. She herself was experiencing medical problems, which were further exacerbated by having to care for her elderly mother. Mr. S. had passed away several years before, and Mrs. S. had moved in with Jan, who was twice divorced. In the last year, Mrs. S. had become increasingly difficult to manage. This took a heavy emotional toll on the daughter, who had never been close to her mother. Consequently, the thought of some respite care was a relief for Jan. Mrs. S. was admitted into the day-care program and attended fairly regularly. When Jan would drop her mother off in the morning, the staff often noticed conflict between the two. They considered the daughter to be physically rough with her mother, who had difficulty following simple directions and struggled to move from the car to a walker. On several occasions, staff also noted bruises on Mrs. S., which she was unable to explain due to her mental condition. The daughter justified her mother's bruises as resulting from accidental falls or bumping into objects. "You know how clumsy and unsteady my mother can be, especially now that she is older and not thinking clearly." The staff grew worried about possible elder abuse. They decided to consult with Adult Protective Services; an abuse report was deemed appropriate. A social worker was assigned to investigate further and found that there was physical as well as emotional abuse by Jan. It became clear that Jan could no longer care for her elderly, disabled mother. Mrs. S. was placed in a board and care home, where ongoing care could be provided and properly monitored. Jan was referred to counseling. She was reluctant at first but was eventually able to begin dealing with her unresolved anger toward her mother for the abuse she herself had sustained through the years.

THINKING THINGS THROUGH

Based on your knowledge of elder abuse, discuss the preceding scenario. How do the respective profiles of victim and perpetrator fit? What family issues are in evidence? How would you have handled it, were you the director of the adult day-care center?

Abuse of Dependent Adults

Reporting abuse of dependent adults is also mandated in many states. Many helping professionals are unaware of this relatively new requirement. Adults who require ongoing care to survive are susceptible to abuse because of their powerless position. Reporting procedures in such cases resemble those for elder abuse.

SCENARIO

Claire is a 38-year-old, recently separated mother of two—a son age 11 and a daughter age 8. After her separation, she had to return to work after many years away from the job market. Her skills were outdated; the only jobs available were low paying. Because of a bitter custody battle with her ex-husband, child support payments were not reliable. Claire found herself in a financial and emotional crisis. Her older brother, Ken (disabled with multiple sclerosis), had been living independently. When he approached her for help, she welcomed the idea. Ken had become progressively more disabled and was at the point where he could no longer live independently without the help of an aide. Ken feared that he would have to be placed in a nursing home, which he did not want. Because he knew of Claire's predicament, he thought they could help each other out. He proposed that Claire and the children move in with him, rent free. In return, she could be responsible for his care. She could also apply to be his aide and receive funding from the state. Both found this arrangement to be mutually agreeable.

As time passed, however, Claire began to resent being tied down to her brother, whose condition had deteriorated such that he needed almost full-time care. She would express her frustrations verbally, was excessively rough with him, would neglect certain aspects of his care, and used his disability checks for herself and her children. One day a close friend witnessed some of this abuse. She later shared her concerns with her own counselor, who suspected abuse of a dependent adult.

THINKING THINGS THROUGH

What was your reaction to this scenario? How would you have responded if you were the sister? If you were Ken? As a helper, what interventions would you consider?

In sum, Leslie (1989, p. 37) states, "All licensees, interns and trainees should be aware of their duty to report child abuse, elder abuse, and dependent adult abuse." Failure to make such a report can result in the conviction for a crime, liability in a civil lawsuit for damages, and loss of license or registration. These obligations to report came about because legislators decided that certain public safety interests took precedence over the principle of confidentiality.

In addition to the cases of abuse we have considered, exceptions to confidentiality also include cases where there is a danger to oneself or an intent to harm others. The ultimate harm to self (suicide) and to others (homicide) are two of the most anxiety-producing situations facing human service professionals. As with child abuse, cases involving suicide or homicide are often crisis situations. Suicide and homicide have legal and ethical implications that must be taken into consideration. The varying roles of human service workers and licensed clinicians may dictate different levels of intervention and responsibility, but an understanding of the issues is necessary for all.

Suicide

Throughout time, for various reasons, some people have opted to take their own lives. In the past three decades, suicide prevention centers have been made available through federal funding (Roberts, 1990, p. 398). The status of these centers is uncertain because of budget cuts and limited resources. Unfortunately, however, suicide attempts will not cease.

Suicide is among the ten leading causes of death in the United States, and the third leading cause of death for young people between the ages of 15 and 24 (National Center for Health Statistics, 1998). Gilliland and James (1990, p. 129) state, "Suicide can strike any family, it cuts across all segments of the population, and it is prevalent in all age and racial/ethnic groups." Statistics show that the two groups for which suicide rates have skyrocketed are the elderly and adolescents. Shea (1998) and Shea (1999) report more than 25,000 suicides in the United States each year with older men, adolescents, and persons suffering from severe major depression or psychosis with religious preoccupation at the most lethal risk of killing themselves. Women over 65 commit suicide twice as often as the population in general, and men of that age kill themselves at a rate that is four times the national norm (Janosik, 1984; National Center for Health Statistics, 1998). These staggering statistics reflect the plight of the elderly. Similarly, adolescent suicides have shown a dramatic increase. Some estimates suggest that among people 15–24 years old, suicide is the third leading cause of death, following accidents and homicide. Estimates show that each year half a million young people between ages 15 and 24 attempt and nearly 5,000 children and youth (age 5 to 24) kill themselves (Peters & Murphy, 1998). In fact, suicide is the second leading cause of death for young people between the ages of 15 and 19 (American Association of Suicidology, 1997). These statistics bear witness to the stark reality that adolescence is a dangerous time of life in our society. Clearly, some teens are in crisis and require our involvement. Similarly, we must not close our eyes to children who also can suffer from depression and attempt as well as succeed in ending their lives.

Hopelessness, desperation, loneliness, low self-esteem, depression, and guilt are feelings that we may all experience to some degree from time to time. It is also not uncommon for people to think about suicide at some point but not to develop a specific plan or follow through with an attempt to take their lives. Anyone contemplating suicide should be taken seriously and certainly not be condemned.

Roberts (2000, p. 134) developed a "Crises Intervention—Seven Stage Model" that is applicable to a wide range of crises. He advocates using the following steps to help individuals in crisis work through their feelings: (1) assessing lethality and safety needs, (2) establishing rapport and communication, (3) identifying the major problems, (4) dealing with feelings and providing support, (5) exploring possible alternative, (6) formulating an action plan, and (7) providing follow-up. Let's elaborate further on assessment guidelines and risk factors.

Guidelines for Assessing Suicidal Intention When clients verbalize their desire to commit suicide, they are likely expressing their feelings of hopelessness, helplessness, loneliness, and other painful emotions. In a sensitive and empathic manner, helpers

need to evaluate the following hierarchy of risk factors: intent, plan, and means (Faiver, Eisengart, & Colonna, 2004). The following risk factors are important to determine:

- The person has indicated by words or actions an intent to commit suicide or inflict bodily harm upon himself or herself.
- The person has a specific plan and the means and ability to harm or kill himself or herself.
- A third party has reported credible information indicating that the person has a plan, the means, and the ability to harm or kill himself or herself.
- Questions to ask the person, a family member, or another person familiar with the client's behavior include the following:

  Does the person intend to kill or harm himself or herself?

  How does the person intend to proceed? Assess whether the person has access to weapons, pills, has thought of leaving gas left on, or other evidence of a plan and the means to carry it out.

  Has the person ever attempted suicide or self-harm in the past? What was the attempt at that time?

  Is there a past history of psychiatric evaluations or hospitalizations? When and where?

Once it is ascertained that the client is indeed contemplating suicide, the next step is to conduct a risk assessment or make a referral to someone who is qualified to do so. The literature offers an abundance of valuable information regarding risk factors for suicide:

- Family history of suicide. Risk increases when "family members, friends, classmates, or co-workers have committed suicide, either in the past or more recently. Anniversary dates of those events may be especially difficult" (Faiver, Eisengart, & Colonna, 2004, p. 143).
- History of previous suicide attempts (80 percent of suicides were preceded by an attempt). "The risk and dangerousness increase significantly when there is any previous history or suicidal ideation, gestures, and particularly previous attempts. Risk increases if a previous attempt was recent, if a potentially lethal method was used, or if an effort was made to avoid rescue" (Jobes, Berman, & Martin, 2000). Previous suicide attempts are one of the best single predictors of lethality (Corey, Corey, & Callanan, 2003).
- Direct verbalization about the desire, intent, and plan to commit suicide increase the risk. If a client has actually verbalized an intent, this signifies that suicidal thoughts are present. It is critical for helpers to assess the psychological intention, purpose, motive, or goal of the suicidal individual—what does the option of suicide mean to the client? "For some, this means seeking relief through death; for others, relief may be sought through changes in others' behavior, effected by gambling with life" (Jobes, Berman, & Martin, 2000, p. 140). In fact, direct verbal warnings are considered one of the most useful single predictors of a suicide (Corey, Corey, & Callanan, 2003, p. 222).
- A specific plan (the more definite the plan, the more serious the situation; do not be afraid to ask directly about intent). Berman and Jobes (1991) advocate assessing for the suicide plan. "The plan reflects both the desired and expected consequence of suicidal behavior and therefore provides one of the best indicators of what may actually come to pass" (Jobes, Berman, & Martin, 2000, p. 140). Research indicates that suicidal risk increases when there is evidence of a carefully thought-through and articulated self-harm strategy. Further, when an individual is knowledgeable about lethal means (e.g., gun or medications), this also increases the risk. In sum, the lethality and availability of the proposed method and the specificity of the self-harm plan must be carefully evaluated by the helper. Also,

the probability of a postattempt discovery and whether efforts could reverse the impact of the potential suicide act (e.g., pumping the stomach after an overdose) must be assessed. It has been found that "assessment of the various aspects of a potential plan is perhaps the best means of evaluating suicidal intent and imminent risk of self-harm" (Jobes, Berman, & Martin, 2000, p. 341).

- Serious depression (sleep disturbance is a key sign). Rush and Beck (1978) have identified a profound sense of hopelessness and helplessness about oneself, others, and the future as signs of depression and suicide. When sleep deprivation is present, this is key. Also, those who suffer from clinical depression have a 20 times greater suicide rate than that of the general population (Corey, Corey, & Callanan, 2003, p. 222).

- Severe anxiety and panic attacks can become so difficult to manage that suicide becomes the only logical outlet.

- When clients who have been taking antidepressants for their mood disorder start to feel better, they may be at greater risk because they have more energy to follow through with their suicidal thoughts and plans (Favier, Eisengart, & Colonna, 2004, p. 143).

- Previous psychiatric history and hospitalization will increase suicidal risk. "Clients who have been hospitalized for an emotional disorder are more likely to be inclined to suicide" (Corey, Corey, & Callanan, 2003, p. 222).

- History of severe alcohol or drug abuse greatly increases suicidal risk. "Alcohol is a contributing factor in one fourth to one third of all suicides" (Corey, Corey, & Callanan, 2003, p. 222).

- Lack of support systems or a belief that one lacks social supports will increase suicidal risk. Being alone can lead to isolation and despair.

- Other clinical risk factors identified by Jobes, Berman, and Martin (2000) include "dramatic and inexplicable affective changes; affect that is depressed, flat, or blunted, the experience of recent negative environmental changes (losses); feelings of isolation or emptiness; and experience of extreme stress, free-floating rage, agitation, and fatigue" (p. 142).

Human service professionals may feel especially anxious when working with suicidal clients. Becoming more knowledgeable about appropriate assessment and treatment of suicidal clients will not eliminate all your fears, but it can help reduce your level of anxiety. If you are not in a clinical position, your role will differ. You will not be expected to engage in the thorough assessment needed to judge the risk of suicide. Nonetheless, you may be the first helper to meet with a suicidal client, so you need to understand the underlying dynamics and the appropriate referrals. Undergraduates may not yet have use for an in-depth discussion of suicide; graduate students or helpers working in agencies, where the potential of suicide and homicide is higher, will need to study the following material more carefully.

Gilliland and James (1990) consider three areas of assessment when evaluating a potentially suicidal client: risk factors, suicidal clues, and cries for help (p. 133). We will follow their format in discussing assessment.

Risk factors Regardless of how you become aware of a client's suicidal intent, certain procedures must be followed. Before exploring risk factors, directly broach the issue of suicide with the client. Jobes, Berman, and Martin (2000, p. 138) state, "The assessment of suicide risk fundamentally requires direct inquiries about suicidal thoughts and feelings. Simply stated, vague suicidal comments should always elicit a direct question from the clinician." The fear that talking about suicide might push the client toward it often makes helpers hesitate to bring it up. For instance, "helpers may fear that a direct inquiry might introduce a dangerous new option not previously considered by the patient [client] (i.e., planting a seed for suicide in the patient's mind)" (Jobes, Berman, & Martin, 2000, p. 138). Faiver, Eisengart, and Colonna (2004) stress, "Asking about suicidal thoughts will not encourage your clients to commit suicide" (p. 143). Helpers may also feel angry at the client or take excessive responsibility for their feelings and behavior. However, only when you have explored the client's latent suicidal tendencies can you take precautions to ensure his or her safety. If you are not in a position to conduct a thorough assessment, seek appropriate assistance from a supervisor, clinician, or crisis team, and make a referral.

In assessing the potential for suicide, carefully observe and listen to the client. If potential suicide is suspected, a qualified helper must be willing to ask certain pertinent questions about the client's suicidal thoughts. It is essential for helpers to address their concerns about the client's suicidal ideation through careful and direct questions. Because about two-thirds of individuals who mention suicidal intent eventually do kill themselves (Gabbard, 2001), you must take all suicidal statements seriously and make sure the client is thoroughly evaluated. In fact, some clients may actually welcome such direct inquiry because it gives them permission to discuss feelings they may have otherwise felt they could not discuss, and it can offer a relief to clients who want to explore these feelings further, but in a safe, supportive, and accepting climate.

Clues Suicidal clients are often fraught with ambivalence about their plans to kill themselves. According to experts in the field, individuals who are serious about suicide leave clues to their imminent action. Sometimes there are vague hints or subtle changes in behavior. Clients who suddenly appear calm and speak of feeling relieved over finding a solution may be ready to take their lives in an effort to end their suffering. Some clients may give away possessions or make arrangements concerning legal or financial affairs. Although suicide is often an impulsive act, there has usually been some premeditation. Being aware of possible clues can help in suicide prevention.

Cries for help Cries for help can be clearly evident rather than neatly disguised. No person is 100 percent suicidal; those with the strongest death wish can also be ambivalent, confused, and grasping for life. Cries for help may take the form of direct verbalization of suicidal intent, calls to suicide hotlines, actual attempts at suicide, or subtle warnings such as self-destructive behavior or an apathetic view of life.

Disguised forms of suicide Related to suicide are self-abusive behaviors such as anorexia, bulimia, self-mutilation, and substance abuse. These are often seen as self-contained disorders or conditions, but one should not discount the possibility that suicidal intent may be present. Failing to pay attention to the suicidal content of these forms of self-abuse can result in lives lost.

To recap, general factors associated with high risk of suicide in adults recently identified by Hepworth, Rooney, and Larsen (2002) include:

- Feelings of despair and hopelessness
- Previous suicidal attempts
- Concrete, available, and lethal plans to commit suicide (when, where, and how)
- Family history of suicide
- Perseveration about suicide
- Lack of support systems
- Feelings of worthlessness
- Beliefs others would be better off if one were dead
- Advanced age (especially for Caucasian males)
- Substance abuse

Patterson, Dohn, Bird, and Patterson (1983) designed the SAD PERSONS scale to help human service professionals determine the risk of suicide. It is an acronym:

Sex (males commit most suicides)
Age (older clients)
Depression

Previous attempt
Ethanol (alcohol abuse)
Rational thinking loss
Social support system lacking (lonely, isolated clients)
Organized plan
No spouse
Sickness (particularly chronic or terminal illness)

As a quick guide to assessing suicide risk, the SAD PERSONS instrument is helpful and also easy to remember. Numerous other intervention strategies exist. Additional scales identified in the literature as useful in assessing suicide include the Hopelessness Scale (Beck, Resnik, & Lettieri, 1974), the Scale for Suicide Ideation (Beck, Kovacs, & Weissman, 1979), and the Suicide Probability Scale (Cull & Gill, 1991).

Adolescents and children also are subject to depression and crises that can result in suicidal feelings and intent. The developmental differences between the various age groups must be taken into consideration. Werner (1994) has found that there is not a marked difference between childhood and adolescent depression: "Behavior manifestations and intensity of feelings are similar, once developmental differences are taken into consideration." Here are common symptoms of depression in adolescents:

- Anhedonia
- Depressed mood
- Significant weight loss or gain
- Insomnia or hypersomnia
- Psychomotor agitation or retardation
- Fatigue or loss of energy
- Feelings of hopelessness, worthlessness, guilt, and self-reproach
- Indecisiveness or decreased ability to concentrate
- Suicidal ideation, threats, attempts
- Recurring thoughts of death

Additional behavioral manifestations identified by the American Association of Suicidology (1999) include:

- Deterioration in personal habits
- Decline in school achievement
- Marked increase in sadness, moodiness, and sudden tearful reactions
- Loss of appetite
- Use of drugs or alcohol
- Talk of death or dying (even in a joking manner)
- Withdrawal from friends and family
- Making final arrangements, such as giving away valued possessions
- Sudden or unexplained departure from past behaviors (from shy to thrill seeking or from outgoing to sullen or withdrawn)

Because adolescence can be a turbulent period when hormonal changes, peer pressures, and identity issues are prevalent, it is important to avoid overreacting. When the preceding warning signs exist for more that 10 days, then there is a potentially serious risk of suicide. Additional red flags are when a child or adolescent has recently experienced the death of a loved one, had severe conflict with parents, lost a close relationship with a key peer or sweetheart, and has no other support system. "Finally, conflicts with others, especially with parents, has been reported as a clear precipitant for both attempted and completed suicides" (Hepworth, Rooney & Larsen, 2002, p. 240)

When working with adults, adolescents, or children, if an assessment has determined the client to be a potentially high suicidal risk, it is then most apropos to mobilize client support systems and arrange for psychiatric evaluation and/or hospitalization if needed.

Helpful Interventions

Be ready to make a referral and willing to accept that dealing with the client's issues may be beyond your competence.

Recognize the high level of stress inherent in working with suicidal clients. Be aware of your limits and the toll stress can take on you personally.

Remain active in your learning about the assessment and treatment of suicidal behavior patterns.

Maintain ongoing training for suicide prevention and crisis intervention. Stay up to date with current research, theory, and practice.

Familiarize yourself with the legal standards that relate to your practice. Be aware of the requirements of your agency, including whether involuntary hospitalization may be necessary.

Be open to consulting with an attorney who has expertise in this area. Most agencies have access to counsel for such purposes.

Always be open to consultation with your supervisors or other appropriate personnel in your agency as these cases can be difficult and complex.

Encourage clients and their support networks to get involved in the treatment process. Avoid taking on full responsibility for clients' welfare. Keep in mind that you can only guide them; they must want to get better.

Maintain documentation as a priority to minimize the risk of future liability.

Make every effort to maintain personal wellness when working with suicidal clients, especially in the event that a client actually does commit suicide. Be sure to have a support system available to process your feelings. Counseling may be appropriate to help you cope with unresolved feelings about the loss and your role as a helper.

Ethical and legal issues Human service professionals are legally and morally bound to protect clients from harming themselves. It is our duty to protect suicidal clients, so we must breach confidentiality when there is sufficient cause to suspect suicidal behavior. The decision whether to breech confidentiality when one deems a client to be potentially suicidal is often a difficult one. According to Remley and Herlihy (2001), practitioners could be accused of malpractice for neglecting to take action to prevent harm when a client is potentially suicidal, yet they can also be liable for overreacting by taking actions that violate a client's privacy. Although helpers may fear reprimand if they breach confidentiality when intervening to prevent a client's suicide, reports show otherwise. According to Vandecreek, Knapp, and Herzog (1988), lawsuits that are specifically related to breach of confidentiality to protect suicidal clients are usually not successful. Whereas, in contrast, "[l]iability for wrongful death can be established if appropriate and sufficient action to prevent suicide is not taken" (Houston-Vegal et al., 1997, p. 105).

According to Szasz (1986), the failure to prevent suicide is now one of the leading allegations in successful malpractice suits against mental health professionals and institutions. This underscores the value of proper assessment and intervention and the need to maintain ongoing communication with your supervisor.

SCENARIO: Working with a Child

Tommy, age 10, is brought into a mental health center by his parents, given their concern about Tommy's negativity and persistent anger. Tommy presents visisbly sad and upset about being at the center. He is very negative in his response to questions and appears doubtful that the clinician really cares about him. In further assessing Tommy's apparent depression, the clinician inquires if

Tommy has ever wanted to hurt himself. Tommy hesitates, but eventually admits that he often thinks about dying, yet has never developed a plan.

SCENARIO: Working with an Adolescent

Becky is a 15-year-old, Caucasian female who lives with both parents and two older brothers. She has felt suicidal on several occasions and has been hospitalized in an adolescent psychiatric facility most recently for attempting to harm herself by taking a variety of pills she found in her parents' medicine cabinet. When evaluating Becky it is clear that she is quite upset with her parents for not allowing her the freedom she believes she should have. She argues that all her friends are allowed to do much more than she can and that "it's unfair." On the various occasions when Becky acted on her suicidal thoughts, it was after a fight with her parents over not being allowed to go somewhere with friends. Becky's parents seem angry as they feel that Becky is often manipulative, yet they still worry that she is capable of harming herself as a way to get back at them. They seek help on how to help Becky learn more adaptive coping.

SCENARIO: Working with an Adult

A 40-year-old, divorced mother of three comes to the drug and alcohol treatment agency where you are an intern. The client says that she is not adjusting well to being on her own after her 15-year marriage, and that she has been feeling depressed. Her ex-husband has not been paying child support as the court ordered him to do, making it difficult for her to care for the children properly. The client says that her husband was using drugs for several years before she left him—mostly marijuana and some cocaine. She is disillusioned with the men she is dating: "They're just like my ex-husband." She reveals that her ex-husband hit her on several occasions. She was recently laid off from her secretarial position at a large insurance company where she had worked for seven years with positive reviews. She has been having trouble sleeping, is extremely tired, has no appetite, and feels helpless and hopeless now that she is without a job. When you ask her about suicidal ideas, she reveals that she has had recurring thoughts of suicide for the last few days. Mostly, she thinks about taking pills or driving her car off a cliff. She has never had these kind of thoughts before, nor has she ever been as depressed as she is now. She recognizes that taking her life would be terrible for her children; she is ambivalent about wanting to kill herself. In relating her history, the client reveals that she was "a good child, obedient and quiet." She says that her mother was often depressed and that her father used to beat her mother. The client would then hide in the other room and has never been able to share her feelings of anger, fear, and sadness. Feelings were seldom expressed openly in her family, and she continues to struggle with the expression of her emotions.

THINKING THINGS THROUGH

Can you identify with any of these scenarios in any way? Have you known a client, acquaintance, family member, or friend who has been suicidal? How did you react? In these cases, how might you intervene? What might some of your feelings be?

If you are not in a clinical setting, how would you deal with these client? How would you proceed? Let's say you are a caseworker at a vocational rehabilitation center in the community. While taking an application, you encounter this mother, who is applying for assistance. During the intake interview, she shares her suicidal feelings. How would you proceed, given that this is not a clinical setting? What might you consider doing first?

Homicidal Behavior

Dixon (1987) uses the phrase *homicidal behavior* to mean potential for or threat to commit physical harm to another individual. Morales and Sheafor (1995, p. 434) define *homicide* as follows: "Death due to injuries purposefully inflicted by another person or persons, not including death caused by law enforcement officers or legal execution by federal or state government."

Human service providers are apt to encounter some form of homicidal behavior in their work. Here, we will discuss the legal and ethical aspects, predictive factors and risk assessment; and appropriate interventions. As in the case of suicide, be cautious about engaging in actual treatment unless you are appropriately trained and licensed.

Ethical and legal implications As with suicide, confidentiality must be upheld, with certain exceptions. If clients inform human service professionals of their intention to harm others, confidentiality may be breached. Corey, Corey, and Callanan (2003, p. 209) capture the essence about working with dangerous clients and the duty to protect potential victims:

> Although practitioners are not generally legally liable for their failure to render perfect predictions of violent behavior of a client, a professionally inadequate assessment of client dangerousness can result in liability for the therapist, harm to third parties, and inappropriate breaches of client confidentiality. Therapists faced with potentially dangerous clients should take specific steps to protect the public and to minimize their own liability. They should take careful histories, advise clients of the limits of confidentiality, keep accurate notes of threats and other client statements, seek consultation, and record steps they have taken to protect others.

If the worker is not a licensed clinician, he or she must make an immediate consultation or referral; both the client and the intended victim must be protected. We advocate consulting any available authority, including supervisors, law enforcement personnel, and mental health agencies.

In most states, helpers are permitted or required to breach confidentiality to warn or protect victims.

Duty to warn This originated in California in the landmark court decision of *Tarasoff* v. *Regents of University of California* in 1976. A therapist at the University of California Counseling Center at Berkeley in 1969 was told by a client, a university graduate student, that he intended to kill his girlfriend, Ms. Tarasoff. Upon learning that this client had purchased a gun, the therapist proceeded to inform the campus police. When the police spoke to the client, he assured them that he would stay away from Ms. Tarasoff, but two months later he stabbed her to death. Her parents brought suit against the university and the therapist. After several appeals, the courts found negligence because neither the therapist, his supervisor, nor the campus police warned the victim or her family (Costa & Alterkruse, 1994; Reamer, 1994).

The court ruled that three conditions make the duty to warn necessary:

1. A "special relationship" exists between therapist and client. In this milieu, the therapist has a duty to control the conduct of the client.
2. There must be a clear need to control the conduct of the client—that is, a clear threat. In this case, the client not only directly verbalized his intent but also had a gun with which to fulfill his threat.
3. There must be a foreseeable victim, even if not specifically named. In this case, the identity of Ms. Tarasoff could be determined (Benke, 1998; Kopels & Kagle, 1993; Mills, Sullivan & Eth, 1987; Reamer, 1994).

Although initially a California decision, *Tarasoff* became a source of controversy nationwide. Many *Tarasoff*-like cases have led to further significant court decisions, including the *Bradley* case, the *Jablonski* case, and the *Hedlund* case (Corey, Corey, & Callanan, 2003, p. 212).

What was initially called *a duty to warn* is now commonly referred to as *a duty to protect*. The *Tarasoff* case has led to the passage of state laws defining mental health professionals' responsibility to protect the public if serious threats are made (Neukrug, 2002). Many states now have laws stipulating that if a human service professional has a suspicion that someone is seriously intent on harming another person, he is required to take all the necessary steps to warn the intended victim, as well as alert the responsible authorities, namely the police (Baird, 2002). Although there continues to be varying interpretations in subsequent cases and in resulting state laws, the two principles that remain consistent are (1) that the helper perceives a foreseeable and imminent threat to (2) an identifiable potential victim. When these two principles are present, the helper "should act to warn that victim or take other precautions (such as notifying police or placing the client in a secure facility) to protect others from harm" (Hepworth, Rooney, & Larsen, 2002, p. 72).

Ethically, the human service professional is entitled to breach confidentiality to protect the victim and the public. The National Organization for Human Service Education (1994) highlights this in Statement 4: "If it is suspected that danger may occur to the client or to others as a result of client's behavior, the human service professional acts in an appropriate and professional manner to protect the safety of those individuals. This may involve seeking consultation, supervision, or breaking the confidence of the relationship."

Predictors According to the literature, there seems to be no single effective way to assess a client's dangerousness or to predict homicidal intent. However, factors such as history of violence, poor impulse control, or previous attempts at homicide increase the likelihood of homicidal behavior. Hoff (1989, p. 288) lists the following predictors:

History of homicidal threats
History of assault
Current homicidal threats and plan
Possession of lethal weapons
Use or abuse of alcohol or other drugs
Conflict in significant social relationships, such as infidelity or threat of divorce

Other predictive factors may include psychotic behavior, need for money, desire for revenge, political concerns, involvement in organized crime, or pathological fascination with murder.

Menninger and Modlin (1972) also view certain childhood history patterns and experiences as "predictive clues to homicide." For instance, they have found that some individuals who experience severe emotional deprivation and rejection early in childhood will resort to violence in a crisis situation. Corey, Corey, and Callanan (2003), Hoff (1989), and Knapp and VandeCreek (1982) stress the importance of identifying the precipitating events or cause of the violence. MacDonald (1967) and Knapp and VandeCreek (1982) state that the risk for homicidal behavior is usually greater when the reason appears more illogical or unjustified.

Risk assessment Ask the following questions of the client, a family member, or someone else who is very familiar with the person's behavior.

Is the person actively engaged in violent or dangerous behavior?
Is the person placing others in peril?
Does the person state intentions to carry out violent or dangerous behavior?
How does the person intend to harm others? (Ask about access to weapons or other evidence of a plan and the means and ability to carry it out.)
Does the person have a background of violent or dangerous behavior?
Is there a past history of psychiatric evaluations or hospitalization? When and where? Does the individual abuse alcohol or drugs? Is this person under the influence of any substance?

As previously noted, the conditions stipulated in *Tarasoff* are the existence of a special relationship, a reasonable determination that violence is likely, and the existence of a foreseeable victim. Monahan (1993) considers the following tasks to be necessary for an adequate risk assessment.

The helper must be educated about the risks for homicidal behavior and the necessary information to be gathered.
The helper must proceed to gather such information.

This information is to be used to estimate risk.

If the helper is not the ultimate decision maker, he or she must give this information to the people who are to be involved in the decision-making process.

Helpful Interventions

Ascertain whether *Tarasoff* issues are applicable in your jurisdiction. Find out what legalities apply where you intern or work.

Make a referral if the case is beyond your level of competence.

Give clients information regarding informed consent and the limits of confidentiality. It is advisable to convey very clearly what information you can and cannot hold in confidence and to alert the client of this before intervention begins.

As the need arises, remind clients of the limits of confidentiality. Clients may not recall the specifics; remind them, especially if they begin a discussion touching on suicide, homicide, or abuse.

Be aware of when *Tarasoff*-like situations may emerge. If a client makes a concrete threat, you are bound, legally and ethically, to take action. Depending on the statutes of your state, you may have to notify the intended victim or family.

Inform clients of your responsibilities to protect them and any intended victims. Explore options with your clients regarding warning the victim and possibly having to hospitalize the clients for their protection and the safety of the potential victims. You must first determine when to report and then what to report and to whom (Corey, Corey, & Callanan, 2003; Gilliland & James, 1990; Knapp & VandeCreek, 1982).

It may be advisable to seek guidance from counsel representing your agency. The *Tarasoff* decision is ambiguous as to whether the duty to warn takes precedence over confidentiality.

If the potential for harm is unclear, consult with peers or supervisors. Consultation can provide support, feedback, and clarification of legal and ethical concerns. Document carefully every step of the way.

Take care of yourself! Make an effort not to be swayed by guilt or threats of legal reprisal by the client. Seek needed support and validation for your interventions.

Working with homicidal clients is almost always a stressful and crisis-oriented experience. This work can be all the more difficult when you are dealing with such a client in a noncounseling capacity. Understand the underlying dynamics and be aware of appropriate interventions. Seeking support during and after the experience can help you deal with your own feelings.

SCENARIO

You are interning in the Office of Academic Counseling at a university. Your role is mainly to advise students on their coursework. Some students reveal personal concerns to you at times. A new student, in the course of discussing his academic program, reveals that his grades have been falling since his father got out of the mental hospital and came to live with him and his mother. The client states that he was severely abused by his father, physically and emotionally. As he tells you the story, he becomes more detached in his manner. He reveals a plan to shoot his father and says he can borrow a friend's gun to do it. As you try to assess his true intent, he becomes even more angry and volatile and repeats his threat to kill his father. When encouraged to consider options and explore the consequences of his behavior, he replies, "I don't care about anything anymore anyway! Maybe I'll shoot myself, too."

THINKING THINGS THROUGH

In the above scenario, how might you proceed with risk assessment and intervention? How might you feel while working with such a potentially homicidal client? Because you are a student adviser, not a counselor or clinician, do you feel you should make a referral?

Gravely Disabled Individuals

In some jurisdictions, confidentiality must also be breached when an individual is gravely disabled. In California law, for instance, there are two separate definitions of *gravely disabled:* one for adults and another for children and adolescents. See Table 6.1 (adapted from Lurie, 1985).

In the case of mental illness that affects peoples' ability to care for themselves, we must make sure they are not in danger of harming themselves or others. In such a case, the individual's identity can be disclosed to secure appropriate care and treatment (often voluntary or involuntary 72-hour hospitalization). You may be uncertain how to proceed with such a client; if so, consult with your supervisor and possibly with law enforcement or county mental health authorities.

SCENARIO

Alex is a 35-year-old whose heritage is part Latino and part Anglo. In his early adulthood, he was diagnosed with schizophrenia, undifferentiated type. Since then he has been hospitalized numerous times. He is unable to maintain a job or live in a stable environment, often preferring to live on the streets. After many attempts to help him, his family has basically given up.

This winter, Alex lives at a homeless shelter whenever there is a space available. While there, he has become increasingly volatile and unable to care for himself. He seems to be delusional, and paranoia is beginning to surface. The shelter manager is fearful that Alex may lose control. She seeks guidance from the local mental health center, where Alex has previously been a patient.

THINKING THINGS THROUGH

In the preceding scenario, imagine that you are an intake worker at the mental health center. You receive a call from the shelter manager about Alex. How might you proceed? What might you recommend to the shelter manager? Who should intervene?

Legal Proceedings

Exceptions to confidentiality also apply when legal proceedings and court orders take precedence. This may involve a request for records or court testimony via subpoenas. Clients who are involved in a court case may request that records be released, either to their attorneys or to the court. Before releasing any client information, have the client sign a release giving you permission to do so. If the client refuses, you must make certain that there is a court order asking specifically that you release the particular information. Without a written court order, a request by an attorney is not sufficient to warrant any release of information.

As with court orders requesting records, subpoenas must also be obeyed (Baird, 2002). However, unless the subpoena so specifies, you should not release client records unless asked to do so in writing by your client or by the court. This does not mean that you do not have to testify in court; you are still required to answer specific questions asked of you. This matter involves legalities that you may not be sure of, so always consult with an attorney or discuss the specifics with your supervisor. Agencies often have a designated legal counsel, assigned to deal with any legal matters involving the agency. If you are an intern working with a client, your supervisor must also be involved because he or she is ultimately responsible for overseeing your work. For many people, the word *subpoena* elicits anxiety, fear, and confusion, and some amount of anxiety is certainly appropriate. Take time to evaluate the situation, and seek legal advice if you are still unclear about your legal and ethical responsibilities.

The legalities involved in confidentiality, privileged communication, and subpoenas are complex and constantly changing. Keep abreast of the statutes and literature in this area, and remain open to seeking support and consultation from your supervisor

■ **Table 6.1**

The Gravely Disabled Adult	The Gravely Disabled Child or Teen
An adult who cannot take care of his or her basic needs for food, clothing, and shelter *due to mental illness,* and who cannot or will not accept third-party assistance to meet those needs. For example:	A minor (under 18) who is unable to use or accept the elements of life, including food, shelter, and clothing provided to meet basic needs, *due to mental illness.* For example:
Despite having money, a person is unable to use it to obtain the food or shelter needed to sustain himself or herself	A minor refuses to eat even though food is provided
A person eats inedible items because he or she cannot tell the difference between food and things that are not fit to be eaten	A minor eats inedible items, such as feces, dirt, or staples
Inappropriate display of nudity, refusal to wear protective clothing in severe weather; wearing several layers of heavy clothing in very hot weather	Dresses inappropriately despite a caretaker's efforts to dress him or her appropriately
History of frequent psychiatric hospitalizations	

and legal counsel. We human service workers must allow legal experts to assist us in an area we are not familiar with, just as an attorney might seek professional counseling to deal with his or her personal difficulties or request professional expertise regarding human behavior.

■ SPECIAL ETHICAL ISSUES

The prevalence of emotional abuse, spousal abuse, and AIDS in our society makes it clear that human service professionals must be knowledgeable about these issues and skilled in providing appropriate interventions. Also, because our society is so diverse, helpers must be sensitive to cultural issues and able to provide services to all groups. Dual relationships also require close attention, because helpers can so easily become involved in conflicting roles.

Emotional Abuse

We have addressed abuse of children, the elderly, and dependent adults. In all three, emotional abuse was seen to have a major impact on victims. However, the reporting of emotional abuse is not mandatory in many jurisdictions. Nonetheless, disclosure is considered permissible—according to the laws of confidentiality, helpers can legally disclose indications of emotional abuse to the proper authorities. Usually, a report of emotional abuse by itself will not result in immediate action by the authorities. Although this type of abuse is difficult to prove, this should not deter you from reporting it. Other forms of abuse often exist where emotional abuse is in evidence. State laws vary regarding reporting of emotional abuse (when reporting is mandated and when it is merely permissible). It is important to know the specifics of applicable state laws.

There are two main categories of emotional maltreatment: emotional assault/abuse and emotional deprivation. Emotional assault includes verbal assaults such as belittling, blaming, or using excessive sarcasm with a child; inconsistency in the parent's behavior; persistent negative moods; constant family discord; and mixed messages (Office of the Attorney General of California, 1988).

A parent engaging in such emotional abuse can scar or incapacitate a child for life. It is clear that emotional maltreatment can impede a child emotionally, behaviorally, and intellectually, even though the scars may not be readily visible. Severe psychological disorders have been traced to distorted parental attitudes and actions. Many children who display emotional or behavioral problems have experienced emotional abuse by their caretakers (Office of the Attorney General of California, 1988).

Emotional deprivation, the other category, can be defined as denying children normal experiences, such that they do not feel loved, wanted, secure, or worthy. If parents ignore their children, whether intentionally or not, the result can be emotional starvation, which may be linked with delinquent behavior in adolescents and criminal behavior in adults. Such parents may be preoccupied with their alcohol or drug use, overwhelmed by their own personal and emotional problems, caught up with outside community affairs, or themselves the victims of emotional abuse.

Just as a child's physical needs must be met, so must the child's emotional needs. For normal development, children need emotional involvement from their parents.

SCENARIO

Juliette is a 20-year-old, U.S.-born, French-American college student. She is seeking guidance counseling about career issues. She begins to discuss the pressures she experienced at home as a child. At first she is hesitant to share these painful memories, but as she becomes more comfortable with her counselor, her emotions seem to flood out. Enormously high expectations were placed on her by her parents, from a very young age. They demanded perfection at school, at home, and socially and would chastise her for the slightest slip-up. She would be criticized for not "being good enough" or "not being the best." Frequently, she was told that she wouldn't amount to anything. Often her mother would taunt her with long-winded lectures about how she was destined to fail. Her father would follow suit, saying, "No good man will ever want you if you don't do better." This maltreatment never turned physical, but it severely harmed Juliette. Now, as a young adult, she constantly questions her abilities, is plagued with self-doubt and low self-esteem, and suffers from much anxiety and fear. Juliette worries about her performance almost every waking minute. She rejects any intimate involvement with other people, for fear of being ridiculed or not meeting their standards. She is quite isolated. Currently preoccupied with her career choice, she is afraid of choosing the wrong career and failing in any profession she pursues.

THINKING THINGS THROUGH

Think about any experience you have had, either directly or indirectly, with emotional abuse. Do you see it in a different light now? In the preceding scenario, how would you proceed in working with this client?

Spousal/Partner Abuse or "Intimate Partner Violence"

Spousal/partner abuse, recently called "intimate partner violence," or IPV, is also quite prevalent in our society. As in the case of child abuse and elder abuse, the number of reported cases is estimated to be much lower than the actual incidence. The FBI estimates that domestic violence is the most underreported crime in the United States (Jacobs, 1989). The following are important facts regarding prevalence, incidence, and risk of intimate partner violence found in the literature (Lenahan, 2004; Roberts, 1998; Roberts & Roberts, 2000; Tjaden & Thoennes, 1998; Valentine, Roberts, & Burgess, 1998):

- Incidence and prevalence of spousal/partner abuse or intimate partner abuse is great. Battering women is considered one of the most life-threatening, traumatic, and harmful public health and social problems in society today.
- In the United States, at least one woman is battered every 15 seconds.
- Nearly 25 percent of women who are battered are revictimized within six months.
- Recent estimates reflect that approximately 8.7 million women have been victims of some form of assault by their partners.

- The prevalence of intimate partner abuse among the gay and lesbian community is approximately 35 percent.
- Women who suffer the most severe injuries are often seen in hospital emergency rooms or hospital trauma centers. Injuries related to domestic violence account for an estimated 35 percent of women who require emergency room treatment.
- Women are more likely to be murdered as a result of intimate partner abuse.
- Intimate partner violence occurs in approximately one of three pregnancies. Battering during pregnancy is dangerous to both the woman and the fetus and has certain risks.
- 70 percent of intimate partner violence homicide victims are women
- 80 percent of women who were stalked by a current or former partner were also physically abused.

Most reported cases involve physical abuse to women by their spouses or partners. This is not to say that men are not abused by their wives or partners, but this is not as frequently reported or as prevalent. Therefore, we will focus primarily on battered women.

The history of the terms *spousal/partner abuse* and *intimate partner violence* reflects the changing times. For the later half of the 21st century, *battered wives* was the term used to describe spousal violence. With the shift in living arrangements from the traditional nuclear family came the term "domestic violence," which includes violence between intimates, including significant others. Around 1994, the more inclusive term of "spousal/partner abuse" emerged; it encompasses physical and sexual assaults, as well as economic and psychological assaults.

"Kicking, biting, pushing, slapping, punching, shocking, sexual assault, or assault with weapons" are all considered forms of spousal/partner abuse (Straus, Gelles, & Steinmetz, 1980). In addition to physical abuse, verbal and emotional abuse are included. Jacobs (1989, p. 130) states, "Contrary to popular opinion, battering is not a black and blue issue. Emotional, mental and psychological abuse typically inflict deeper and more lasting scars than skin-deep or bone-breaking battering." Lenore Walker agrees that humiliation and verbal torment may be even more traumatic than actual physical abuse or the threat of physical violence (Walker, 1979, 1984). Morales and Sheafor (1995, p. 307) state, "Most women are reluctant to report their abusers due to shame, guilt, and obstacles created by the legal, health, and social service systems."

By 1999, both the American Medical Association (AMA) and the Centers for Disease Control (CDC) started collecting and reporting instances of spousal/partner abuse in the following four categories:

1. *Physical abuse,* which may include pushing, spitting, biting, shaking, scratching, throwing, shoving, slapping, choking, punching, burning, and/or restraining a person against his or her will.
2. *Sexual abuse* that encompasses a wide range of behaviors such as pressured sex, coerced sex by threat or manipulation, sexual assault accompanied by violence, and physically forced sex or rape.
3. *Economic abuse* where the perpetrator abuses the victim by controlling his or her financial resources with respect to food, clothing, shelter, transportation, communication devices, insurance, and money. Included are threats to leave the family and withdraw financial resources.
4. *Psychological abuse* that includes threats, intimidation, and isolation. Often, the perpetrator will threaten harmful acts against themselves (suicide) or the partner and/or the children (homicide), as well as threats to kidnap the children if the victim leaves the abuser. Often the perpetrator will intimidate the victim by making accusations against the victims' character and parenting skills. Far too often the children are used as pawns to control the partner or to force the child to take sides in an argument. Isolation includes keeping the victim imprisoned at home, prohibiting her or him from contact with friends or family, and preventing use of the phone or family vehicle (Mansergh, 2004).

Most recently, the CDC has defined intimate partner violence as "the threatened or actual use of physical force against an intimate partner that either results in or has the potential to result in death, injury, or harm. IPV includes physical and sexual violence, both of which are often accompanied by psychological or emotional abuse" (Lenahan, 2004, p. 1).

According to the California Penal Code: "The physical injury that is committed by a spouse, former spouse, cohabitant, person with whom the victim has a child, or dating or engagement relationship" (CPC 13700, Lenahan, 2001) constitutes IPV. Interestingly, the definition of intimate partner violence was expanded to include former partners and cohabitants because of the inherent risks associated with both present and former violent relationships. Current behavioral definitions of intimate partner violence are similar to those discussed earlier under spousal/partner abuse. These include physical and sexual abuse along with emotional and psychological abuse that includes threats, intimidation, and isolation. More specifically, verbal and emotional abuse also involve name calling, accusing the victim of various things (e.g., seeing someone else, turning the children against the batterer), or making humiliating comments designated to erode the victim's self-esteem. Examples are statements such as "You're so fat [skinny, ugly], no one else would want you." Other forms of emotional abuse involve withholding behaviors such as failing to provide attention to the victim, disrespecting the victim's opinions, and yelling, screaming, or swearing at the victim (Lenahan, 2004).

In terms of psychological abuse, the batterer attempts to manipulate the victim through guilt and shame, capitalizes on the victim's fears, and subjects the victim to intimidation and threats. Isolation is encouraged whereby the victim is prevented from seeing family or friends. At times, the victim is even prevented from using the phone, her calls are monitored, she cannot have a cell phone, her time at the store is limited, she can only go out with the batterer, and/or the batterer may call home several times a day.

Economic abuse by the batterer often coexists by withholding money, failure to contribute sufficient funds to meet family needs, taking out loans without the victim's consent, running up the victim's credit, refusing to put the victim's name on property deeds, interfering with the victim's ability to work, preventing the victim from working outside the home, and refusing to pay for transportation to work.

Sexual abuse can take the form of spousal rape when the batterer forces the partner to have sex or to engage in sexual activities that she doesn't want to engage in. It can include debasing or degrading experiences, forcing the victim to have sex with children, animals, or other individuals, and spousal jealousy. The batterer may engage in other sexual relationships and/or infect victims with sexually transmitted diseases or expose them to HIV.

Prestalking behaviors may occur in the early stages of a relationship when the perpetrator becomes excessively attentive, flattering, and obsessive. He may follow the victim and wait for her outside of work or home. This becomes stalking usually when the victim attempts to leave. The batterer may use stalking to persuade the victim to return home, to end a subsequent relationship, or to discontinue legal actions. Stalking becomes another form of power and control over the victim. Stalkers are likely to follow their victims on social occasions, including dates or visits with friends, to destroy the victim's social system. As a means of regaining control over the victim, the stalker may threaten to take the children away. Former intimate partners are more likely to be stalked than those currently in relationships. Stalkers represent a serious threat to the life of the victim. In the year before their murders, 89 percent of murder victims experienced physical abuse in combination with stalking (Lenahan, 2004, p. 10) Cyberstalking has been defined as "the use of the Internet, e-mail, or other forms of electronic communication to stalk or harass an individual on an ongoing basis" (Lenahan, 2004, p. 10). Delusional stalking is when the stalker pursues an individual he or she doesn't know.

Patterns

As with other forms of abuse, spousal/partner abuse or intimate partner violence is not confined to any one racial, ethnic, or socioeconomic group. However, reports of battering appear to be more frequent among couples of lower socioeconomic status, especially where the spouse is unemployed or underemployed (Margolin, Sibner, & Gleberman, 1988). These statistics may be skewed because middle- and upper-class families have more resources available to keep any violent interactions a private matter.

Intimate partner violence is a multifaceted matter, involving conflicts over cultural values and the division of labor and often creating a climate of oppression. Intimate partner violence occurs in both heterosexual and homosexual relationships. Both short- and long-term relationships can fall victim to such abuse, which can range in age from adolescents to senior citizens. The CDC has reported that although alcohol/substance abuse may be a contributing factor, there are many instances when the perpetrator is "clean and sober." According to Mansergh (2004, p. 2):

> The only consistent factors for intimate partner violence are:
>
> - Perpetrators have a continuum of access to their victims' vulnerabilities, routines, and daily lives.
> - In addition to dealing with the "controlling" behavior that often goes along with the perpetrator's profile, the victim must deal with the intricacies of an intimate relationship and the sense of entitlement that the assailant frequently displays.

Additionally, there are "patterns in terms of assaultive behavior, patterned attacks and tactics, and patterned types of relationships" (Mansergh, 2004, p. 2). Understanding these patterns is paramount to providing effective assessment, detection, and intervention. Such assessment includes being able to identify the perpetrator's warning signs, the cycle of violent behaviors, and the function those behaviors serve.

The Cycle of Violence

According to Walker (1979), the cycle of violence in domestic violence cases has three distinct phases: the tension-building stage, or the "build-up"; the explosive or acute battering incident, or the "blow-up"; and the calm, loving response, or the "honeymoon." The length of each phase and cycle varies with couples and circumstances. The build-up phase tends to be the longest, while the blow-up state rarely lasts longer than a day or two. Table 6.2 provides details on the cycle of violence.

Deschner (1984) presents a more detailed breakdown of the stages of spousal violence: (1) mutual dependency, (2) noxious event, (3) coercions exchanged, (4) "last straw" decision, (5) primitive rage, (6) reinforcement for battering, and (7) repentance.

Understanding the cycle of violence is crucial to comprehending abusive relationships. In our society, it is often easier to blame the victims of such abuse rather than trying to understand the underlying dynamics. Human service workers must be sensitive to be effective in working with battered women. Any pressure to have them resolve the problems by leaving their abusive partner may create further conflicts. Despite good intentions, such interventions can further alienate some women, who may wind up feeling more misunderstood and alone. Many women's shelters now give their staff intensive training on domestic violence and crisis intervention. Such workshops aim to enhance the helper's understanding of the cycle of violence and the empowerment of the survivor (Jacobs, 1989).

Predictive Factors

There is no set "profile" of a batterer or a battered woman. However, various factors, when found together, may indicate a risk of becoming a perpetrator or victim. Table 6.3 summarizes research regarding these factors.

Interestingly, many of the predictive factors are similar in both the perpetrator and the victim. Perhaps this may reveal something about how men and women are socialized

■ **Table 6.2 Cycle of Violence**

PHASE ONE: TENSION-BUILDING PHASE

Step One: Victims have a sense of "walking on eggshells"

- Minor battering occurs
- Victim's denial of anger helps her to cope
- Victim blames outside factors

Step Two: Batterers don't want behavior made public, causing fear that the victim will tell, thus increasing the oppression

- Batterer's brutality keeps victim captive
- Learned helplessness syndrome

Step Three: As the tension increases, the anger escalates, the victim realizes that phase two is coming, and works hard to control external situations (e.g., keeping the children quiet, no phone calls)

- Soon coping techniques fail

Step Four: Batterer increases possessive smothering and brutality

- Victim is less able to defend against the pain and hurt
- Victim withdraws, batterer moves in more oppressively
- As the tension builds and becomes unbearable, victims sometimes trigger phase two to break the tension and to "just get it over with"
- Batterer characteristics: jealousy, isolation of the victim, rule changing, name calling, dominating, threats
- Victim characteristics: fear, minimizing, anger suppression, fatigue, confusion, self-doubt, withdrawal

PHASE TWO: THE ACUTE BATTERING INCIDENT

The Acute Battering Incident

- Lack of predictability
- Acute battering with major destructiveness
- Lasts usually from 2 to 24 hours, with some reports of a week or more of terror
- Only batterers can end this phase
- Victims express extreme futility of trying to escape
- Victims suffer emotional collapse 24 to 48 hours after the acute battering
- Victims will seek isolation
- Extreme sexual abuse is possible during this time
- Batterer characteristics: anger, uncontrolled tension, assault on the victim, exhaustion
- Victim characteristics: fear, anger, may call police, may flee

PHASE THREE: HONEYMOON PHASE

- Unusual period of calm
- Batterer is extremely loving and contrite, saying he is sorry and promising that it will never happen again
- The batterer believes he can maintain control
- He believes he has taught the victim a lesson
- He promises to give up drinking (or drugs, gambling, other women, etc.)
- He convinces the victim she's needed, makes the victim feel guilty for leaving, makes the victim feel responsible for what happened
- He promises they will get help if the victim stays
- The victim sees the batterer being sincere and loving
- The victim chooses to believe this is what the batterer is really like, everything she wanted in a partner
- This is when almost all of the rewards of being married or partnered take place, thus making it difficult for the victim to leave or end the relationship
- Batterer characteristics: apologizes, promises change, loving, dependency
- Victim characteristics: guilt, hope, loneliness, may return home, low self-esteem, dependency, deceived feeling

Sources: Cappell and Heiner (1990); Gilliland and James (1990); Jacobs (1989); Lenahan (2004); McLeer (1988); Roberts and Roberts (2000); Walker (1989).

■ **Table 6.3 Predictive Risk Factors**

The Batterer/Perpetrator	The Battered Woman/Victim
1. History of family violence	1. History of family violence
2. Aggressive coping style—use of force or violence	2. Passive coping style
3. Abuse of alcohol or drugs	3. Use or abuse of alcohol or drugs, or codependency
4. Low self-esteem	4. Low self-esteem and feelings of inadequacy that seem to mount as the cycle of violence is repeated
5. Strong traditional view of male–female roles	5. Belief in traditional family values: male–female roles
6. Excessive jealousy	6. Feels deserving of abuse, may feel guilty for not doing enough or believe she is responsible for the battery
7. Use of or access to weapons (guns, knives, or others)	7. Feeling trapped and without hope
8. Authoritarian toward wife	8. Submissive toward spouse; may be quite assertive or even aggressive with others
9. Mood swings	9. High and low moods, according to the cycle of violence
10. Poor impulse control; uncontrolled anger	10. May have difficulty expressing anger
11. Early in relationship, treated girlfriend/spouse roughly	11. May be blind to the early signs of abuse
12. Threatening demeanor at times	12. Feels weak and easily dominated
13. In denial—doesn't believe he has a problem	13. Does not believe anyone can or will want to help

Sources: Bersani and Chen (1988); Cappell and Heiner (1990); Gilliland and James (1990); Hughes (1988); Jaffe, Wolfe, Wilson, and Zak (1986); O'Leary (1988); Margolin, Sibner, and Gleberman (1988); Lenahan (2004); McLeer (1988); Roberts and Roberts (2000); Schechter (1982).

and about the violent nature of our society. Sadly, the United States has been called the most violent society in the Western world.

Reporting Responsibilities

There has been much controversy regarding reporting responsibilities. AB 1642 (1993) in California, for instance updates and expands the law requiring any health-care professional employed in a health-care clinic, physician's office, local or state public health department to report all patients whom the practitioner reasonably suspects is suffering from any wounds or other physical injuries caused by firearms and assaultive or abusive behaviors. The law further states that "any person suffering from any wound or other physical injury inflicted upon the person where the injury is a result of assaultive or abusive conduct must be reported to law enforcement The law requires that the health-care practitioner report any injury, whether it is minor or major. Verbal reports must be made immediately or as soon as possible and written reports must be sent "to a local law enforcement agency within two working days of receiving the information regarding the person." The California law recommends that any comments by the injured person regarding past domestic violence or the name of the person who allegedly inflicted injury be included in the report (Penal Code 11162, 11161.9, 11162.5). Finally, Section 11161.9 holds the reporting party immune from liability as a result of any report that was mandated by law. Because the law is specific regarding the reporter to be a health-care professional, one must be careful about the legalities about reporting if working in another type of human service setting (e.g., mental health or counseling). We recommend that beginning helpers unsure of their reporting responsibilities seek consultation and/or legal counsel regarding this matter.

If the helper is not considered a "mandated reporter," then reporting could be viewed as a breach of confidentiality.

Some states have similar laws, and others have no legislation regarding this issue. In many jurisdictions, battering is a now a punishable crime, and batterers are subject to arrest for their violent behavior. Probation may be granted, and participation in a diversion program may be mandated. Psychoeducation groups are becoming more prominent, and anger management treatment programs for batterers seem to be on the rise. Ethically, the issues are difficult. There is a temptation to encourage a woman to get out of the battering situation, without regard to her particular situation.

Treatment for the Battered Woman

In working with the abuse victim, various levels of intervention are commonly used: emergency response, crisis intervention, education, advocacy and referrals, and short- and long-term counseling.

Hotline counseling The goal is to listen and ensure the safety of people at risk for physical, verbal, or psychological harm. "Hotline workers distinguish between a 'crisis' call—one in which the woman is in imminent danger or has just been beaten—and other types of calls in which the individual is not in immediate danger but is anxious or distressed and is seeking information or someone to talk to" (Roberts & Roberts, 2000, p. 193).

Crisis intervention This is aimed at empowering the survivor. The woman is helped to deal with her crisis and to develop more adaptive coping. Children can also be profoundly affected by the violence that pervades the home. They too need to receive crisis treatment, support, and reassurance. The Abuse Counseling and Treatment (ACT) program used in Fort Myers, Florida, is an example of crisis intervention used to helper battered women. "It is referred to as the A-B-C process of crisis management—the A referring to 'achieving contact,' the B to 'boiling down the problem, and the C to 'coping with the problem'" (Roberts & Roberts, 2000, p. 193).

Advocacy services This involves information and referral to shelters, tips on police protection, assistance in obtaining legal help, and referral to specialists in abuse counseling. It is important to encourage use of resources. Assess the client's contact with health care providers, mental health providers, law enforcement, the legal system, community resources (shelters, victim witness, social service agencies), and clergy or religious leaders.

Educational services Women are taught about the cycle of violence and the role they play in it. They are helped to pinpoint the precipitators of abuse in their situation and to assess the part they play. Their self-esteem is enhanced, as is their repertoire of coping skills. Assertiveness training is crucial in helping them begin to stand up for themselves. Neighbors and McNeair (1996) say that battered women should also participate in psychoeducational programs; they believe that the cycle of violence will not stop if only the batterer engages in treatment.

Safety planning with victim Help client obtain bank account information, copies of Social Security numbers, health records for self/children, copy of last income tax return, proof of custody, license information for car, and marriage license. The client may also need extra cash, car keys, clothes, food, diapers, medicine, a safe place to store a suitcase and/or important papers, and phone numbers for doctors, lawyers, and other significant resources. You need to help the client determine what needs to be done to leave quickly. When the client has safely left, help the parent create a safety plan with the children when they are not with the abused parent. Help client screen incoming phone calls, change door locks and add locks and safely devices, obtain a post office box to receive mail, keep shelter and hotline numbers handy, devise a safety plan for work, travel to/from work, and alert the children's school or day care about who has permission to pick up the children (Lenahan, 2004; Roberts & Roberts, 2000).

Short- or Long-Term Counseling   Women are encouraged to take a closer look at the abusive relationship and to make plans for their lives. The emphasis is on helping survivors break through the isolation and work on rebuilding their self-esteem. At times, psychotropic medication may be required for extreme anxiety or depression. Appropriate referrals are made to psychiatrists and others in the medical profession.

Screening for intimate partner violence   Use gender-neutral questions, avoid use of terms such as *abused* or *domestic violence*, provide examples of what being hurt or threatened may mean (e.g., Has anyone hit, slapped, kicked, choked you? Has anyone said he was going to kill you or someone in your family?). Use active listening, offer hope to indicate the victim is not alone, and validate the victim's feelings, fears, and concerns. Riley (1994) has found that art therapy can be used as part of a comprehensive treatment approach in shelters for battered women and their children to help victims communicate their painful experiences in a nonverbal manner that is less threatening than traditional talk therapy.

Treatment for the Batterer

Treatment for batterers covers a wide spectrum of modalities: individual counseling, group therapy, marital therapy, couples' group therapy, and anger management groups. Some feel that psychoeducational groups may not be the most appropriate treatment for batterers, because social reinforcement for violence against women may minimize their effectiveness (Gondolf & Russell, 1986). Alternative treatments include resocialization and theme-centered programs.

Treatment for Children

Recently, there has been increased concern for the child who lives in a home where domestic violence occurs. The child is clearly a victim of emotional abuse and is certainly at risk for being physically abused as well. The emotional effects of witnessing domestic violence are very similar to the psychological trauma associated with being a victim of child abuse. Child victims have been found to suffer a variety of emotional, behavioral, academic, social, and developmental problems that may require treatment (Osofsky, 1995; Roberts & Roberts, 2004). Individual and play therapy, art therapy, group counseling, and education are a few interventions we recommend. Give the child the opportunity to vent feelings, fears, concerns, and anger in a safe environment. Also, helping children learn healthier ways to manage their own emotions must be a component of any treatment program.

SCENARIO

Juana is a 25-year-old Latina who has been married to John for eight years. They met in high school and were married shortly after she became pregnant when both were 17 years old. Neither finished high school. Juana became a homemaker and had three more children. John has worked on and off in the construction industry. There have always been conflicts over John's excessive drinking and absence from the home. He is also irresponsible with their finances and has poor impulse control. Through the years, Juana has sustained numerous beatings. He tells her she could not survive without him. With four children, all under the age of 8, and no marketable skills to speak of, Juana feels trapped. She hesitates to tell her parents of her predicament; they warned her about John and were never in favor of their relationship. She also realizes that her father was often just as abusive to her mother. Through the years, her father has mellowed. Now he and Juana's mother tend to get along much better.

Socially, Juana feels isolated. John dislikes her having friendships or even a close relationship with extended family. He repeatedly tells her that her place is in the home, to care for him and their children. "That's the way it was when I grew up." He comes from a traditional home, where the father was the primary breadwinner and the dominant authority figure. It is unclear whether John's father physically abused his mother, although there certainly was verbal and emotional abuse. His father also went on drinking binges with buddies and would not return home for days. John remembers this with some anger but tends to minimize its impact.

Juana is finally at the end of her rope. As the children grow, the financial demands do as well. Furthermore, Juana is aware that the children are beginning to notice the abuse. When she overheard them acting out the abuse while playing "house," Juana realized she needed to seek help. Apprehensively, she contacts a mental health center where she is received on a crisis basis.

THINKING THINGS THROUGH

Can you recall a personal, professional, or social situation in which you have known that spousal abuse was occurring? Knowing what you know now, would you react any differently? In the preceding scenario, what are some of the salient issues? How would you intervene as a human service worker?

Let's assume that instead of seeking help at a mental health center, Juana instead calls the crisis hotline where you are doing your internship. You receive her frantic call for help. How would you proceed?

AIDS

The issues involved in the AIDS crisis are complex. As a helper, you are likely to encounter clients who are HIV positive or afflicted with AIDS. Historically, human service workers have been called upon to help those with special needs, and AIDS sufferers are one of today's current target populations.

AIDS (acquired immune deficiency syndrome) is a contagious, presently incurable disease that targets the body's immune system. AIDS is caused by the human immunodeficiency virus (HIV) and can be transmitted from one person to another during sexual contact or through sharing of intravenous needles and syringes (Zastrow, 1990).

> Acquired immunodeficiency virus (AIDS) is the last stage of a disease caused by the human immunodeficiency virus (HIV), which attacks and weakens the body's natural immune system. Without a working immune system, the body gets infections and cancers that it would normally be able to fight off. By the time an individual develops AIDS, the virus has damaged the body's defenses (immune system). People who have AIDS are vulnerable to serious illnesses that would not be a threat to anyone whose immune system was functioning normally. These illnesses are referred to as "opportunistic" infections or diseases. AIDS weakens the body's immune system against invasive agents so that other diseases can prey on the body (Corey & Corey, 2003).

Victims of this disease are left with little recourse when the HIV virus attacks their immune systems, which become unable to fight off other diseases. This physically vulnerable position also has major emotional and psychological ramifications. Once people are infected with HIV, there are several potential outcomes:

- Some remain well and show no signs of the virus, but they are *carriers*. They can transmit the virus to others, primarily through unprotected sex and use of infected syringes.
- Others develop AIDS-related complex (ARC), which is less severe than full-blown AIDS.
- Many develop AIDS. They cannot fight off opportunistic diseases such as pneumocystis carinii pneumonia and Kaposi's sarcoma (a rare form of cancer) (Martelli, 1987; Zastrow, 1990; Corey & Corey, 2003). Because of greater susceptibility to infections and other illnesses, coupled with an inability to fight off these infections, many eventually die as a result.

The term *epidemic* applies to the extent of AIDS in the United States and in the world today. Statistics show a dramatic rise in the death rate from AIDS. Statistics reflect that AIDS remains a "contemporary crisis" because "between 800,000 and 900,000 Americans are HIV-positive, and almost one-half million have died of the disease (Centers for Disease Control, 2002). Despite the bleak prognosis for many, newer medications are now available to treat the opportunistic infections that in the past would kill people with AIDS, and many HIV-positive individuals are living long and relatively symptom-free lives for many years.

Risk Factors

Anal Intercourse. This form of intercourse is considered most high-risk because it can result in tearing of the lining of the rectum, which allows infected semen to get into the bloodstream.

Multiple Sex Partners Without Use of Safe Sex Practices. Logically, the risk of infection increases according to the number of sexual partners, male or female.

Blood Transfusions. In a comparatively low percentage of cases, people have contacted the HIV virus during blood transfusions (e.g., during surgery). In these cases, the blood used was contaminated. Increased precautions now minimize this possibility.

Reuse of Contaminated Syringes. Sharing of intravenous needles with an individual who is carrying the virus can be quite dangerous, because "a small amount of the previous user's blood is often drawn into the needle and then injected directly into the blood-stream of the next user" (Zastrow & Kirst-Ashman, 1994, p. 399).

The complexity of this medical condition, the unfavorable prognosis for its victims, and the ease with which it can be transmitted heighten people's anxiety and encourages stigmatization and alienation.

The Plight of People Infected with HIV

Undoubtedly, almost anyone would feel overwhelmed upon being diagnosed with HIV infection or, even worse, with AIDS. People with HIV or AIDS often have to contend with stigma, prejudice, discrimination, alienation, rejection, and abandonment. Although AIDS is similar to other terminal illnesses, such as cancer, the reactions of others are markedly different—and often negative. People suffering from the AIDS virus must contend with attitudes and fears of people around them (Corey & Corey, 2003; Martelli, 1987; Neukrug, 2002; Popple & Leighninger, 1990; Zastrow, 1990):

The stigma felt by the individual suffering from this disease.

Prejudice against those who are infected.

Fear of infection is prevalent because AIDS is contagious.

Revelations are often necessary when one is diagnosed with AIDS. Many people who are homosexual or intravenous drug users have kept this a private matter; with the diagnosis of AIDS, they may have to reveal certain private aspects of their lives.

Alienation and abandonment are not only common fears but also actual possibilities for people who suffer from AIDS. Many people will consciously or unconsciously alienate themselves from an infected person.

Such defense mechanisms as denial and avoidance are common among people who are close to individuals diagnosed with AIDS.

Anxiety and depressive disorders and progressive dementias can affect AIDS patients. Lack of support from family, friends, employers, insurance companies, government agencies, and the medical profession further exacerbate the predicament of those afflicted with AIDS

For people with AIDS and their significant others, "a complex web of psychological issues, mainly related to some of our deepest fears, comes into play" (Martelli, 1987, p. 50). Among them are fear of mortality, fear of contagion, fear of differing lifestyles, and fear of helplessness.

Role of Human Service Professionals

In this new millennium, human service professionals continue to be an important component in the education and prevention of this disease, deliverers of counseling services for those who are already infected, and supervisors of those many volunteers who are giving of their time and humanity to assist in the caretaking of those infected, their families, and their friends.

The functions of helpers are likely to include educator, counselor, supervisor, advocate, caretaker, and case manager, among others (Neukrug, 2004). Buckingham and Van Gorp (1988), professionals active in this area, cite various interventions that have been useful in their work with the AIDS population:

> Problem solving with everyday concerns and difficulties
> Estate planning
> Decreasing the level of hypochondriacal preoccupation
> Guidance in designing adequate structure and limits for activities of daily living
> Assisting with family conflict

Neukrug (2002) and Shannon and Woods (1995) offer certain recommendations for working with HIV:

1. *Know the cultural background of client:* Besides working with the issues that are unique to HIV-positive clients, awareness of cross-cultural issues may also be necessary.
2. *Know about the disease and combat myths:* To avoid perpetuating the myths or discriminating against HIV clients, helpers must themselves be knowledgeable about the disease and its process.
3. *Be prepared to take on numerous helper roles:* Various roles, including advocate, counselor, caretaker, and resource person, may be necessary in working with clients who are HIV positive.
4. *Be prepared to deal with a number of unique treatment issues:* These include likely loss of friends and loved ones from this high-risk group, depression and feelings of hopelessness given declining health, uncertainty about relationships and future, and anxiety over loss of income or high cost of medical expenses. To be effective in implementing interventions, helpers must be sensitive, caring, and knowledgeable about AIDS. They must also be aware of the ethical dilemmas and legal issues involved.

Ethical and Legal Issues

Ethical concerns Human service workers must examine their own value systems, underlying attitudes, biases, and misperceptions, as well as possible fears about working directly with AIDS clients. If you remain open to your own issues, you can minimize the risk of their spilling over into your work with clients. Having any of these feelings does not make you an uncaring person, unfit to be a helper—it just makes you human.

SCENARIO

You are a new assistant medical social worker assigned to assist social workers throughout the hospital as needed. After a week, you are beginning to have doubts about this career choice. You feel very uncomfortable when asked to work with HIV patients. Whenever you are in contact with an AIDS patient, you feel that you have been contaminated by even having been in the same room. You find yourself dreading future contact; you begin to make excuses about not being able to follow through with their cases. You ask yourself whether you will be able to continue to treat this population. Most are homosexual or ex-drug users, and you tend to regard them as responsible for their plight, because of their "unhealthy" lifestyles.

How would you feel if you were in a position like this?

SCENARIO

Your new position as a drug rehabilitation counselor entails working in both inpatient and outpatient units. You are looking forward to the variety this will allow. While conducting inpatient groups, you become aware that many of the patients are potential HIV carriers. Many have been intravenous drug users and/or sexually promiscuous. When the topic of safe sex is broached in the group, many participants scoff at the suggestion. They are especially critical of abstinence. After the group sessions, you often feel angry, disgusted, and confused. You are concerned that you will eventually lose all empathy for people who do contract the HIV virus. You wonder whether this is the population you should be helping.

THINKING THINGS THROUGH

How do you suspect you might feel if you encountered a client with AIDS in your field placement? Do you believe you could remain objective? How might you proceed if you discover that you have some of the same prejudiced attitudes indicated in the scenario?

Legal implications With regard to AIDS, there are no clear legal obligations for the human service professional. However, two familiar ethical dilemmas could also involve legal considerations: breaching of confidentiality and duty to warn. When an HIV-infected client is involved in unsafe sexual practices with others, is it our duty to warn the potential partners, as *Tarasoff* might suggest? If so, would this be permissible under the laws of confidentiality? Legally, no clear decision has yet been issued to guide helpers caught in this dilemma; it remains unclear whether *Tarasoff* applies to AIDS cases. According to Corey, Corey, and Callanan (2003, p. 236), "[u]p to now, courts have not applied the duty to warn to cases involving HIV infection, and therapists' legal responsibility in protecting sexual partners of HIV-positive clients remains unclear. Practitioners have few legal guidelines to help them determine when or how to inform a potential victim of the threat of HIV transmission." In addition to differing state laws regarding HIV and limits of confidentiality, deciding whether to warn others about an HIV-positive client is also controversial and emotion laden, which clearly complicates one's ethical and legal obligations. Human service workers must be attentive to new legislation and ethical or legal guidelines in this area. For further specifics on guidelines that some authors provide, please refer to ACA Ethical Standards, B.1.d (ACA, 1995), and Corey, Corey, and Callanan (2003, pp. 236–239).

Remain open to consultation, supervision, and support if you encounter moral conflicts, ethical dilemmas, or emotional difficulties in working with AIDS clients. Just as the AIDS disease is a human disease, helpers are only human and will surely experience struggles in their work and may need guidance and support.

SCENARIO

Jim is a 30-year-old Caucasian who was born and raised in a small Midwestern town, to a middle-class, religious family. As the oldest of six and the first son, Jim was expected to excel in all areas. He was told he had to be a good role model to his younger siblings. While Jim was growing up, his father would often insist that he accompany him to Bible study groups. His mother was also actively involved in church functions. In fact, most of the family activities were church related.

During high school, Jim dedicated himself to his studies. A younger brother who was heavily involved in sports would often tease Jim for his reluctance to participate in athletic activities and would jokingly call him a "sissy." Jim was also heckled for his shyness. He was reluctant to date and basically kept to himself. It wasn't until Jim moved away to college that he felt free to find himself. In his new environment, Jim eventually had to face the realization that he was attracted to males. He could no longer deny his homosexuality.

Jim developed a supportive network of friends who accepted him for who he was. In time, he began dating other gay men. Eventually he became seriously involved with his first lover, Mike.

When Mike tested HIV-positive, Jim recognized that he too was at risk. Later, Mike died of AIDS. Jim tested positive about that time and felt alone as never before. He knew that he could never tell his family, who didn't even know of his homosexuality, let alone his HIV status. Because of their religious beliefs, he could envision their negative, critical, and unforgiving views of him. Jim opted to remain distant from his family and seek the support of the people he could relate to the most. Still, he felt saddened that his family would never know who he really was.

THINKING THINGS THROUGH

How did you feel when reading this about Jim? How would you work with Jim? What interventions would you use?

Ethnic-Sensitive Practice: A Multicultural Perspective

In our increasingly diverse society, we must maintain ethnic sensitivity in our work with people. In this section, we elaborate an integrated approach that we have developed from the literature and from our own experience. We will also consider ethical aspects of ethnic diversity. Various authors have offered significant insights into ethnic sensitivity. In an effort to provide a comprehensive view, we will integrate these as well as our own suggestions.

An Integrated Approach

Know yourself Human service workers must be willing to engage in self-definition and introspection. Gilman and Koverola (1995) list the following ways to work on knowing who you are.

Having a clear sense of your own ethnic and cultural background—getting in touch with your roots

Being clear about your religious and cultural belief systems

Examining your own value system

Remaining open to examining prejudices and negative stereotypes that you may associate with certain cultural groups

Accepting yourself, with your strengths and limitations; being open to examining your underlying motivations for working with others

SCENARIO

You are a new social work intern at a county mental health center. The client population is ethnically diverse. You were raised in a very limited environment; your parents and extended family believe other cultures are inferior. Despite this, you take pride in working with various ethnic populations and believe you are not prejudiced. However, as you work with certain ethnic backgrounds, you find yourself reacting as your parents would. You are embarrassed to admit this; you prefer to deny the existence of these thoughts and feelings. Is this an appropriate response? Is this accepting oneself and working to remain open and objective? How would you proceed in this situation?

Empathy versus sympathy "Oppressed clients do not need sympathy. In fact, sympathetic [helping] is simply another form of oppression" (Gilman & Koverola, 1995). Being empathetic involves sensitivity to clients' circumstances, but does not include feeling pity or claiming you understand how they feel.

SCENARIO

As a graduate intern on a neurology unit, Lupe once worked with a young mother who was bedridden and paralyzed by a neurological disorder. In a sincere effort to help, she said, "I'm so sorry for you. I can understand how you feel." The client became agitated, answering, "How can you understand? You can get up and walk out of the room. I can't." This experience taught Lupe the difference between empathy and sympathy. She learned never to assume that she *knows* how a person feels. Instead, she *imagines* how difficult it must be to be in another person's circumstances.

Being yourself Allow yourself your own mannerisms, style of dress (keeping it appropriate), and sense of humor.

SCENARIO

Randy has found that he prefers to dress casually and interact with others in an informal manner. Similarly, his office decor reflects his interest in Native American culture and the Southwest. He has found that when he is comfortable, he can be more effective with others. They notice his genuineness.

Do you feel comfortable being truly yourself? Describe ways in which you behave congruently with your personhood. In contrast, discuss ways in which your behavior may be incongruent.

Avoid stereotyping Refrain from making assumptions that may be incorrect or even prejudiced. By staying aware, you will be able to keep yourself from perpetuating any stereotypical beliefs that your assumptions may be based on. Be aware of the policies of the agency you represent.

SCENARIO

Several of your classmates are placed at a homeless shelter for their first graduate internships. In your practicum, they share their feelings and struggles in working with some resistant and manipulative clients at the shelter. Based on their statements in class, they seem to have categorized all homeless people negatively. You have also worked with homeless individuals, including some who truly want to improve their situations and are not resistant or manipulative.

Do you believe that the other students have developed stereotypes about the homeless? Can you describe situations where you have caught yourself entertaining stereotypical views of certain individuals or population groups? Please explain.

Know your clients Be willing to learn about your clients in order to be understanding and sensitive to their special needs.

Recognize that as human beings we all possess the capacity for thoughts, feelings, and behavior.

Develop an appreciation for different cultures. According to Brill and Levine (2002) it is essential for the helper to know about the history of the particular group he is working with as well as be sensitive to the significance of the present status and situation of the ethnic group they belong to.

Be knowledgeable about different cultures. Avail yourself of any opportunity to learn more about people who are ethnically different from yourself.

Be sensitive to the different communication styles of the cultures you will be working with.

Learn from your experiences. Allow your clients to teach you about their culture.

SCENARIO

You have been placed in an agency that works with developmentally disabled children and their families. This is your first experience working with this population; many of the diagnoses, interventions, and issues are foreign to you. After your first week, you feel completely lost and question your ability to continue to work in this setting. Your supervisor reassures you that your anxieties are common. He invites you to do some additional reading on some issues of people who are developmentally disabled. He also lets you borrow some teaching videos that he uses in his parenting groups because these might help orient you better to this population. You feel less overwhelmed. Taking the time to learn about this population from the literature, the videos, and the clients and their families themselves, you feel much more confident. Although you still have much to learn, you at least have some basic knowledge to guide you.

If you were in a similar placement, where you felt overwhelmed by your lack of knowledge regarding the client population, how might you proceed? Have you ever experienced this?

Appropriate assessment of your clients' needs Individualize each case by learning to take into consideration the clients' personal circumstances, their culture, and their value systems. Be aware that not all individuals from a particular ethnic group are the same; differences exist even within the same ethnic group (Brill & Levine, 2002).

SCENARIO

You are a young, single, Caucasian woman who is quite independent and successful. One of your first clients is a 24-year-old single Latina woman who still lives at home and is struggling with mounting conflicts with her parents' old-fashioned values. You identify with this client and unknowingly push her toward separating from her family. You have a hard time accepting her reluctance to do so, expecting her to be as independent as you were at 24.

Would this be a fair expectation? Why or why not? What cultural issues might be involved? Can you think of a situation where you might make an erroneous assessment because you were not considering the client's culture?

Ethical Issues Related to Culture and Ethnicity

We cannot meet our responsibility to help others if we ourselves are unwilling or reluctant to work with certain people because of their race, culture, or ethnicity. Let's look at several relevant professional codes of ethics and ethical standards.

Ethical Standards of Human Service Workers

The Ethical Standards of Human Service Professionals (NOHSE, 2000), under the heading of "The Human Service Professional's Responsibility to the Community and Society," lists six standards relevant to human diversity that human service providers must adhere to. These include:

1. Human service professionals advocate for the rights of all society, particularly those who are members of minorities and groups at which discriminatory practices have historically been directed.
2. Human service professionals provide services without discrimination or preference based on age, ethnicity, culture, race, disability, gender, religion, sexual orientation or socioeconomic status.
3. Human service professionals are knowledgeable about the cultures and communities within which they practice. They are aware of multiculturalism in society and its impact on the community as well as individuals within the community. They respect individuals and groups, their cultures and beliefs.

■ **Table 6.4 Code of Ethics Related to Cultural Diversity**

NASW Code of Ethics	ACA Code of Ethics
The Social Worker's Ethical Responsibility to Society	*The Counselor's Ethical Responsibility to Diversity*
Promoting the General Welfare	Respecting Diversity
• The social worker should act to prevent and eliminate discrimination against any person or group on the basis of race, color, sex, sexual orientation, age, religion, national origin, marital status, political belief, mental or physical handicap, or any other preference or personal characteristic, condition, or status.	• Nondiscrimination. Counselors do not condone or engage in discrimination based on age, color, culture, disability, ethnic group, gender, race, religion, sexual orientation, marital status, or socioeconomic status.
• The social worker should act to ensure that all persons have access to the resources, services and opportunities which they require.	Respecting Differences
• The social worker should act to expand choice and opportunity for all persons, with special regard to the disadvantaged or oppressed groups and persons.	• Counselors will actively attempt to understand the diverse cultural backgrounds of the clients with whom they work. This includes, but is not limited to, learning how the counselor's own cultural/ethnic/racial identity impacts his/her values and beliefs about the counseling process.
• The social worker should promote conditions that encourage respect for the diversity of cultures which constitute American society.	

4. Human service professionals are aware of their own cultural backgrounds, beliefs, and values, recognizing the potential for impact on their relationships with others.
5. Human service professionals are aware of sociopolitical issues that differentially affect clients from diverse backgrounds.
6. Human service professionals seek the training, experience, education and supervision necessary to ensure their effectiveness in working with culturally diverse client populations.

Similarly, the codes of ethics for social workers (National Association of Social Workers, 1999) and for counselors (American Counseling Association, 1995) include special sections regarding cultural diversity. Table 6.4 compares them.

Additional revisions of the NASW Code of Ethics in 1999 reflect a continued commitment to ensuring cultural competence and recognition of social diversity:

• Social workers should have a knowledge base of their clients' cultures and be able to demonstrate competence in the provision of services that are sensitive to clients' cultures and to differences among people and cultural groups [1.05.b].
• Social workers should obtain education about and seek to understand the nature of social diversity and oppression with respect to race, ethnicity, national origin, color, sex, sexual orientation, age, marital status, political belief, religion and mental or physical disability [1.05.c].

In all of these codes, the focus is on encouraging ethnic-sensitive practice and a multicultural perspective. As a human service professional, it will ultimately be up to you to put into practice the codes of your particular profession.

SCENARIO

Joe is a 14-year-old African American. The court has ordered him to engage in counseling because of several problems with the law related to gang activities. Joe comes from a lower-class family. He is the middle child; his brother is 17 and his sister is 10. His father abandoned the family when Joe was 5, and there has been no contact since. His mother eventually remarried but is recently divorced from her second husband, due to numerous domestic disputes. Joe had been close to his stepfather and seems angry at his mother for the breakup. His brother is no longer in the home; he was recently placed in a group home after his third stay at juvenile hall. Joe misses his brother, with whom he used to spend a great deal of time. The mother is rarely home; she had to start working two jobs after the divorce. The sister is cared for by extended family. Joe refuses to be "babysat," because he believes he is too old. He spends most of his time out on the streets. He is not currently in a gang, but he does hang out with known gang members in his neighborhood. When his mother questions him about this, he angrily replies, "At least they are there for me, like a family should be."

SCENARIO

Farrah is a 35-year-old married woman who recently immigrated from the Middle East. She was first seen at an emergency room for chest pains caused by panic attacks. She was referred to a mental heath center for follow-up. After several cancellations, she reluctantly attends a session. She appears withdrawn and hesitant to say much about her home life or family history. Her primary role has been that of a homemaker and mother of four children. Due to financial struggles, her spouse has told her that she must also find outside employment to help him. She shares her fears of working outside the home. She is worried about not being able to "do it all." She knows little English, is not yet familiar with American culture, and is concerned about the care of her children. She describes feeling different from others and seems afraid of not fitting in. Farrah also talks of being ridiculed by family members if they learn of her counseling appointment.

SCENARIO

A family is referred to Child Protective Services by an anonymous reporter, who suspected abuse in the home. The family consists of foreign-born parents from Guatemala and their two teen daughters, Anna and Carmen, who are born Americans. There have been conflicts between the teens and their parents about house rules. According to both Anna and Carmen, their parents are very old-fashioned and too strict compared to their friends' parents. For instance, their friends get to stay out late on weekends, are allowed to go to school functions, and can spend the night at friends' homes—none of which they are allowed to do. The parents argue that the girls are too young to be staying out late and that they have no business staying at someone else's house when they have their own. In their home country, such things were never allowed. Anna and Carmen answer that this is the United States and the new millenium. Their father is especially offended by what he feels is blatant disrespect of his authority. On occasion, he has slapped or used a belt on them, out of frustration. This seems to have further alienated his daughters. He feels extreme anger at having been anonymously reported to CPS and blames the girls. They deny having told anyone. The mother, who is more passive, seems to have become depressed and withdrawn as a result of the report. When they arrive at the counseling center, their animosity toward each other is evident. It becomes very clear that these parents have come to the appointment only because of outside pressures. In essence, these clients are involuntarily seeking assistance.

SCENARIO

Than, a 30-year-old Cambodian refugee, has only been in the United States a short while. In his country, he was a professor and was well respected. Now, however, his job opportunities are limited due to poor English and minimal marketable skills. He has three children and a wife who is pregnant with a fourth child. She has been very ill during this pregnancy. Than is especially stressed because all the extended family on both sides remains in Cambodia, unavailable to help out. He is struggling to find gainful employment and help care for the young children. Besides not being accustomed to doing such "womanly duties," he feels humiliated at the low-paying jobs he has been offered. He reluctantly talks of his struggles to a relief worker at the family services center, where he has been referred for help until he can locate a stable job. His shame and embarrassment are obvious, but his appreciation is great.

THINKING THINGS THROUGH

Identify the multicultural issues in each of the preceding scenarios. For each case, specify the issues involved. How you would proceed if working with these people in the helping capacity? Also consider examples from your own personal and agency experience.

Dual Relationships

The relationship between helper and client is an important factor in providing effective treatment. However, dual relationships may arise when human service workers mix professional and personal relationship with their clients. Engaging in any other type of relationship with a client with whom you are professionally involved with can be considered an unethical *dual relationship* (Corey, Corey, & Callanan, 2003).

A dual relationship is a particular type of boundary breakdown. By ethical standards, helping professionals must be aware of their potentially influential position in relation to clients. They must avoid exploiting the trust and dependence of their clients. "Dual or multiple relationships occur when professionals assume two or more roles at the same time or sequentially with a client. This may involve assuming more than one professional role (such as instructor or therapist) or blending a professional and nonprofessional relationship (such as counselor and friend or counselor and business partner" (Corey, Corey, & Callanan, p. 2003, p. 246). "Dual (or multiple) relationships may be sexual or nonsexual and may occur when (helpers) simultaneously or sequentially assume two (or more) roles with a client" (Herlihy & Corey, 1993, 1997a). Welfel (2002, p. 155) adds, "Whenever [helpers] have other connections with a client in addition to the therapist-client relationship, a dual or multiple role relationship exists." Cormier and Hackney (2005) and Welfel (2002) believe it is imperative for the helper to honor a division or a boundary between the professional and personal lives of helper and client. "In instances when this boundary is not honored, there is a blurring of roles and what can be referred to as boundary crossing or boundary violations." (Cormier & Hackney, 2005, p. 177) Various authors (Corey & Herlihy, 1997; Cormier & Hackney, 2005; Welfel, 2002), along with the various professional codes of ethics, recommend avoiding dual- or multiple-role relationships when possible. Three basic reasons include:

1. "A dual relationship implies that the counselor is vulnerable to other interests that compete with promoting the welfare of the client" (Welfel, 2002, p. 160).
2. "When a counselor has another role in the client's life, the client's emotional reaction is confused" (Welfel, 2002, p. 161).
3. "There is a power differential in all helping relationships, meaning that the practitioner holds more power than the client, resulting in the fact that the client who feels powerless may feel coerced to participate in a dual- or multiple-role relationship" (Cormier & Hackney, 2005, pp. 177–178).

This complex ethical issue affects all human service professionals, regardless of their work setting or client population. There is much controversy over what is ethical and what is unethical. Clearly, helpers must avoid, without exception, sexual relationships with their clients. However, other dual relationships may seem to be appropriate or are unavoidable. Exactly what is acceptable and what is not? Among professional groups, it was concluded that nonsexual dual relationships should also be avoided: "No matter how harmless a dual relationship seems, a conflict of interest almost always exists, and any professional counselor's judgment is likely to be affected" (St. Germaine, 1993, p. 27).

The ethics code of the American Counseling Association (1995) states that dual relationships are to be avoided when possible. The relevant code emphasizes that by engaging in any type of dual relationship, there is a risk of impaired professional judgement. Helpers "are aware of their influential positions with respect to clients, and they avoid exploiting the trust and dependency of clients. Helpers make every effort to avoid dual relationships with clients that could impair professional judgment or increase the risk of harm to clients. When a dual relationship cannot be avoided, counselors take appropriate professional precautions such as informed consent, consultation, supervision, and documentation to ensure that judgment is not impaired and no exploitation occurs" (A.6.a.).

The ethical standards and codes of ethics of both human service providers and social workers offer similar stipulations. NOHSE's ethical standards of 2000 stress the power and status differential that exists between clients and helpers (p. 2):

> Human service professionals are aware that in their relationships with clients power and status are unequal. Therefore they recognize that dual or multiple relationships may increase the risk of harm to, or exploitation of, clients, and may impair their professional judgment. However, in some communities and situations it may not be feasible to avoid social or other nonprofessional contact with clients. Human service professionals support the trust implicit in the helping relationship by avoiding dual relationships that may impair professional judgment, increase the risk of harm to clients or lead to exploitation.

The NASW's Code of Ethics (1999) cites similar concerns regarding the risk of exploitation or potential harm to clients:

> Social workers should not engage in dual or multiple relationships with clients or former clients in which there is a risk of exploitation or potential harm to the client. In instances when dual or multiple relationships are unavoidable, social workers should take steps to protect clients and are responsible for setting clear, appropriate, and culturally sensitive boundaries [1.06.c].

Furthermore, the social worker should avoid relationships or commitments that conflict with the interests of the clients. "The social worker should under no circumstance engage in sexual activities with clients" (NASW Code of Ethics, 1999, p. 5).

All three of these professional ethical codes reflect a concern that the helper could potentially harm a client while engaging in a dual relationship, especially if there is sexual contact. Any exploitation of a sexual nature is prohibited, including sexual intercourse, sexual contact, or sexual intimacy with a client, a client's spouse, or a client's partner (Leslie, 1993).

Other types of dual relationships to be avoided "include, but are not limited to, familial, social, financial, business, or close personal relationship with clients" (American Counseling Association, 1995). Welfel (2002, p. 165) provides a helpful list of nonsexual dual or multiple relationships:

- Accepting a friend or relative as a client
- Providing counseling to an employee
- Employing a client
- Going into business with a current and/or former client
- Providing counseling to a student and/or a supervisee
- Allowing a client to enroll in a course taught by the counselor
- Inviting a client to a party or going to a social event with a client
- Selling something to a client

Although not as severe as with sexual intimacy, the inherent danger in such relationships is the impaired objectivity and judgment of the professional, which would be likely to result from the blending of two types of relationships.

In some situations, boundaries are difficult to distinguish and therefore are easily crossed. Leslie (1993) indicates that there are many gray areas that deserve careful attention. Herlihy and Corey (1997) contend that it is important to be self-aware of the ethical issues tied to dual relationships so as to be able to deal appropriately with these gray areas.

Dual Relationships with Unclear Boundaries

Social Relationships: Client and helper attend the same clubs, social activities, religious functions, and so on.

Business Relationships: Client and helper barter for services or engage in some joint business venture.

Combining the Roles of Teacher and Therapist: Educator (teacher, professor, etc.) agrees to engage in a therapeutic relationship with a current student. Sometimes this isn't even openly agreed on, but some form of therapeutic counseling is being provided.

Combining Roles of Supervisor and Therapist: A supervisor crosses the line and becomes a therapist to his student/employee, or an actual therapeutic relationship develops from the supervisory relationship.

Providing Therapy to a Relative or a Friend's Relative: Such a dual relationship can create unforeseen difficulties because it is almost impossible for the helper to be completely objective.

How can such seemingly innocent relationships be considered as dangerous as the flagrant abuse involved in sexual contact? In reviewing the literature on dual relationships, Herlihy and Corey (1997) noted a wide range of perspectives. Conservative practitioners such as Pope (1985a) and Pope and Vasquez (1998) advocate stricter ethical codes and enforcement. Others, including Herlihy and Corey (1997), Hedges (1993), and Tomm (1993), believe that some ethical standards are too rigid and inhibiting, and that dual relationships cannot always be avoided.

How then do you, a beginning professional, determine what is an unethical dual relationship? The one guideline we have found useful is to focus on the potential for exploitation.

Exploitation means taking advantage of another person in some way. Some exploitation (e.g., sexual) is quite obvious; an ethical practitioner would never do so with a client. Other situations, not so clearly unethical, could nonetheless become problematic for you and the client.

SCENARIO

The helper agrees to engage in a special relationship with clients in an effort to help them. Examples include bartering for services, giving clients extras, going out of his or her way for a client, or attending a social function in the guise of supporting the client. In any of these examples, what exploitation do you believe could occur? How might this affect the helping relationship?

SCENARIO

The client offers to do something for you as the helper. It may seem totally innocent: using his connections to acquire tickets for you, recommending a friend or relative who will give you a break on a service, investing in a business venture you have, or donating to a special fund you are endorsing. How might any of these lead to exploitation? How could this affect the helping relationship?

The issue of gift giving often comes up between client and helper. Many clients show their appreciation for the helper by giving presents—baked goods, produce from their backyard, or elegant and expensive gifts. Whether to accept such gifts must be considered on a case-by-case basis. The most sound ethical practice would seem to be to refrain from accepting monetary or expensive gifts. Accepting tokens of appreciation such as food, handmade gifts, or souvenirs from a client's home country cannot be considered unethical per se, but each case must be weighed individually and with great care. Let clients know about your policy early on, perhaps when you discuss informed consent issues. Some clients respond well to such limits, whereas others may take the refusal of a gift as an insult. Approach each situation with sensitivity and be aware that any suggestion of a dual relationship must be avoided.

When helper–client boundaries are crossed or some form of exploitation occurs, whether intentional or not, problems are sure to arise. Engaging in dual relationships may be unavoidable for the human service worker who play many roles that overlap. Here we delineate some guidelines (adapted from St. Germaine, 1993, and Herlihy & Corey, 1997) to minimize the risks involved in dual relationships.

Minimizing the Risk of Dual Relationships

Set healthy boundaries from the outset.
Discuss with clients any potentially problematic relationship.
Seek consultation and/or supervision.
Make a referral when indicated.
Fully inform clients about any potential risks.
Disclose and clarify areas of concern.
Consult with other professionals periodically if you are engaged in dual relationships.
Document discussions about any dual relationships and relevant steps.

Dealing with dual relationships requires personal and professional maturity. Awareness of your own needs in relation to your clients' is crucial. Avoid any interaction in which your own financial, social, or emotional needs take precedence.

SCENARIO

Alan is a 32-year-old, married Caucasian man who owns a successful insurance business. He seeks counseling for marital problems and seems to benefit from the therapy. He bonds quickly with his male therapist; they seem to have much in common and relate easily to one another. At one session, the counselor mentions in passing that he would like to buy a new car, were it not for skyrocketing insurance costs. Alan immediately suggests that he come see him for a great insurance deal. The counselor initially rejects this notion but later agrees to an estimate. The estimate is indeed lower than all the others he has received, leading him to contemplate switching insurance companies. He wonders whether this would be inappropriate but concludes that it shouldn't affect their therapeutic relationship. He purchases insurance from Alan. Everything seems to be going well with this arrangement, until there is a conflict over insurance coverage on a claim the counselor submits. Alan is angry at the counselor for being so demanding, while the counselor is upset with Alan and questions his integrity. The counselor becomes increasingly uncomfortable with Alan. Their therapeutic relationship begins to deteriorate, to the point where the treatment goals are compromised. After a difficult session, the counselor engages in soul searching and realizes that the dual relationship is the source of their conflict. In retrospect, he regrets ever agreeing to purchase insurance from his client. This has clearly impaired their relationship and affected the helping process.

THINKING THINGS THROUGH

Discuss a case where you have unknowingly (or knowingly) engaged in a dual relationship or were close to doing so. Would you now change your approach? Why or why not? Please elaborate. Also consider the preceding scenario. How would you act if you were the human service worker engaged in a client–helper relationship with this client?

▪ CONCLUDING EXERCISES

1. Of the ethical and legal issues discussed in this chapter, which one affected you the most? What feelings does this issue bring up for you? What does it say about your values?

2. Imagine that one of your best friends, or a valued coworker, revealed that he or she had violated ethical standards with regard to the issue you chose in Exercise 1. What do you imagine your reaction would be? How might you choose to deal with the situation? How would you reveal your values to this person?

3. Let's say that you have moved to another country, where the concept of professional helping is nonexistent. However, you notice that many people who are struggling have developed their own resources. For instance, they may seek help from a priest, palm reader, herbalist, medicine man, or elder. Such consultation seems useful to many people, but you become concerned about cases where vulnerable individuals are hurt by the advice they are given. You wonder about the dangers inherent in practicing without sufficient knowledge or skill. A colleague, also from the States, approaches you with a plan about opening a community center to provide help to people in need, at minimal cost. You are enthusiastic. Of course, for this to become a reality, funding is needed. You both decide to pursue this dream by writing a grant proposal and presenting the idea to community leaders who can authorize fund-raising events.

   You realize that ethical codes and legal stipulations will be necessary, to protect you as well as the clients. Taking into account the culture at hand, how would you proceed? What might you tell community leaders and other professionals about the importance of ethics? Specifically, tell how you might begin the process of developing some basic ethical and legal guidelines. Elaborate on a few of these guidelines that you believe would be the most crucial.

CHAPTER SEVEN

The Challenges of Working in Human Services

AIM OF THE CHAPTER

In this chapter, we explore the challenges that human service workers face, both professionally and personally. Being an ethical, skillful, and caring helper requires that you learn to face a multitude of challenges—some obvious and easily dealt with, others elusive and difficult to confront. In your fieldwork experiences, you will no doubt encounter some of these challenges.

The challenges are grouped into the following categories: personal, organizational, and environmental. Helpful coping strategies will also be discussed to help you recognize the many paths for dealing with adversity. Ultimately, you will develop your own coping style for dealing with the ongoing challenges.

PERSONAL CHALLENGES

Personal involvement is one of the greatest assets human service workers possess, allowing us to understand others' pain and struggles as well as their joys and excitement. The ability to empathize contributes to developing trust and rapport with clients. Similarly, we can serve as role models for those we work with. In this personal involvement, there is a risk of becoming overwhelmed or disillusioned. The sections that follow highlight some key personal challenges, including those we have faced in our professional careers. We will also include challenges encountered by our students and colleagues.

Fears Associated with Being a Student

A student placed in an agency setting may feel anxious about the unknown, overwhelmed by lack of knowledge, and pressured by the demands of the agency and the educational institution. The process of becoming a professional "is not without its challenges, particularly early in the journey. Typically, beginning helpers struggle with anxiety about competency and evaluation" (Cormier & Hackney, 2005, p. 170). If you are in this position, maybe you are telling yourself that you should know it all and be able to help everyone you come into contact with. Teyber (2000, p. 53) shares a similar view:

> Too often, novice therapists [helpers] are thinking about where the interview should be going next, wondering how best to phrase what they are going to say next, preparing advice, or worrying about what their supervisors would expect them to do at this point in the interview. Such self-critical monitoring often immobilizes student therapists [helpers], blocks their own creativity, and, sadly, keeps them from enjoying this rewarding work. Further, it prevents therapists [helpers] from listening and understanding as fully as they could.

Do not be overly hard on yourself or create undue stress. It is not only common to have some fears and self-doubt, but also to be expected. Not even the most experienced professionals can help all people. In fact, seasoned helpers realize that they are only able to help those who are ready, able, and willing to change. We believe that the helper's desire and skill must be paired with the client's willingness if change and growth are to occur. We concur with Teyber (2000, p. 53) that "beginning therapists [helpers] are encouraged to slow down their inner monitoring process and to focus more on the moment, on their current interaction with the client, and on the meaning this particular experience seems to hold for the client right now." Acknowledge your limitations and assertively seek guidance whenever necessary. Experience will help you develop a broad knowledge base and allow you to refine your skills. Remember, "The more you know, the more you realize how much you don't know." "Our willingness to recognize and deal with these anxieties as opposed to denying them by pretenses, is a mark of courage. That we have self-doubts seems perfectly normal; it is how we deal with them that counts" (Corey, 2005, p. 29).

SCENARIO: Anxiety Faced by a New Student

In her first year of graduate internship, Lupe was placed at Rancho Los Amigos Hospital. Her supervisor was a social worker on the pediatric unit. He had a play area in his office, where children's chairs and a table were situated. Lupe, unaware of her high anxiety level, would sit in one of the children's chairs during her supervision hour. After several weeks, her supervisor finally made her aware of how odd this was. There followed an important breakthrough in their relationship: The issue of her feeling inferior due to lack of knowledge was challenged. Although this was embarrassing, it was also a potent learning experience. It taught Lupe not to feel that she should know it all. Years later, she continues to give herself this message when she begins to place too many expectations on herself.

THINKING THINGS THROUGH

Consider the various feelings you had upon entering your fieldwork placement. Can you describe them? Did you feel some anxiety, trepidation, or fear? How did you cope?

Challenges Faced by Beginning Professionals

Clearly, it is irrational to believe that you are solely responsible for your clients' progress. Unfortunately, this is a common fallacy that beginning helpers hold. If not dispelled, it will lead to much self-criticism and disappointment. According to Corey and Corey (2003), many students and beginning practitioners develop an unrealistically high expectation that they must be the "perfect helper," which creates undue stress and a natural fear of making mistakes. Attempt to remember that this need to be the "all-knowing" or "perfect" counselor can significantly hinder the helping process (Baird, 2000). One way to avoid this pitfall is to be clear from the outset about your responsibilities and those of your clients. These will vary depending on the agency, the type of clients, and your overall helper role.

SCENARIO

Beginning professionals, having earned a bachelor's or master's degree, may feel that they should not have to ask for help. However, the beginning professional faces a new role and setting, and so should expect to need help. Being new, he or she should be curious and questioning. Because he or she is well educated and appears professional, staff and supervisors may assume that the helper can function independently. Although this can be taken as a compliment, don't lull yourself into believing that you can function without input from others. Surely there will come a time when you will not know how to carry out a certain procedure, how to deal with a specific client, or how to network with the community. If you have closed the door to asking for help early on, and developed a style that isolates you from others, then it will become difficult to seek help when you most need it.

Have you ever hesitated to ask for help because you believed that you should already know the answer? How can this approach be detrimental to your overall learning?

SCENARIO

The opposite situation can also be a problem: Staff and supervisors may assume that the new professional knows very little and will need constant guidance and supervision. A professional who has already worked in agencies during internship and field placements may find this constraining. You can maintain a balance by being inquisitive and assertive. Once you have settled into your new position and feel comfortable, let your coworkers and supervisor know that you appreciate their concern and help but believe you need the opportunity to work more independently.

How were you treated by staff when you first started at your agency? Were they supportive? Did they expect too much or too little from you? If you were uncomfortable, how did you cope?

SCENARIO

When Randy was a community worker in a community mental health center, he was not allowed to co-lead therapy groups with licensed professionals. The reason given was his status as a bachelor's-level professional. In fact, the supervisory staff were uncomfortable because they had never had a B.A.-level professional work in that capacity. Randy's persistence, dedication, and assertiveness eventually paid off. He was finally allowed to co-lead several groups within the center. This experience laid the groundwork for Randy to create other groups within the agency. This opportunity would not have existed were it not for his belief in himself and his persistence in seeking a professional position.

A transition occurs when you leave the safety of your educational setting for work in the real world. Even though you are looking forward to graduation, be aware that it can be difficult to switch from a structured and somewhat sheltered life to the real world. When you are truly on your own, with no teachers to guide you, you may feel frightened, disappointed, and frustrated. If you have a job lined up, ending school may seem straightforward, but you must still change from being structured externally by the educational program to having to provide your own structure from within. Woodside and McClam (1994, p. 171) state:

> In a classroom, the student is expected to work within the guidelines of an instructor's outline or syllabus and may not deviate from these expectations. Workers, although expected to follow the rules and regulations of the agency, are constantly in situations that require creativity, initiative, and judgement. Not all actions are predetermined (especially those involving clients), and not all decisions that need to be made are clear cut.

Overall, remain realistic about your role as a beginning helper. A career path in the helping professions is a lifelong process. Allow yourself the time to settle into your new role, and recognize that even as a professional, you may need guidance and support.

SCENARIO

While in graduate school, Lupe had a professor who told the class that it would take at least five years for graduating students to become even mediocre clinicians. Most students were skeptical and felt that this statement was meant to intimidate them. Twenty-five years later, Lupe sees the wisdom of his comment. Only in the last ten years has she felt that her clinical skills are sound, and even now they are still developing.

The transition from a student or intern role to that of a full-time professional helper can often be traumatic. Caseloads can quadruple, the demands triple, and the expectations double. Students who had time to give 100 percent to each case may now find themselves having to focus primarily on the most pressing cases and less on the least urgent ones. This can be overwhelming and frustrating for people who have always taken pride in giving their all to every client. Time management skills and prioritization are key to survival. It is not humanly possible to give 100 percent to every case. Setting limits and being realistic can help you make this transition a smooth one.

SCENARIO

Lupe, during her internship as a medical/rehabilitative social worker on a neurology unit, learned a substantial amount about the particular neurological conditions her clients faced. She gained experience in working with a variety of unusual cases, which she could study in depth. Her caseload averaged eight patients at any given time during her internship. Upon graduation, she was hired full-time at this facility, as the social worker in the spinal cord injury unit. Her caseload increased dramatically—to about 40 patients, with many in the critical stages of rehabilitation. No longer was a preceptor or supervisor frequently available, nor did she have the time to dedicate 100 percent to every case. By talking to others about her situation, and seeking guidance, support, and encouragement, she survived this difficult transition.

If you are a beginning professional, have you experienced similar difficulties with regard to having less time and guidance?

Woodside and McClam (1994, p. 170) highlight the stressors involved in the transition from student to professional:

It is typical for entry-level professionals to be totally consumed by their jobs, spending long hours at work and worrying about their clients during their off-time. It is also common for beginning professionals to start work with high job satisfaction, which decreases when realistic assessments of the work replace earlier dreams of what the job would be like. If the transition is successful, beginning professionals emerge with a more stable sense of professionalism.

THINKING THINGS THROUGH

Can you describe some of your fears as you begin your professional career? How have you handled them? Have you had any problems coping with the transition from intern to staff member?

Balancing Family Life and Career

On the surface, balancing family life with your professional career may seem simple enough. However, this balancing act is often far more complex than meets the eye. Every professional has a personal style and boundaries about how and when career is allowed to influence personal life. Visualize a continuum: On one end of the spectrum, you find professionals who keep their careers totally separate from their family life. At the other end of the spectrum are professionals who live their professional lives both at work and at home. In between are people who maintain more of a balance between their professional and personal lives. The scale below illustrates the spectrum:

Totally Separate   Some Commingling   No Separation

|—————————————————|—————————————————|

Certainly, individuals differ in their comfort level and attitude in this regard. Depending on the circumstances, many human service workers fluctuate in terms of how much they allow their careers to be a part of their family life. Some believe it possible to involve the family in their work life, if this is done carefully and constructively.

In our personal experience, we have found that striking a balance is most suited to our personalities. Our case may be unusual, because both of us are clinicians and university professors in the same field. Also, because our work involves helping others, we have sought to instill this as a strong value in our own children. We often include them in general discussions about our work and ask for their input on certain social ills. We encourage their participation in charitable events and seek creative ways to involve them in our work, taking care not to allow any direct client involvement. For instance, we have asked our children to draw pictures for our classes and offices, and we consider them expert resources in understanding childhood and adolescence.

Let us not mislead you. Learning to integrate career with family is no easy task. At times we have not set appropriate limits or managed our time wisely, resulting in family conflicts and frustrations. When our children make negative statements regarding our work, or our marriage begins to suffer, we are forced to reevaluate our stance. It is sometimes difficult to do our jobs professionally and still remain active, involved, and committed parents.

SCENARIO

In writing this text, we found it especially difficult to balance family with career. One specific incident comes to mind. Our son and daughter were outside playing and kept banging at the door (which was unlocked), to come in for a drink or a toy or to tattle on each other or a playmate. We became very frustrated as the day wore on. These interruptions affected our concentration and motivation and depleted our stamina. Later, when we discussed the situation, we realized that our children were only acting normally. Our expectations had been too high. We expected them to understand our situation and be totally supportive. From this experience, we learned to pace ourselves better, be more realistic regarding our children's demands, and try to schedule time for writing at night or away from home.

THINKING THINGS THROUGH

Consider ways in which you balance your personal life with your career or education. Are you content with your style of time management, or have you considered changing?

Keeping Personal and Work Issues Separate

Keeping personal issues apart from work with clients is often a true challenge. Entering work, we are still the same individuals we were earlier that morning—when we might have received a speeding ticket, had a flat tire, argued with our children about their attire, or gotten up an hour late. We may subconsciously bring such frustrations into our work.

SCENARIO

You are a new intern at a community center, where you aspire to work in an administrative capacity some day. You enjoy going to work and find your administrative assistant role to be quite gratifying. One morning, however, you felt especially sluggish because of a lack of sleep the night before. You and your spouse were arguing over parenting and household responsibilities and didn't seem to resolve these issues as readily as in the past. Feeling upset, you struggled to get to sleep. The next day, your usual level of motivation and ambition was not evident when you arrived at work. You missed an important business lunch, were curt with office staff, and failed to meet a deadline on a news release. The next day, when you met with your supervisor, she noticed that your mind wasn't at work. You told her that you hadn't had much sleep and were upset about family matters. She suggested that you work on keeping outside stressors separate from work as much as possible. You agreed to work on this.

How might you have handled this situation? Would you have decided to stay home? Did the supervisor act appropriately? How would you have responded to the supervisor?

THINKING THINGS THROUGH

Have you found it easy to keep personal issues separate from work? If so, how? If not, why not? Consider ways to help you with this endeavor. Can you foresee any problems if you do not keep them separate?

Similarly, it can be a challenge to keep from taking your professional concerns home. You may bring excessive paperwork home and find yourself consumed by it rather than interacting with your family or engaging in leisure activities. (You students may have forgotten the meaning of *leisure*.) Agency demands may at times make it inevitable that work will affect your home life. It may be helpful to set specific times to engage in paperwork and give yourself a deadline.

It is also a common experience to feel physically exhausted after a long day of helping others. Ongoing emotional investment can be physically draining. You may need to develop some healthy after-work activities to reenergize yourself. You may find yourself worrying about a particular case and allowing it to affect your home life. Before re-entering the world of home life, find a supportive colleague, spouse, or significant other to talk with. This can minimize the tendency to brood and may keep your worries or concerns from interfering with your personal life. Sometimes the drive home from work can be a nice transition time, the very bridge you need to help you "change hats." You might listen to your favorite music or plan what to do when you get home, which can help take your mind off work.

SCENARIO

Years ago, we worked together at a mental health center, where we each coordinated separate programs and saw difficult clients almost daily. Often, when we car-pooled, the ride home was dominated by talk of our individual frustrations and conflicts. Upon reaching home, we were often exhausted, moody, and unable to enjoy the evening or each other. The solution that worked best for us was to set some time limit on venting and then make an effort to refocus on the positives of our day and our lives. Also, planning outside activities seemed to be a great source of relief and help.

Can you relate? Are you successful at keeping your professional life separate from your personal life?

THINKING THINGS THROUGH

Consider some of the problems that are likely to arise if you are not successful at keeping clear boundaries and maintaining firm limits. What suggestions might you have to help keep such clear boundaries and limits?

Caretaking and Rescuing

The human service worker is often naturally adept at caretaking and rescuing. Caretaking is certainly a natural and healthy aspect of helping, a strength that allows the helper to be empathetic and motivated. However, it has a counterproductive side. The worker may be trying to avoid painful encounters by caring for clients in ways that minimize their experiencing the full range of their emotions. Clients often seek help with the assumption that the helper will take care of them by giving them advice or pointing them in a specific direction. Depending on the situation, this may or may not be appropriate. Long-term gains may be sacrificed when a short-term solution is simply handed to the client. Be aware of your role and the goals of your work.

Rescuing is a form of caretaking. Some types of rescuing are obvious, such as wanting to take an abandoned child home. Subtler forms include discouraging clients from expressing sadness, pain, hurt, anger, or sexuality and avoiding discussion of such topics. The reason for such avoidance may be the helper's own difficulties in expressing these emotions. (See the section on countertransference later in this chapter.) In fact, change usually occurs through people exploring and expressing their painful emotions and deep feelings, not by shutting them off. When painful emotions are suppressed, depression may develop. When feelings are left unexpressed, they may later surface in a pathological form.

If you find yourself experiencing difficulty when encountering pain or anger in those you work with, or if you struggle with sexual issues, perhaps you need to engage in self-exploration. Seeking supervision or therapy is often advised at such a juncture.

SCENARIO

A client seeks your advice regarding her struggles with her current boyfriend. You are greatly affected by the emotional abuse she describes and the tremendous power differential between the two. You encourage her to leave him, telling her that she deserves better and shouldn't tolerate his abuse any longer. When she returns, she announces that she has left him. Months later, the client returns and shares her current struggle with another man who is just as abusive as the first. You become aware of the error you made during your first intervention: rushing to the rescue but overlooking the client's need to understand the underlying dynamics that affect her choice of mates and lead to her pattern of abusive relationships. Had you been more willing to help her explore these issues, she might not have returned to your office later with a similar problem. The second time around, you learn from your mistake and work on helping the client in a different way that doesn't involve caretaking.

THINKING THINGS THROUGH

Can you think of situations where you acted as a caretaker or rescuer? Are you comfortable with this? Would you considering changing? If so, how?

Wanting to Be Liked

A most difficult challenge is helpers' need to be liked by their clients and coworkers. In and of itself, wanting to be liked is not negative but there is an inherent danger. Wanting to be liked, a helper may neglect to confront a client when appropriate, or may discourage the expression of painful or difficult feelings. Certainly, alienating a client or coworker is not recommended, but successful human service work does not mean being popular with everyone. Our work at times requires that we deal with raw emotions and people in crisis. Clients may feel vulnerable, leading them to express strong feelings they might normally have kept subdued. A crisis can be a dangerous event as well as an opportunity for growth (Hoff, 1989). The crisis state may be most conducive to change, because of the clients' vulnerability. They may be more receptive to looking at themselves and moving toward positive change than they were before the crisis. Although this process may have a positive outcome, pain and other strong emotions are likely to be expressed. The helper is often the catalyst that encourages the expression of these painful feelings, so he or she may not receive accolades from the people who are undergoing this process. Refrain from simply pleasing clients; risk giving your honest assessment of their situation. If pleasing is your major goal, both you and the client are cheated of honest and realistic feedback. Our role is not merely to put out fires or to please everyone we come into contact with; we must act as change agents. Those clients who may not respond favorably to our interventions at the time may experience benefits years later. We often plant seeds that may not flourish until much later and often without our knowledge.

SCENARIO

Lupe recalls working with a very rebellious adolescent, whom she had to confront numerous times about her acting-out behavior. The client often resisted attending the sessions and complained that she didn't feel the therapy was helpful. Several years later, much to Lupe's surprise, she encountered the former client one day. At first, she did not recognize the client, who had matured considerably. She was further surprised by the feedback this person gave her. The former client had learned a lot from her treatment and was in college, pursuing studies in counseling. She was able to share her appreciation for Lupe's earlier efforts.

What do you think would have happened if Lupe had not confronted the client but merely tried to be liked? Do you think this person would have benefited as much?

THINKING THINGS THROUGH

Discuss a situation in which your desire to be liked prevailed over your work with the client. What were the results? If you could do it over again, would you change your approach?

Transference and Countertransference

Transference and countertransference are issues you will undoubtedly face as you begin fieldwork or agency employment. Initially analyzed in Freudian psychoanalytic theory (Corey, Corey, & Callanan, 2003; Nugent, 1990), transference and countertransference are likely to exist in any helping relationship. In fact, many authorities say they are universal in all relationships, social or therapeutic (Gelso & Carter, 1985; Brammer & Shostrom, 1982). "To varying degrees, transference occurs in probably all relationships. The therapeutic situation, however, with its emphasis on a kind of controlled help-giving, magnifies and intensifies this natural reaction" (Gelso & Carter, 1985, p. 169). Similarly, Faiver, Eisengart, and Colonna (2004, p. 145) add, "These concepts constitute central dynamics in psychoanalytic therapy; however, an appreciation of the importance of transference and countertransference is pertinent to your understanding of all interpersonal interactions, including all therapeutic relationships."

Transference

Definitions of transference are many. Faiver, Esengardt, and Colonna (2004) describe it as the "transfer of feelings, attitudes, fears, wishes, desires, and perceptions that belonged to our past relationships onto our current relationships, so that the people in our present lives become the focus of these thoughts and emotions from long ago" (p. 145). Transference is considered to be an unconscious process "whereby clients project onto their therapist past feelings or attitudes they had toward significant people in their life" (Corey & Corey, 2003, p. 143). The feelings typically involved in such a transference range from love, lust, high praise, and regard to anger, ambivalence, hate, and dependence.

In our role as social workers and clinicians, we can certainly attest to the existence of transference in therapeutic relationships with our clients. We have also witnessed transference in nonclinical settings. By a broader definition of transference, the client's feelings and attitudes are directed toward the helper regardless of whether or not they are rooted in early childhood. Teyber (2000) considers this more contemporary definition of transference to be "broader and more useful. It refers to all of the feelings, perceptions, and reactions that the client has toward the therapist—both realistic and distorted" (p. 222). All helpers must recognize the impact transference can have on the helping relationship. Furthermore, it is important that helpers be open to distinguishing when clients' reactions are transference based or legitimate and justified feelings toward the helper.

Examples of Transference

>As a caseworker in a adult day-care center, you are faced with several seniors who treat you as a daughter and tend to infantilize you.
>
>You are working in a child-care agency where several children are especially attached to you and often call you "Mom."
>
>As a probation officer at juvenile hall, you have encountered several teenagers who look up to you as a father figure. They alternately test your limits and ask you for fatherly advice.
>
>You are a career counselor for women re-entering the workforce. You often come across women who identify closely with you because of your history of returning to school in your 40s. Some seem to idolize you and see you as a perfect person.

THINKING THINGS THROUGH

Can you think of any examples of transference from your own experiences working in agencies? Were you aware of the transference at the time? How did you deal with it?

There is no consensus as to appropriate responses in cases of transference. Generally, the interpretation and analysis of transference depends on the nature of your role and the setting. For instance, if you are in a clinical position, then it may be considered therapeutically necessary to interpret, analyze, and work through transference. By recognizing and working with this phenomenon, helpers can gain a better understanding of their clients' conflicts and therefore develop a tailored approach to helping.

SCENARIO

You are a recent graduate, embarking on a career in clinical practice. One of your first long-term counseling clients is a woman who becomes very attached to you and eventually says that she is in love with you. At first, you are shocked by this revelation. Your supervisor helps you understand the significance behind the strong feelings. It seems that this client is still grieving the loss of her father, who is a political prisoner in a far-away country. You are the first male to provide her with the warmth, understanding, and nurturing she lost when she left her father. In your work with this client, you are careful to respect her feelings, but you also help her explore their symbolism. Eventually she is successful in working on the grief issues that are at the core of her feelings for you.

How would you have handled this case? Why?

By contrast, if you are in a helping position where there is not a therapist-client relationship, it may be necessary only to understand the transference issues. In fact, in some settings, it is considered both unethical and inappropriate to explore the nature of transference with clients. Such interpretations may lie beyond the scope of the services being offered and the competencies of the helper.

SCENARIO

You are an intern at a teen shelter. The girls you work with have either run away from home or are victims of abuse and have been removed from their homes. You spend entire weekends with them and have become very close and integrally involved in their lives. You are not their counselor or social worker; your role is to be their cottage manager. Overall, you are well liked by most of the girls and staff. However, there is one particular 15-year-old who is cold and aloof whenever you are in her presence. You make an effort to engage in conversation with her and seek her out whenever possible. She avoids you and is increasingly distant. You consult with the social worker who works individually with the girls. She suggests a possible reason for her behavior: The girl's mother resembles you and has abandoned her for her series of boyfriends. You ask whether you should bring this up with the girl. The supervisor tells you not to, because you are not involved in a clinical, therapeutic relationship with the client. You are not to take it personally, but rather to continue to do your best to relate to her.

Have you ever experienced this type of negative transference? If so, how did you feel? How did you respond?

Regardless of whether you are a clinician, case manager, crisis or outreach worker, or group home caseworker, you are not immune to clients' transference. Deciding how to proceed with transference requires careful consideration.

Countertransference

Countertransference is considered "the other side of transference" (Corey & Corey, 2003) or "the counterpart of transference" (Hepworth, Rooney, & Larsen, 2002). It occurs when the therapist or helper projects unresolved conflicts onto the client (Corey, Corey, & Callanan, 2003; Nugent, 1990). Some see countertransference in broad terms, as any projection that could conceivably interfere with helping a client. Examples may include "counselor anxiety, the need to be perfect, or the need to solve a client's problems" (Corey, Corey, & Callanan, 2003). Another viewpoint contends that countertransference can involve any of the helper's feelings about his or her clients. We tend to favor this perspective, in which the term *countertransference* encompasses all of a helper's thoughts and feelings in reaction to clients, whether prompted by the clients themselves or by events in the helper's own life.

SCENARIO

You are an intern at a college women's center, where you co-lead a support group for older women who desire additional support during their transition back to college. Last week, a new group member was abrasive, loud, and insensitive to the others' issues. As you plan for this week's session, you hope that this client does not show up, as you find yourself not particularly liking her.

How should you cope with these feelings?

SCENARIO

You are working at a health-care clinic. Your role is to interview incoming patients and orient them to the hospital. In one orientation group you lead, a male patient is very inquisitive about the cardiac rehabilitation unit. He is very appreciative of your time and compliments you on your knowledge and informative approach. You also appreciate his positive outlook, find yourself very much at ease, and enjoy spending time with this patient.

Is this okay? Why or why not? Why might you be careful about giving rein to your positive feelings?

Countertransference may not always pose a problem, yet it can "interfere with objective perceptions and block productive interaction with clients. Countertransference contaminates helping relationships by producing distorted perceptions, blind spots, wishes, and antitherapeutic emotional reactions and behavior" (Hepworth, Rooney, & Larsen, 2002, p. 576). Helpers who have many unresolved issues with anger, rejection, loss, trust, or authority figures or who have excessive needs to be loved and admired are at a greater risk of engaging in countertransference. Working with clients often sets off our own unresolved feelings. Old wounds that have not yet healed may be perfect triggers (Corey & Corey, 2003), consciously or subconsciously. If you are unable or unwilling to explore your own issues, or if you avoid certain feelings, some facet of your work is likely to be affected. Certain themes—such as separation, loss, dealing with anger or assertiveness, or denial of sexuality—may surface through your work with clients. Learning to recognize these themes and risking introspection is clearly advantageous, for both your client's well-being and your own.

SCENARIO

In your first week at a mental health center, your encounter with one particular client has left you tense. At first you cannot understand why she affected you this way. As you explore your feelings more closely, your reactions seem to make more sense. This client reminds you of your "critical parent"—an aunt who cared for you when you were young. This client's appearance, demeanor, and presenting problems all remind you of this aunt, whom you have never really confronted or come to peace with.

How will you keep these negative feelings in check? Do you think you'll be able to continue to treat this client effectively?

The Identification Process

According to Watkins (1985), an identification process occurs between helper and client and influences the development of countertransference reactions. Helpers' ability to identify with clients in their experience is essential to a positive working relationship. When there is optimal identification, the helper can better understand the client yet still maintain an appropriate distance. Conversely, when such identification does not take place, the relationship may be compromised and the helper's ability to empathize minimized. A helper's reactions or feelings toward clients are therefore affected by their status in the identification process. Several factors help determine the extent of this identification:

Values: Are your values similar to the client's?
Demeanor: Is there a similarity in your attitudes and beliefs?
Language: Do you share a similar language?
Physical Appearance: Is there a similarity in physical appearance between you and your client?
Expectations: Do you share similar goals and expectations?

THINKING THINGS THROUGH

Think of a relationship with a specific client. Using the factors listed above, name areas in which there is strong identification and those in which there is not. If great differences exist, do you think it is still possible to work successfully with the client? What countertransference issues may be worth examining?

When helpers find themselves at either extreme of this identification process, destructive countertransference patterns can develop, significantly impairing the helping relationship. Harmful effects include the erosion of trust and a "unidirectional interaction" between helper and client. Various ways to minimize potential problems between helpers and clients include self-analysis, personal counseling, supervision, genuineness, self-disclosure, and referral when appropriate.

Positive and Negative Countertransference

As noted, both positive (constructive) and negative (destructive) feelings can be projected onto clients or felt by the helper (Corey, Corey, & Callanan, 2003; Nugent, 1990). The following list gives some examples:

Positive/Constructive

> You look forward to seeing certain clients.
> You feel overly attached to certain clients.
> You enjoy being with a specific client.

Negative/Destructive

> You feel unexplained anger and resentment toward a client.
> You dislike a certain client.
> You feel annoyance toward a client.

Clearly, no one is exempt when it comes to countertransference. It is normal and expected to have feelings toward clients. Hepworth, Rooney, and Larsen (2002, pp. 576-577) cite the following as possible indicators of countertransference:

Typical Manifestions of Countertransference

- Being unduly concerned about a client
- Having persistent erotic fantasies or dreams about a client
- Dreading or pleasurably anticipating sessions with clients
- Being consistently tardy or forgetting appointments with certain clients
- Feeling protective of or uncomfortable about discussing certain problems with a client
- Feeling hostile toward or unable to empathize with a client
- Blaming others exclusively for a client's difficulties
- Feeling persistently bored or drowsy or tuning out a client
- Consistently ending sessions early or permitting them to extend beyond designated ending points
- Trying to impress or being unduly impressed by a client
- Being overly concerned about losing a client
- Arguing or feeling defensive or hurt by a client's criticisms or accusations
- Being overly solicitous and performing tasks for clients that they are capable of performing
- Being unusually curious about a client's sex life
- Having difficulties accepting or liking certain types of clients (which may also be reality based)

Experiencing any of the above is natural. Countertransference can occur in any type of helping relationship, whether clinical or nonclinical, so you must remain aware of your feelings. Asking yourself the following questions may give you added insight into how you can best deal with your feelings.

> How strong are these feelings? If they persist, it may be necessary to take a look at underlying dynamics and seek supervision and consultation.
> Is your objectivity impaired? Countertransference becomes a problem when the helper's own reactions stand in the way of dealing with the client's problems.
> How are these feelings and reactions dealt with? Are they handled in accordance with strict ethical and legal guidelines?

When feelings are not dealt with appropriately, problems are sure to arise. Those helpers who continue to experience countertransference reactions may have long-standing and unresolved conflicts that interfere with the helping relationship. Introspection is essential for helpers to be able to reach a realistic perspective in their relationships with clients.

We have found it useful to have a set of specific guidelines for coping with these feelings, as follows:

Guidelines for Working with Countertransference

> Work toward acceptance of your feelings. They may contain significant information about your clients or about yourself.
>
> Be careful to not judge or disregard your own feelings.
>
> Learn to use these feelings and reactions appropriately to enhance the helping process. Gain insight by examining your countertransference feelings. This can help you understand the client and develop more appropriate interventions.
>
> Consult with professional colleagues if your feelings interfere with your work with clients. A more objective viewpoint is helpful and at times necessary to guide you in the right direction. Sometimes a decision to refer a client may be the most appropriate action.
>
> Engage in counseling or therapy of your own if extreme countertransference reactions persist.

Resistant clients may very well be the ones who elicit the most negative countertransference in helpers. Next, we address some salient issues related to working with such clients.

Resistance

The term *resistance* conjures up negative, resentful, and anxious feelings. Brammer (1988, p. 52) considers the term to denote clients' "conscious or unconscious reluctance to begin a helping relationship as well as their covert thwarting of the goals of the interview once the process is under way." Resistance can thus occur before and during the helping relationship. Corey (2005, p. 72) defines *resistance* more broadly: "anything that works against the progress of [treatment]. *Resistance* refers to any idea, attitude, feelings, or action (conscious or unconscious) that fosters the status quo and gets in the way of change."

In terms of human services, *resistance* is defined by Meier and Davis (1993, p. 15) as an "obstacle presented by the client that blocks treatment." It can take on different forms, from open hostility to passive resistive behaviors. *Reluctance* is another term for resistance used in the literature. Egan (2002) states, "Reluctance refers to the ambiguity clients feel when they know that managing their lives better is going to exact a price. Clients are not sure whether they want to pay that price. Incentives for not changing often drive out or stand in the way of incentives for changing" (p. 163).

Next we examine the most common reasons for resistance and possible interventions in each case. In each accompanying scenario, we present one intervention that we believe might be helpful. This is not to say that other alternatives do not exist, or that the interventions presented will always be the most effective ones.

Reasons for Resistance

Fear of change Simply put, "Clients are ambivalent about change, uncomfortable about the need to disclose painful or shameful things about themselves, or afraid to face what the [helper] is trying to help them face" (Nugent, 1990, p. 76). Egan (2002) echoes that "some people are afraid of taking stock of themselves because they know, however subconsciously, that if they do, they will have to change, that is, surrender comfortable but unproductive patterns of living, work more diligently, suffer the pain of loss, acquire skills needed to live more effectively, and so on" (p. 165). From a psychodynamic viewpoint, resistance is viewed as an attempt to prevent anxiety-provoking material from entering awareness (Hansen, Stevic, & Warner, 1986). The social learning viewpoint considers resistance to be related to clients' fears of change and the consequences that might result (Shertzer & Stone, 1980). Being aware of this fear can make us more empathetic and perhaps less resistant ourselves.

SCENARIO

A client enters treatment because of unhappiness in her marriage. Her spouse is alcoholic and she is unable to tolerate his drinking, verbal abuse, and irresponsibility. She is aware of her co-dependent behaviors but is reluctant to make the changes that would allow her to feel satisfied. She is therefore resistant to any recommendations you make.

Effective intervention Rather than focusing strictly on her co-dependent behaviors, or making an effort to convince her to change, take a more humanistic approach. Encourage her to express her mixed emotions, to help her recognize that change may be necessary for her to feel more satisfied with her life.

SCENARIO

You are a caseworker in a special probation unit for repeat offenders who have drug and alcohol addictions. In one of your educational meetings, you are amazed at the clients' anger and resistance. They seem uncomfortable and even fearful of considering an abstinent lifestyle. What might be the basis for their discomfort or fear?

Effective intervention There might be a temptation to lecture these clients about the evils of drugs, but this might lead to further resistance. All you can do is give them factual information about the impact of drugs and alcohol and the difficulties commonly experienced when making changes in one's life.

Resistance as a defense mechanism Resistance can be viewed as a "defense against anxiety and as a protective coping mechanism" (Faiver, Eisengart, & Colonna, 1995). Brammer (1988, p. 53) states, "Fear of confronting our feelings is a strong deterrent for seeking help. The pain must be great in most of us before we will admit that we need a helping relationship. We may try all kinds of substitutes like drugs, for example, first. It is important for helpers to recognize this ambivalent or mixed feeling in all helpees." Resistance can also be viewed as an adaptive process. People have developed certain coping skills that have contributed to their survival. These may be internalized to such an extent that asking clients to change their coping may not be welcomed (Teyber, 2000).

SCENARIO

A client seeking treatment for depression brings up issues that lead you to suspect that she may have been sexually abused when young. You carefully broach this issue with her and are surprised to be met with fierce resistance. Later, when she is more receptive, she shares having felt extremely anxious at the very thought of exploring this issue for fear of what might be uncovered.

Would you have persisted? What if the client refused even to discuss this possibility?

Effective intervention In a situation like this, it is essential to respect the client's wishes and be sensitive to her feelings. A slow-paced, gentle approach might be more helpful. Insisting that the client deal with unresolved past issues, against her will, might be viewed as another violation.

SCENARIO

A middle-aged man seeks help at a vocational rehabilitation center. He is no longer able to work in the aerospace industry, due to a work injury. He is reluctant to consider most positions that involve his intellectual capacity. When you pursue this, it becomes obvious that he is scared to face underlying fears of failure and inadequacy. He felt competent and comfortable in his previous employment and now feels doubtful of his abilities to work in any other capacity.

Effective intervention Begin by allowing this client his feelings with regard to the loss of his previous occupation and his struggle to begin a new career. Encourage him to talk about past job experiences, in hope that he will recognize his strengths and learn to transfer the skills he has developed into a new career.

Previous negative experiences Unfortunately, some clients have valid reasons for their resistance: a previous negative encounter with a professional helper or agency (Corey & Corey, 1993). Certainly, this makes sense; it is instinctive to avoid repeating a prior negative experience.

SCENARIO

Years ago, a client sought help after the death of her mother. A helper told her that such a loss is a normal event that did not meet the criteria for ongoing treatment. After this interaction, the client felt alone with her grief and hesitant to pursue any further help. She is currently going through a divorce and is contemplating seeking counseling, but she is reluctant because of her previous negative experience.

Effective intervention It is important to recognize clients' hesitancy and explore possible underlying reasons. In this case, it would be helpful to allow the client to express her unresolved feelings regarding her previous experience with a helper. Then reassure her that her current feelings of loss are valid and worthy of intervention.

SCENARIO

A middle-aged female college student hesitates about seeking out a guidance counselor. She vividly recalls a negative experience years ago. She recollects that the helper questioned her abilities and deterred her from pursuing her original goals, which she is now returning to college to fulfill.

Can you blame this client for being hesitant? If you were her new guidance counselor, how would you approach her?

Effective intervention In working with this client, be sensitive to her fears and approach her in a supportive, caring manner.

Cultural factors Often, depending on the cultural perspective of the client, seeking help may be viewed as weakness, incompetence, disrespect, or disloyalty.

SCENARIO

An Asian college student is feeling overwhelmed with his heavy workload at school and at his job. He feels depressed and apathetic but is hesitant to seek help. He finally makes an appointment with a school counselor to discuss his academic struggles.

Effective intervention Taking into account the cultural issues, respect this client's hesitancy and recognize the difficulty he has in even coming to see you. Make an effort to stay at the client's level rather than rushing to pursue deeper feelings he may not be ready to entertain. Learn more about the culture and its view of seeking help outside the family.

SCENARIO

A middle-aged Latina female is referred for a health screening by the physician who recently delivered her third child. He is concerned about her use of herbal healing to treat a possible ulcer. You are a bilingual social worker in the hospital who is asked to help with the screening.

What cultural factors would you consider in working with this client?

Effective intervention By recognizing the cultural aspects of herbal healing, you can be supportive rather than critical. This will help you build rapport, which can allow you to suggest additional healing through a recommended diet and possible medication.

Authority issues For various reasons, some people resist authority. They may therefore rebel or act out in a helping relationship where the helper is seen as an authority figure. Many clients also have a need to be in total control and view the helping process as a loss of control. (This may be a defense mechanism that has developed for a valid reason.) Some clients may actually feel more powerful if they refuse or resist help.

SCENARIO

A common fear of people who enter counseling is that they will completely lose control as they begin the process of exploring their feelings. Especially vulnerable are those who customarily exert control and authority as a way to hide their underlying insecurities. For example, a man owns his own car repair business, is the primary breadwinner, and is the major authority figure both at home and at work. Surprised by his wife's desire for a divorce, he seeks counseling. His focus in sessions is on having the counselor help him explore ways to get his wife to come back to him—in essence to change her behavior. When the counselor offers feedback, the client shoots it down, because he is accustomed to being in control.

Effective intervention It is important not to engage in the power struggle. Educate the client about the process of helping and the fact that he can only change himself.

SCENARIO

A 50-year-old executive has just learned that his job may be eliminated due to downsizing in his industry. He seeks vocational rehabilitation but finds it very difficult to consider any of the suggestions he is given regarding other job options.

Is this resistance? Why might it be difficult for him to accept help from others?

Effective intervention Be understanding of the client's lost role and the grief, anger, and fear he may be experiencing. Give him information about available resources, but respect his current resistance. He may not be ready to find another job. Furthermore, he may feel that his identity as an executive is being overlooked.

Manipulation Certain clients "are incapable of healthy relationships, and may use them as a medium for manipulations and power struggles. . . . All other people are adversaries" (Brill, 1990, p. 242). Clients who are needy may develop a dependent relationship with a helper. They may not really want help or change, but simply the safety and nurturance of the relationship. When asked to work on their issues, to change, or to consider termination or a referral, they may resist rigorously.

SCENARIO

A divorced couple is referred for court-ordered counseling in connection with a bitter custody battle over their 6-year-old son, who lives with the mother. Initially they are seen separately because both of them want to tell their own side of the story, which differs dramatically. The court has stipulated conjoint counseling with both spouses, so the counselor tries to set up an appointment. On various occasions, the session is canceled, rescheduled, or only attended by one party. When the counselor attempts to address this, he is met with resistance and passive aggressiveness.

Effective intervention The helper should remain focused and neutral and should directly confront the clients about their resistance to the conjoint sessions. The emphasis must be on their son's emotional well-being. Referral to another clinician may also be indicated if countertransference begins to interfere with the treatment.

SCENARIO

You are organizing job-training programs at a homeless shelter as part of your internship at a community services center, where you are specializing in "macro" practice. Shelter clients must attend these programs weekly if they are to remain at the shelter. Many are eager to look for employment, but others just come to classes and are resistant to any suggestions you offer. It becomes clear that they attend only to be able to stay at the shelter longer.

Effective intervention To minimize this problem, clarify the need for participants to be motivated and receptive from the outset. With manipulative clients, point out the discrepancy between their stated desire to seek employment and their resistance to do so in fact. Also, their inappropriate behavior in the training seminar may need to be confronted directly, on an individual basis.

Depression What appears to be mere resistance may in reality be clinical depression. Symptoms may include a negative, pessimistic outlook; apathy; lack of energy or desire; "paralyzed" behavior; a sullen attitude; self-deprecation; and inability to trust.

SCENARIO

You are a new social worker in Child Protective Services. You have assumed the caseload of a transferring social worker. Treatment plans have been established by the court, and it is your responsibility to make sure the clients are in compliance. Several of your first cases involve child abuse in single-parent families. As you review the case notes, you are struck by the apparent lack of compliance by the parents. In home visits with these mothers, you find that they appear very apathetic and hopeless. Your initial reaction is to become critical of their lack of follow-through, but you soon realize that you have been seeing symptoms of depression in these women.

Effective intervention Refrain from blaming, criticizing, or demanding because this would only worsen the depressed state and increase the women's feelings of low self-worth. Instead, attempt to provide support and educate them about the state of depression they may be experiencing. If possible, refer them to mental health professionals for further evaluation and treatment. Also, it may be helpful to give them structured tasks with tight time lines. Start small to avoid overwhelming them. Give them positive feedback even for small successes. If at all possible, provide some case management services that can help empower them. Above all, remain sensitive to their plight and recognize that the depression may be a realistic response to their situation.

Involuntary clients These are clients who do not voluntarily seek the services of the helper but are court ordered, coerced by parents or spouse, or manipulated by threats of divorce by the spouse or of grounding by the parents. Such clients are not likely to welcome your interventions with open arms. They may express their resistance through hostile behavior, through indifference or apathy, or by pretending acceptance (Brill, 1990).

SCENARIO

A male client who is struggling with marital issues convinces his wife to attend a marital therapy session. She goes only because he wants her to. She is not aware of her part in the conflicts and has no desire to change at this point. In the session, he talks about his issues and asks for her input. She is obviously angry and conveys her view that all the problems stem from his behavior. She is clearly unwilling to consider any viewpoints other than her own.

Effective intervention Be direct and honest with your impressions of their relationship. Tell how you see them both with regard to the marriage. Educating them about marital therapy is also vital if future treatment is to be effective.

SCENARIO

You lead drug-education groups for clients who have been ordered by the court to attend because of arrests for driving under the influence. Some clients in your group create havoc by talking among themselves, falling asleep, asking inappropriate questions, clowning around, and not taking the process seriously.

How would you handle this group? Would you confront the members? Do you think their involuntary status is contributing to their resistance?

Effective intervention You may approach this situation in a variety of ways, depending on your style and personality. Some helpers would directly confront the acting out, presenting the clients with the choice of either participating or being terminated from the program. Others might consider working with the resistance, by talking about how frustrating it is to be at this class. When you have gained the clients' trust, they may feel safer about exploring pertinent issues that may be painful and difficult to discuss openly.

Additional reasons for resistance or reluctance include fear of intensity, lack of trust, and fear of disorganization (Egan, 2002). Attempting to understand why a client might be resistant is essential in establishing rapport and developing an appropriate treatment plan.

Ellis (2002) and Ellis and MacLaren (1998) highlight forms of resistance that are also helpful to be aware of in dealing with resistance. These include:

- Healthy resistance is common for anyone embarking on change.
- Resistance motivated by a client-helper mismatch often occurs despite "good intentions."
- Resistance caused by counselors' [helpers'] relationship problems that are related to countertransference issues.
- Resistance related to counselors' [helpers'] moralistic attitudes.
- Resistance created by fear of discomfort "low frustration tolerance" in clients who demand immediate relief.
- Resistance due to secondary gain.

- Resistance stemming from feelings of hopelessness.
- Resistance motivated by fear of change or fear of success.

THINKING THINGS THROUGH

Reflect on your interaction with resistant clients. Describe the circumstances and how you dealt with the resistance. How might you handle a similar situation in the future?

Besides the effective interventions suggested for the above scenarios, the following general guidelines for dealing with resistance may also be useful.

Guidelines for Dealing with Resistance

Be accepting and strive to develop a respect for resistance, as it may serve a valuable purpose and may be normal given the circumstances.

Examine your own issues with regard to resistance; be willing to explore your own resistance.

Be willing to explore the significance of the resistance and its inherent dynamics. Emphasize the development of a trusting relationship. Little will be accomplished without building a foundation.

Address the resistance directly and empathetically.

Examine the quality of your interventions.

Be willing to use constructive confrontation, according to the following criteria. It must possess the elements of caring and concern, be descriptive, be specific, and be well timed. Point out resistive behaviors to the client, as well as other incongruent behaviors. Some may not even be aware of these incongruencies.

Consider resistance as an opportunity for positive change and growth. Often, addressing and exploring the nature of the resistance can lead to important revelations.

Respect the fact that change is not universally accepted or rigidly defined. Not everyone wishes to alter or modify his or her life as you might. Refrain from pushing your own agenda and expectations.

Recognize that resistance may simply be a result of poor communication.

Be realistic, flexible, creative, and willing to work with the resistance; help the client search for incentives for moving beyond resistance.

Settle back and listen to the client; this can open up new insights and awareness.

Difficult Clients

Hand in hand with resistance is the challenge of working with difficult clients. There is no universal definition of a difficult client; each of us will experience clients differently. Generally, however, certain types of clients tend to make the helping process difficult. Table 7.1 is an adapted version of Corey and Corey's (2003) analysis of types of difficult clients, along with suggested interventions.

Examples of Difficult Clients

"Yes but" clients may elicit feedback on how to cope with a particular situation. When you provide them with suggestions and alternatives, they say, "Yes, but. . ." and always have some reason for not following through.

Blaming clients are often encountered in work with couples or families. One member of the couple or family tends to blame the other for whatever problems they are having. This creates anger, resentment, and helplessness and hinders effective problem resolution.

Overly dependent clients are very needy and seem to want you to direct their lives for them. They may hang on every word you say and are not content unless you instruct them on how to proceed. They may call you frequently for advice and seem unable or unwilling to make decisions on their own.

■ **TABLE 7.1 Difficult Types of Clients**

Type of Client	Suggested Intervention
Involuntary clients	Be careful not to become apologetic or defensive. Place the responsibility on the client.
Silent or withdrawn clients	Be aware of the possible purposes of the silence: protection, lack of knowledge regarding the helping interaction, cultural issues, or intimidation (the client may see you as an authority or feel dumb). Explore the nature of the silence.
Talkative clients	Be aware that they might feel intimidated, angry, or anxious. Point out body language. Discuss your own reactions in a nonjudgmental way.
Overwhelmed clients	Resist becoming overwhelmed yourself. Explore with the client the ways in which he or she becomes overwhelmed.
"Yes but" clients	Such clients are seldom satisfied, so you may find yourself working too hard. If so, place the responsibility back on the client. Point out the behavior.
Blaming clients	Place the responsibility and focus for change back on the client.
Clients who deny needing help	Refrain from trying to convince such clients that they have problems. Encourage them, but refrain from pushing them.
Moralistic clients	Help them recognize how their judgments of others affect their relationships and may distance them from others. Ask them to indulge their tendency to lecture: Have them make up a very stern diatribe. This may reveal how they possibly have incorporated a critical parent.
Overly dependent clients	Explore how you might possibly be fostering dependency. Encourage individuation and separation, in a supportive way.
Passive-aggressive clients	Explore the underlying dynamics. Be aware of how the client's behavior affects you. Share your reactions while avoiding making judgments about the behavior.
Intellectualizing clients	Recognize that intellectualization is often a defense mechanism, developed out of a need to insulate people from their deeper feelings. Refrain from insisting that these clients dig deeply into their feelings. Rather, gradually encourage the expression of feelings.
Emotionalizing clients	Ironically, similar to intellectualizing clients—in the sense that they may also be defending against deeper emotions and have gotten stuck in their pain. Gradually encourage more cognitive reactions.

Source: Adapted from Corey and Corey (2003, pp. 155–162).

Passive-aggressive clients are upset with you but do not directly tell you. Rather, they act out their resentment by coming late to sessions, not following through on homework, complaining about you to your supervisor, and so on. It may be difficult to deal with such clients because they do not directly express their feelings. Perhaps not until afterward will you notice yourself feeling frustrated, annoyed, and angry, too.

THINKING THINGS THROUGH

From the examples above, select which types of difficult clients you would find most challenging to work with. Discuss what feelings and thoughts they may evoke in you. Elaborate on why these particular clients may be more trying to work with. Explain how you might work with them.

Many of the feelings a helper has in working with difficult clients may apply to working within organizations. Next, we take a closer look at organizational challenges.

▪ ORGANIZATIONAL CHALLENGES

Understanding the Organization

In Chapter 2, we established the importance of understanding the agency you work in. Because agencies often function according to informal rules and policies, it may be quite a challenge to understand the entire system. Also, as in families, an agency system changes over time, with the loss and addition of different members. It has been our experience that key personnel in an agency tend to set the tone for the informal functioning of the system, much as a parent would determine the rules within a family structure. Discovering these hidden rules and agendas can take much time, effort, and inquisitiveness.

THINKING THINGS THROUGH

Reflect back on the Thinking Things Through exercises in Chapter 2 that involved the formal and informal lines of command, policies, and norms of agencies. Briefly discuss which policies, rules, or norms you advocate and which you do not. Explain why.

Fitting in

For beginning helpers who are eager to effect change, taking time to fit into an existing agency structure may be frustrating and difficult. Allowing yourself to fit into an agency system is similar in importance to building trust in a helping relationship. Both require that helpers be genuine, patient, and respectful of the other party (the client or system). This process may take more time than anticipated; resist trying to fit in too quickly. Sweitzer and King (2004) stress that learning about the history, mission, goals and objectives, values, strategies, and funding of the organization will help you gain the information you need to begin this process. Also, becoming aware of organizational politics and the unique culture in existence can facilitate fitting in. You need time to orient yourself to the agency system, and the staff will also need time to get to know you. Therefore, a sense of belonging may not arise immediately. Furthermore, issues you are not aware of may affect how staff members feel about the hiring of new personnel. Take time to learn about these nuances.

SCENARIO

You are a recent graduate hired to work at a family services agency where there are both caseworkers and clinicians. Because of your education, you are offered a position as a clinical therapist, working with children and their families. You are excited about this new opportunity, but you encounter some negativity from existing staff, who seem distant and cold. As the weeks go by, you become close to one particular coworker. She tells you the reason for some of the staff's animosity. It seems that they resent your status and are upset about office space assignments. Unbeknownst to you, you were given one of the larger, more desirable offices because of your need to accommodate large family groups. Many caseworkers do not even have an office of their own. When you learn this piece of information, everything begins to make sense. You approach your supervisor with your concerns about wanting to fit in and not alienate coworkers. He agrees to bring up the issue of office space at a staff meeting and explain the reason for certain assignments.

Do you feel this will help create a more cohesive atmosphere? How else could you have handled this situation?

THINKING THINGS THROUGH

Consider your current agency system. Recall how you were first introduced and how receptive others were toward you. Do you recall experiencing any difficulties? Do you currently struggle to fit in? Explore the reasons why.

Working with Other Organizations

In addition to understanding your clients and your organization, it is important to know about the community. "Just as it is impossible to understand a system without understanding all its component parts and relationships, it is impossible to understand that system without understanding its relationships to other systems and to larger societal forces and pressures" (Sweitzer & King, 2004, p. 150). Because it involves new people and places—operating outside of your safety zone—this endeavor can create anxiety and insecurity.

Agencies often have a specified catchment area, which may include a whole city or cities, rural areas, entire counties, or smaller segments of these divisions. Learning about this catchment area can be valuable.

Knowledge of the community can provide important insights into the environmental and cultural aspects affecting the client population that your agency serves. Learning about the service organizations in your community and their relation to your agency allows you to build important ties that will facilitate networking. Also, the history of the relationship between your agency and the community may provide you with useful data on how to proceed in encouraging constructive affiliations. Conflicts between agencies may have previously limited any productive partnership. Knowing about this may prepare you to anticipate problems and help you in devising effective intervention strategies. Clearly, it is helpful to your client when you are able to call on community resources with whom you have developed a positive working relationship. Acting as an advocate for your client requires great skill in communication. By creating positive community connections, you pave the way for such advocacy.

Among the organizations you may interact with are the local law enforcement entity (police or sheriff), the Department of Social Services, Child Protective Services, counseling agencies, charitable organizations, chemical dependency programs, Legal Aid, shelters for battered women and children, and homeless shelters. Cultivating key contacts in other agencies helps you learn the most effective ways to refer or seek assistance on behalf of your client. Minimizing the bureaucratic red tape is obviously beneficial to all concerned.

Also valuable are positive relationships with neighborhood groups, activist groups, and local political organizations. These contacts can be advantageous in fundraising ventures, in obtaining support for your agency's continued services to the community, and possibly in formulating any new programs your agency is considering.

SCENARIO

When Lupe was working in a community mental health center, she found that the Latino population was underserved, despite great needs. By going to community centers serving this population, she built positive rapport and trust. Often she provided short-term parenting classes or women's support groups at these centers, so clients and staff got to know and trust her. Eventually these positive connections made clients feel more comfortable to seek services at the mental health center itself.

How might you have handled this situation? Do you feel this community involvement was helpful?

SCENARIO

Randy, while working at a mental health center, found it helpful to connect with the local community college where students often needed help in crisis situations. Many students were not comfortable seeking services directly at the mental health center, but they were amenable to seeking crisis counseling at the college health center. Randy negotiated an agreement to see students at the college for up to six sessions; if further services were needed, there would be a transition to the mental health center. He found this to be a fruitful venture. Students and the college staff became more comfortable with direct referrals to the mental health center, because their key contact, Randy, was someone they knew and trusted.

Do you feel this community contact was beneficial? How so?

THINKING THINGS THROUGH

Consider your current placement and ways in which you might connect with the community. Which community agencies would be most important to develop a relationship with? Why? Explain how such contacts can benefit you and your clients.

Sometimes, working with the community allows you to be more objective about the strengths and limitations of your own agency. Both are important.

Organizational Limitations

According to Satir (1972), systems can be divided into two kinds: closed and open. "An open system offers choices and depends on successfully meeting reality for its continuing life. A closed system depends on edict and law and order and operates through force, both physically and psychologically." You may experience agencies that clearly fall into either of these categories, but most likely your organization has elements of both. Much or all of your professional life will probably be spent working within an agency, so learning to cope with organizational limitations will be an important asset in your development. No organization is perfect; each organization, like each person, has its strengths and limitations. Systems that are rigid, inflexible, and bureaucratic may engender feelings of frustration in their workers. You might notice this by paying attention to how you feel when you are working at your agency. Are you frustrated and angry or depressed and down after spending time at your placement? Then begin to pay attention to what organizational factors are influencing how you feel. Perhaps you have little chance for input or any voice within the organization. Or certain coworkers may be antagonistic or resentful toward you. Do not ignore such warning signs.

Organizational limitations are most often discussed in terms of the restrictions they place on the provision of services. For instance, many agencies now set rigid limits on services provided, length of treatment, and use of available resources. This can be exasperating. Such limits can conflict with the very values that are inherent in helping. This can be especially difficult for the beginning helper, who is set on making a difference and unprepared to deal with such barriers. Do not give up; there is usually a way to help the client, remain faithful to the ethical codes of your profession, and still abide by agency policy. Flexibility and imagination are key elements in overcoming obstacles.

SCENARIO

A common restriction in service delivery today is being allowed to provide only brief, short-term, or crisis work when extensive, long-term treatment is actually necessary. Altering your style, being more directive, and teaching clients techniques they can continue to work on independently are some ways to cope with this restriction. Spacing out visits may also work; you can continue to see the client on a more ongoing basis, albeit less frequently. Interestingly, some clients may actually benefit from a short-term focus.

How would you deal with a situation where you felt a client required more intensive treatment than you were able to provide because of agency limitations?

SCENARIO

An agency might stipulate that rather than only working with children, the entire family must participate in treatment. This may seem ideal, but it may not be culturally appropriate in every case. Learn to work within the appropriate cultural framework to encourage family participation. Dispel any negative stereotypes that might exist about counseling and family therapy.

How might you handle such a situation? Would you encourage family treatment, attempt to change the policy, or refer the family to another agency that does not stipulate family involvement?

SCENARIO

Interns at a homeless shelter where clients were allowed only a limited stay often found it difficult to perform effective case management. They were frequently unable to complete their work with a particular client because he or she was no longer there the following day.

Have you ever experienced similar feelings caused by organizational limitations?

Agencies often have certain role restrictions that serve as limitations as well. Some agencies may have strict policies specifying what treatments or training activities student interns, bachelor's-level graduates, and nontenured staff may engage in. Although this may not seem just, it is prudent to explore the reasons behind such policies before reacting. If, after such an inquiry, you still feel the policy to be unjustifiable, you can explore appropriate channels to modify the policies. Again, finding out the history behind such policies may give you clues on how to pursue constructive change.

SCENARIO

At an agency, there is a legal mandate that interns may not engage in any clinical service because they are unlicensed and therefore not insurable against any liability if a client is adversely affected. How might you feel if you were not allowed to work with specific clients because of your intern status? How would you handle this?

SCENARIO

In many nonclinical settings, such as group homes, shelters, and school sites, there are onerous requirements for documentation and paperwork. This organizational limitation may exhaust helpers' time and energy. Despite the frustrations involved in paperwork, documentation is done for valid reasons. Understanding the underlying reasons, you may feel less resistant and be able to explore how to conduct such record keeping more efficiently. How do you cope with documentation and paperwork at your placement?

Students may find their agency placement not to be the ideal setting for learning what they want to learn. We advise students to research which placements best fit their learning goals and then interview several of the most suitable ones before accepting any. Just a little extra time and effort can end up saving you from frustration in an agency that is incompatible with your needs. Students who wait until the last minute to find an internship are likely to end up feeling cheated and discontented.

Whether the placement is a mismatch because of poor preparation or did not turn out to be what was expected, students can turn the situation into a positive learning experience. You may learn what population you do not want to work with, what role you do not like, or what helpers you do not want to emulate. To reframe your feelings about your placement by viewing it as a learning experience is certainly more adaptive and healthy.

SCENARIO

Sarah majored in accounting while in college and subsequently worked as a tax accountant for 15 years before suffering burnout. She had aspired to work in human services as a counselor helping adolescents. Returning to school, she decided to make her dream come true and chose an internship working with teens. Initially elated, she quickly realized that this placement was not to her liking. Instead of writing off this internship, she attempted to learn as much as she could about adolescents.

THINKING THINGS THROUGH

Consider your current agency. First, make a list of its strengths and weaknesses. Then discuss ways in which you might cope with the limitations you have noted.

Change in Organizational Life

As with all systems, organizations are subject to influences from within and from outside. Such forces have an unavoidable impact on your position within the system. Living in an organization, like living in a family, is a process of ongoing adjustment and adaptation to change. Brill (1990, p. 109) writes, "Workers are as much a part of many social systems as are their clients, and as such are subject to controls and pressures of the social systems." When systems change, the worker must adapt as well. Sometimes the transition is easy; sometimes it can be an extremely challenging experience.

SCENARIO

In today's mental health system, changes in funding are forcing agencies to shift from long-term treatment to brief or crisis intervention. Many insurance companies will not pay for treatment beyond a certain number of sessions, so providers must change their treatment focus and be more creative in their overall approach. Many are surprised to find that exploring new ways of helping can benefit the client and serve as a learning experience for the practitioner.

As a counseling intern, how might you deal with this restriction? How might you feel?

SCENARIO

Midway through your internship at a small senior community center, you are informed that changes in management have resulted in many staff leaving. You are surprised by this sudden change. In time, you become angry. You feel that this transition has left clients and staff (including yourself) confused. You are especially affected; your supervisor has been changed twice since the shakeup.

How would you feel? How might you react? How might you involve your school fieldwork instructor, liaison, or director?

Considering new options and ways to cope with change is crucial to your survival in an organization. Although it is normal to be upset and disillusioned by unforeseen or unwarranted changes, dwelling on these can be counterproductive. Focus the majority of your time and energy on exploring ways to adapt to these changes or modify them to meet your needs and to fit your style.

THINKING THINGS THROUGH

Discuss changes you have witnessed in your agency. Describe how such changes affected staff and clients, as well as the overall functioning of the agency. Do you feel change was introduced and handled appropriately? Explain ways in which change could have been introduced and dealt with more smoothly.

ENVIRONMENTAL CHALLENGES

Along with individuals, agencies have a network of support systems they interrelate with. Such a network can be considered the agency's immediate environment. Also, the broader fiscal and political environment directly or indirectly affects the overall functioning of the agency. Corey and Corey (2003, p, 281) stress that "the helping process does not take place in a vacuum isolated from the larger sociopolitical influences of society." Next, we address these environmental issues.

Fiscal Realities

Coping with fiscal realities is an ongoing challenge in human service agencies. Many government and private agencies are being affected by budget cuts. The threat of layoffs looms over these agencies. Staff members are often demoralized when they are asked to work more and more while pay and benefits stay the same or even diminish. According to Woodside and McClam (1994, p. 72):

> The trend of decreasing federal funding of human services has affected many state, local, and private, and nonprofit agencies that previously depended on the federal government to support at least some of their services. Ideally, the community and the state would absorb costs and work with the private sector to develop ways of providing the needed services. In reality, all organizations involved in service delivery are struggling to get by.

During tough economic times, vulnerable population groups often seem to be blamed for the problems of the day. This can result in funding cuts for human services, which can greatly affect both the quantity and quality of the care offered. The client may become ineligible to receive help, and the worker's job may be ultimately eliminated. Available funding often fluctuates, depending on factors such as the nation's overall economy and the political climate.

In today's volatile economic climate, most human service agencies are being forced to implement budget cuts. This situation will probably affect you (if it hasn't already), directly or indirectly. Your agency may have had to reduce or limit services. Grant funding and new program development are also likely to diminish. Workers who specialize in community organization may be particularly demoralized by these dismal fiscal realities. How can a new human service worker, zealous to help others and effect change, cope with such environmental stressors? This question has no simple answers. By keeping an open and positive mind and being creative in the use of existing funds, effective treatment is still possible. You may need to be realistic about the limitations you will face and learn to adapt to them to the best of your ability. Also, it never hurts to keep on hoping for things to improve and funding to increase. Many who have worked in the helping field for years have successfully survived repeated budget cuts at their agencies over the years past. It requires working together to weather the storm and creatively developing new ways to cope with demands and expectations that seem unreasonable or even impossible.

SCENARIO

In one agency where funds were scarce, educational and therapeutic groups were developed to meet the needs of the clients in a more cost-efficient manner. Also, increased efforts were made to enhance access, responding more effectively to the client population. Although these changes were warranted economically and allowed clients to be seen more readily, many staff felt overwhelmed and resentful at not being included in the decision-making process.

How might you have felt? How would you have coped? What suggestions might you have for coping with such unpleasant realities?

SCENARIO

As an intern placed in the probation department, you face staff reductions and program cuts because of county budget restraints. You notice a decrease in morale. The staff has less time and energy for clients. Also, fewer supervisors are available; your supervision time is now only once a month.

How would you feel if you were placed in such an internship? What environmental factors are at work? How might you cope with the direct impact this change has had on your supervision?

THINKING THINGS THROUGH

Based on your fieldwork placement, discuss what fiscal realities exist. What impact have they had on the overall functioning of your agency? On the personnel within the agency? On the clients? What might the consequences be? What suggestions do you have to cope with these consequences?

Political Realities

Various political issues are apt to affect organizational life in some form or another. Local, state, and federal politics can impact funding and the way in which agencies are managed. Sometimes, a change in governmental officials will also lead to a change in staffing, philosophy, regulations, and funding distribution (Corey & Corey, 2003; Sweitzer & King, 2004). As human service workers, you may face a hostile political environment, which you will have to learn to manage. We have observed that difficult economic times often go hand in hand with political turmoil. Such is the case today. There seems to be a war on Capitol Hill with regard to federal aid and funding of various social programs. As a consequence, the need or benefit of providing certain services is put in question. Some social expenditures are labeled as wasteful. This attitude leads to reluctance to fund existing programs or allow new ones to be developed.

SCENARIO

Proposition 187 in California, which deters undocumented immigrants from seeking services, is a prime example of how politics can affect service delivery. Many undocumented immigrants are hesitant to seek help and do so only when in dire need. The cost of emergency treatment is often higher than the programs they avoid. If the actual intent was to save money, the proposition may be counterproductive, in terms of both dollars and human costs.

THINKING THINGS THROUGH

Consider how political changes affect agency systems. What recent political changes do you feel have directly or indirectly affected the agency you are currently placed in? Please describe these and offer some suggestions on how to best handle them.

New Technology

A major trend in the past decade has been the technology boom—computers and Internet use, copiers, fax machines, cellular phones, pagers, and modern medical equipment. The benefits of these new technological devices are too numerous to name. Many are thrilled with these advancements and are "naturals" at learning how to use them effectively. Others are hesitant, resistant, or phobic about new devices, especially when the old ones continue to be effective. By staying in sync with technology, human service workers can open communication channels with other disciplines and professions and become more efficient in their work. The benefits of learning new technology usually outweigh the negatives. One way of coping with such changes is to focus on how you can personally benefit. For instance, computer technology facilitates billing, payroll, and budgets; increases one's ability to plan, monitor projects, and remain accountable to others; improves statistical and tracking methods; aids in problem analysis; and simplifies record keeping. Moreover, the amount of time saved by the use of computers is staggering. Use of computers has also been shown to be very effective in training helping professionals (Casey, 1999; Haney & Leibsohn, 1999; Hohenshil, 2000; Neukrug, 1991). According to Neukrug (2004, p. 291), "[m]ental health professionals now use computers and other technologies such as interactive videos and CD-ROMs in case management, record keeping, clinical assessment and diagnosis, personality assessment of clients, comprehensive career counseling, documentation of client records, billing, marketing, and assisting clients in learning new skills (e.g., parenting skills, assertiveness training, and vocational skills for specific jobs)."

SCENARIO

As a newly converted computer user, Lupe can attest to the struggles she endured in reaching the point where she was even willing to consider trying a computer. Because of Randy's interest in computers, the realization that she needed to keep up to date with this new technology, and the fact that both her young children were familiar with the technology, she finally decided to challenge her computer phobia. She started out using a word processor and then eventually moved up to a standard computer. At first it was not easy. There were many frustrating moments when she wished she had never touched a computer key. However, as she become more comfortable with the computer language and functions, Lupe became intrigued with all the possibilities. She still has her word processor as a backup and has learned to share her frustrations with others, who echo some of her remaining fears and frustrations.

THINKING THINGS THROUGH

What if you are not totally comfortable with the use of computers? What if you feel the pressure to keep up with the rapidly changing advances in computers today? How do you cope with your fears and anxieties?

Impact of the Internet on Human Services

As we have already addressed, the use of the Internet is perhaps one of the greatest of technological advances in this new millennium. Sampson, Kolodinsky, and Greeno (1997) and Sampson (2000) report a rapid increase in the number of network-based computer applications in counseling and human services. For instance, list servers and bulletin board systems have become popular and novel ways for helpers to post messages and encourage dialogue between colleagues. Use of e-mail between human service professionals and in helper-to-client interactions has become a common form of communication. Use of websites that are maintained by professional organizations, educational programs, and individual practitioners have become commonplace. Video streaming for professional conferencing, Web-based courses, search engines for research, and downloadable files on just any human service subject imaginable has become fairly routine. Even certain professional journals such as the *Journal of Technology in Counseling* are found online. Lastly, the ever-growing practice of online counseling, although controversial, has seen rapid growth in the past decade.

Besides the ethical and legal issues of privacy discussed earlier (Chapter 6), it is also essential that professional helpers consider the risks and benefits to their clients before they engage in Internet use. Some clients may not have access to computers or feel comfortable in their use, others may prefer direct, personal contact versus online communication.

eTherapy, which refers to the delivery of mental health services online (e.g., e-mail, discussion lists, live chat rooms, or live audio or audiovisual conferencing), has certain benefits as well as certain limitations. Potential benefits include:

- Easy accessibility at any time; providing help at the client's greatest time of need
- Serving clients in remote and underserved regions
- Connecting with those who may be physically unable to leave their homes
- Providing confidentiality for clients in small communities
- Being able to help clients who are hesitant to seek face-to-face services
- Furnishing time for reflection on discussion when using e-mail
- Making it feasible for practitioners to develop more specialized practices
- Allowing practitioners more flexibility in their work schedules

Limitations also exist such as:

- Potential for miscommunication due to lack of nonverbal and paralinguistic cues
- Problems with assessment and diagnosis because of inability of observing client's nonverbal behavior
- Ineffectiveness of online services
- Differential effectiveness because online services may be appropriate and effective for some (e.g., divorce adjustment) and clearly inappropriate for others (e.g., psychotic or paranoid clients)
- Problems ensuring confidentiality
- Inability to immediately intervene in crisis situations (e.g., suicidal clients)
- Computers/Internet inaccessible to certain clients

Caution is necessary when using the Internet. At the very least, if one intends on using the Internet in a professional capacity, additional specialized training is recommended. It is critical for helpers to be knowledgeable about these uses. Knowing about the many computer games, chat rooms, virtual reality rooms, and other features that exist will help us better understand those clients who find themselves addicted to some form of Internet use. We predict that Internet addictions will soon be just as prevalent as substance abuse, eating and sexual disorders.

We live in a very fast-paced and ever-changing world where technological advances are constantly evolving. To remain at the forefront of new technology that can greatly enhance client services, human service practitioners need to possess certain technical competencies. The competencies that helpers should master according to Gladding (2004, p. 22) include "being able to use word processing programs, audiovisual equipment, e-mail, the Internet, list servers, and CD-ROM databases."

THINKING THINGS THROUGH

Have you had experience using the Internet in your work? What was your experience like? What are some precautions you might need to take? What do you believe are some benefits of modern technology? Limitations?

COPING STRATEGIES

By way of review, we next outline some strategies that can help you cope with the challenges outlined in this chapter.

Open Communication

Recognizing and allowing your feelings is the first step in developing useful coping strategies. A second step is being able to communicate effectively with others—clients, workers, supervisors, administrators, and professionals from other agencies. The next chapter will have more to say about communication skills. Here we simply note that maintaining open communication channels will enhance your ability to cope with existing stressors.

Flexibility and Patience

If you are flexible, patient, and willing to reevaluate your expectations and options, you are well suited to become an effective helper. Flexibility is of utmost importance; clients are not always willing, able, or ready to follow the steps helpers think they should follow.

Our clients' lives are often unpredictable, and we must be willing to bend in our approaches and interventions.

Working with others also requires that you be patient and willing to work at their pace. One of the basic concepts of the social work profession, "Start where the client is at," relates to the need to be patient. If the helper rushes the helping process, the outcome may be the loss of the helping relationship. Being flexible and patient also applies to working within an agency and with others in the community. Changes in agency policies and government programs often involve a lengthy process, not completed overnight.

Patience may be easy for some, difficult for others. Beginning helpers may unwittingly push their own agenda rather than work at a pace that is comfortable for the client or the system. Just as fitting into an organization takes time, change in clients or systems is a lengthy process.

Avoiding Negativity

If you hope to avoid burnout, resist being poisoned by others' negativity. It is not uncommon to encounter human service professionals who are negative and critical of their work. Such negativity is often contagious. Catch yourself if you begin to engage in negative thinking. Use cognitive restructuring to replace a negative thought with a more positive one, so that you will not become overly pessimistic. Hope and positive efforts to make things better are the cornerstones of human service work.

■ CONCLUDING EXERCISES

1. Imagine yourself living the ideal balance between work and home life. How would your life be different? Considering your ideal, are there any changes that you could realistically implement right away?

2. Describe some of your greatest fears upon beginning your fieldwork experience. What do you tend to feel the most anxious about? Attempt to capitalize on your anxieties by recognizing them as normal feelings that help you prepare for a new situation. Reframe your anxiety as excitement about beginning a new endeavor. Now restate your fears, but use the word *excitement* in place of *fear*. Is there a different emotional outcome when you view your feelings this way?

3. Imagine that you have just been recommended for a post in curriculum development in a new human services department at a university where you have been a guest lecturer before. You have been asked to be creative in developing a course: The Challenges of Human Services. As a practitioner in the community, you are the most appropriate person to tackle this endeavor. Tell what this course would be about. In outline form, what would be the objectives, goals, and subject matter to be incorporated into the syllabus?

CHAPTER EIGHT

Interpersonal and Professional Relationships

AIM OF THE CHAPTER

Relationships serve a vital function in our lives. Even before birth, we need others for our physical and emotional survival. As we grow, so do our relationships with family members, and we develop additional relationships with extended family, teachers, friends, and other adults. As we mature into adults, our repertoire of relationships expands. In the process of becoming professional helpers, we are exposed to an array of individuals we need to relate to.

Working in human services, you must interact with many people at various levels. Some interactions will be limited in time, others ongoing—relationships with clients, colleagues, agency staff, supervisors, administrators, professionals at other agencies, and the community at large. Of course, working in this field inevitably affects your personal relationships with family and friends as well.

Thus far in the text, we have focused on relationships with clients and their families and with supervisors. In this chapter, we highlight aspects of additional interpersonal and professional relationships you will encounter in human service work. First, we will address personal relationships with significant others and family members because they are affected by our working in human services. We then discuss the interpersonal aspects of relationships with colleagues, who may also become friends. We also examine the professional aspects of relationships with administrators, agency staff, and the community. Our discussion will center on developing and maintaining healthy relationships, both interpersonally and professionally. Key points are taking time to develop healthy relationships, understanding and respecting different styles, the value of assertive communication, and successful conflict resolution. We begin by outlining the importance of assertiveness and conflict resolution skills.

EFFECTIVE COMMUNICATION AND CONFLICT RESOLUTION

Forming positive relationships is a foundation of the helping professions. Clients are more likely to benefit from the helping process if there is a strong working relationship with the helper. This also applies to relationships we have with other people at work and with people we relate to outside of the work environment. Striving to enhance these relationships is equally as important as building strong bonds with clients. In fact, our work with clients can be greatly strengthened if we have constructive relationships at home and with our professional colleagues. Among the benefits of developing constructive relationships are:

- Creation of a support system for venting and help with problem solving
- Increased cohesion, which decreases isolation and feelings of alienation

- Enhanced teamwork, allowing helpers to work together to benefit the client
- Reduced conflict, which counteracts tension and anxiety
- Increased creativity and a greater sense of accomplishment
- Camaraderie and closeness
- Increased ability to consider other options, viewpoints, and perceptions
- Expanded learning opportunities through increased contact with professionals within and outside your field
- Enhanced self-esteem and confidence in one's skills and ability to relate to others
- Heightened self-awareness
- Opportunities to develop skills in communication and conflict resolution

The benefit listed last—communicating effectively and resolving conflicts constructively— is all-important. Relationships can be impaired if communication is poor, indirect, or unclear or if conflicts are not dealt with effectively. To avoid this, we emphasize the need to develop skills in both assertive communication and conflict resolution.

Effective Communication Skills

In earlier chapters, we have discussed the importance of communication, attentive listening, and assertiveness with regard to making the most of your fieldwork experience, your interaction with supervisors, and your work helping others. Here the context is expanded: We will discuss effective communication and assertiveness in developing, maintaining, and enhancing personal and professional relationships.

Attentive listening, sometimes referred to as *active listening,* consists of endeavoring to truly understand what the other person is saying. This requires developing new ways of responding so that people you are listening to feel understood and open to further interaction. There is an important distinction between hearing and listening. *Hearing* "is a word used to describe the physiological sensory processes by which auditory sensations are received by the ears and transmitted to the brain," whereas *listening* "refers to a more psychological procedure involving interpreting and understanding the significance of the sensory experience" (Drakeford, 1967, p. 67).

In essence, we can *hear* what other people are saying without really listening or understanding them. Human service providers must be careful not to succumb to the tendency to hear without listening. In this regard, various skills can contribute to becoming an effective listener. These *attending skills* are maintaining a posture of involvement, using appropriate body language and eye contact, and encouraging an undistracting environment. One must attend with all the senses. According to Bolton, (1979, p. 23), "[a]ttending is giving your physical attention to another person. I sometimes refer to it as listening with the whole body. Attending is nonverbal communication that indicates that you are paying careful attention to the person who is talking." Listening carefully and attentively fosters an atmosphere of caring, concern, and respect, which are at the core of any positive relationship. Here are ten techniques that you may find helpful in sharpening your listening abilities.

> *Give your wholehearted attention to the person.* Temporarily put aside your own interests, preoccupations, and problems; concentrate 100 percent on what the other person is saying.
>
> *Really work at listening intently.* Your heartbeat will increase if you are really listening rather than pretending.
>
> *Show an interest in what the speaker is saying.* Maintain eye contact, demonstrate attentiveness by appropriate facial expressions, and block out any distractions.
>
> *Resist distractions.* Keep outside interruptions to a minimum.
>
> *Show patience.* Listen silently until the person has stopped speaking.
>
> *Keep an open mind.* Refrain from comparing your values and views. Do not assume that your way is better than others.
>
> *Listen for ideas.* Do not become defensive because this may negate the person's message.

Judge the content, not the delivery. Focus primarily on what the person in attempting to tell you, not how correct his or her grammar is or how eloquently he or she speaks.

Hold your fire. Spend time listening to the person rather than figuring out how to respond.

Learn to listen "between the lines." Use your eyes as well as your ears. You may learn more about what clients are saying by what they haven't said and by their body language.

Although these skills are basic, and you probably practice them already, they do merit review from time to time. Even the most seasoned helpers can continue to improve their listening skills.

SCENARIO

A coworker states that he is very upset at another worker, who is telling others how unprofessional he is for not following through on established procedures for referrals. You listen attentively, attempt to consider his viewpoint, and allow him to express his emotions. You try to understand the underlying reasons for his intense feelings. He states that he is so angry that he is considering writing a letter about the other worker and making copies for all the staff.

THINKING THINGS THROUGH

What are the attentive listening skills employed in the above scenario? How might you respond in an attentive manner to the coworker's last comment? Which techniques would you consider to be most useful?

Additional aspects of communication discussed by Sweitzer and King (2004) include understanding the concepts of clarity and openness. Clarity refers to "how well people in the organization listen to one another and how they are helped to do that" while openness applies to individuals who "generally say what they think about issues and about one another, an honest and clear way" (Sweitzer & King, 2004, p. 145). Every organization differs in how it goes about accomplishing this goal; some are more successful, yet others only partially so. Much depends on the specific individuals involved and their respective personalities and views, the management style and history of the organization, and the overall layout. It will be up to beginning professionals to observe and learn as they navigate their way.

Communication is sometimes hampered by barriers such as judging, sending solutions, and avoiding others' concerns (Bolton, 1979).

SCENARIO: Judging

A woman who works with children and adolescents at an agency receives a referral from the intake worker of the day. The worker discovers that the issues involved are not related to the child, but rather to the parents' marital problems. This worker is irate that the intake worker did not do a better job of screening. She feels that this referral has taken up precious time she could have otherwise used to see a new child intake. She attacks the referring helper's character and clinical experience.

Do you feel that this was appropriate? How could it have been handled more effectively?

SCENARIO: Sending Solutions

While working at a child guidance agency, Randy felt upset after an interaction with his supervisor. Upon arriving home, he talked about his anger with his wife. She responded by immediately suggesting how he could resolve this problem and avoid any further conflict. Randy experienced this as frustrating. He wanted to be heard and to have his anger validated, rather than to be given solutions.

How would you have handled this if you were in this situation? What would you do if you wanted your feelings to be heard and had not asked for solutions?

SCENARIO: Avoiding Others' Concern

As an intern, you have been invited to attend the general staff meetings each week. In a meeting, you attempt to express your concerns about confidentiality. You have noticed that staff often talk freely with each other about clients in the hallways, sometimes loud enough that others can overhear. When you voice your concern, the meeting abruptly shifts to another topic and never returns to address your concern. You are frustrated and annoyed at not being heard by the others.

What would you do if this happened to you at your fieldwork placement? How would you handle this if the meeting were still going on when you realized how you felt and what you wanted? What if the meeting had ended and your concerns were never discussed?

THINKING THINGS THROUGH

Give an example of one of these roadblocks to effective communication that you have experienced, personally or professionally. Tell how it came out, how you felt, and how you might handle this differently in the future.

As with attentive listening, assertiveness is necessary to effective communication. Assertiveness (discussed in detail in Chapter 3) involves verbal and nonverbal skills that help "maintain respect for oneself and others; satisfy one's needs; and, defend one's rights (without dominating, manipulating, abusing, or controlling others)" (Bolton, 1979, p. 12). Learning to be assertive in all areas of your life can benefit you personally. It may help keep you from alienating others, and it can allow you to build healthy relationships. Many of the relationship problems helpers face are linked to difficulties in being assertive in their personal and professional lives.

SCENARIO

As a female intern at a girl's club, you begin to feel uncomfortable with your female supervisor when you notice that she is paying special attention to you and making flirtatious comments. She openly shared her lesbian orientation with you at the beginning of your internship, and you had no discomfort then. In a supervision meeting, she invites you out on a date and says that she is attracted to you. You are taken aback by her announcement, but decide that this would be a good chance to practice your assertiveness skills. You say that while you are somewhat flattered by her compliments and invitation, you are uncomfortable with this personal invitation. You respectfully decline her offer and let her know that her personal comments cause you discomfort. You state that you enjoy having her as your supervisor and would like her to keep giving feedback about your performance. However, you make it clear that you are not interested in a personal relationship.

THINKING THINGS THROUGH

In the above situation, how would you have responded differently? If the genders or sexual orientation of the supervisor or intern were different, would the need for an assertive response be any different? Consider a situation in your fieldwork experience where there may be a need to become more assertive. Explain how you might respond to this fieldwork situation in an assertive manner.

An aspect of assertiveness is fully accepting your responsibility in relationships. We define *responsibility* in this context as openly accepting your role in the relationships you are part of. This includes a willingness to focus on your own thoughts, feelings, and actions, as opposed to blaming or attacking others for theirs. Maintaining a responsible stance facilitates positive relationships where you are respected and valued.

SCENARIO

A coworker tells you that she has noticed that you are often sarcastic with your supervisor. She also comments that you have often been late to meetings and to the group you and she co-lead. You are surprised to realize that the feedback she is giving you is true. You own your responsibility in this situation and tell her that you have been feeling frustrated with your supervisor. She is seldom available and she has not given you assignments that are meaningful. You recognize that your feelings have been coming out inappropriately in your behavior and that you need to find more constructive ways to cope with your frustrations.

THINKING THINGS THROUGH

How would you handle this situation? Draft an assertive statement that is respectful but still owns your own responsibility for indirectly expressing your frustration with the supervisory relationship.

Conflict Resolution Skills

Any two or more people in a relationship may experience conflict. No two people are exactly alike in their values, views, opinions, styles, and priorities. In the human service field, and also while in your fieldwork placement, you will no doubt encounter some sort of conflict. Conflict is commonly seen in a negative light—to be avoided or quickly eradicated. Surely, most of us would prefer that there be little conflict in our lives, but conflict is a natural occurrence and is neither negative nor positive in and of itself. Weeks (1992, p. 7) writes, "Conflict is an outgrowth of the diversity that characterizes our thoughts, our attitudes, our beliefs, our perceptions, and our social systems and structures. It is as much a part of our existence as is evolution. Each of us has influence and power over whether or not conflict becomes negative, and that influence and power is found in the way we handle it."

We agree that conflict, if handled properly, can actually be an opportunity for growth in certain circumstances. Managing conflict requires learning conflict resolution skills. These skills can enhance not only your work with clients but also relations with others, both interpersonally and professionally. When conflict is delayed or avoided, the lingering effects may result in strained or failed relationships.

Types of Conflict

Conflict arises for a variety of reasons. Bolton (1979, p. 233) cites three types of conflicts:

Conflict of Emotions. One person may feel saddened by a tragedy while another feels anger. Consider people's reactions to the bombing at Centennial Park in Atlanta in July 1996.

Value Conflict. One person may speak out for gay and lesbian rights, while another considers homosexuality a sin.

Conflict of Needs. A person might have a deep need to live ethically but also be badly in need of food and therefore steals from a supermarket.

Conflicts can arise from differences among individuals. Learning to accept such differences is much more constructive than expecting uniformity. Weeks (1992) identifies seven basic ingredients of conflict:

Diversity and Differences. Perceptions, needs, values, power, desires, goals, and opinions, and many other components of human interaction differ. At least one such difference occurs at the core of every conflict.

Needs. When ignored, obstructed, misunderstood, or incompatible, these needs can lead to conflict.

Perceptions. There may be conflicts over self-perceptions and perceptions of others or of the situation.

Power. There may be conflict over who has power and how it is used.

Values and Principles.

Feelings and Emotions. Most conflicts involve some kind of emotional reaction.

Internal Conflicts. Conflicts within a person, if not dealt with effectively, may have negative effects on relationships with others.

SCENARIO

In a group for adolescents that you are co-leading at your internship placement, the teens ask for someone to explain birth control methods to them. You are comfortable with this and call a Planned Parenthood agency to send a representative. However, your coworker expresses his discomfort and opposition to the idea. Not only would this promote sexuality and acting out in the adolescents, he says, but agency policy prohibits discussing these topics with the teens. You find yourself unsure how to handle this conflict.

THINKING THINGS THROUGH

Which of the seven types of conflict described above fits most closely with the conflict in the scenario? How would you handle this situation? Discuss which of the seven types of conflict you have not considered before and think of an example of such a conflict.

Approaches to Conflict

The primary goal of most approaches to conflict is to work toward a resolution in a cooperative and appropriate manner. Self-awareness is the first step toward developing sound conflict resolution skills. By looking inside ourselves, we can better understand our own part in the conflict and then work on corrective action.

One approach, developed by Bolton (1979) is appropriate for human services and incorporates the basic elements of effective communication. This three-step process aims to help people argue and disagree in a constructive manner that is "systematic, noninjurious, [and] growth-producing" (p. 218):

Treat the other person with respect.
Listen until you "experience the other side."
State your views, needs, and feelings.

THINKING THINGS THROUGH

Consider a recent conflict you have experienced, either at home or at the agency. Apply this three-step process to determine its effectiveness and your ability to resolve the conflict collaboratively.

Conflict may engender feelings of fear and anger that must also be dealt with. A first step in most conflict resolution is to recognize the existence of such feelings. Accept that fear and anger are human emotions we all experience. It is not these emotions that are negative but rather the way they are sometimes inappropriately expressed. Understanding the underlying reasons for the feelings can be helpful in determining how best to respond. Engaging in healthy *anger management* is also essential. This involves learning to work with our angry feelings in ways that lead toward appropriate resolution. Weeks (1992) lists parts of this process:

Learning to be in charge of your own anger
Learning how to receive anger
Learning how to express one's anger constructively

There is no way to completely rid ourselves of our angry feelings; they are a part of who we are. Potter-Efron and Potter-Efron (1995) state that anger is a natural human emotion necessary for survival. Anger may alert us that something wrong deserves our attention. Part of being in charge of our own anger is learning to identify our feelings when we first experience them. Expressing these feelings early on, before they become unmanageable, enables you to stay calm rather than allowing your feelings to be in charge of you.

SCENARIO

You are an intern at an agency that brings together abused children and their parents for eventual reunification as a family. Your role is to observe the parents while they interact with the children. In your first case, you have read the notes that describe the history of the parents' physical abuse of the child. This particular case upsets you, and you ask your supervisor to assign you a different case to start. The next day, you learn that this case has in fact been assigned to you. You become angry at your supervisor—more so when you see the parents in the waiting room. Before meeting with the family, you hastily approach your supervisor in her office. You lash out in frustration and say that she is "insensitive, cold, and uncaring." You then refuse to see the family and walk out of the office, leaving no opportunity to discuss the situation further.

THINKING THINGS THROUGH

What do you think was really going on in the above scenario? Why might you have experienced difficulties with this case? What would be the consequences of handling your anger in this manner? Can you think of a more constructive way to handle this situation? How could you have expressed your feelings in a respectful yet assertive way? Think of a situation in a professional relationship where anger was dealt with inappropriately and one where it was handled more effectively. What do you see as the consequences of each of these?

Clearly, both assertive communication and conflict resolution are fundamental to healthy relationships. Learn to apply these concepts. Opportunities to practice are plentiful in daily interactions with others. Without a doubt, most of our relationships could be strengthened were we to adopt these skills in our daily lives. In organizations, "If there are not some structured, accepted, and supported ways of dealing with conflict in place, more informal and often less effective methods will be used" (Sweitzer & King, 2004, p. 145). In the next sections, we consider how various types of relationships can be enhanced, beginning with personal relationships.

PERSONAL RELATIONSHIPS

Human service workers, in helping clients enhance their overall functioning and make necessary changes in their lives, are likely to witness human suffering, frightening abuse, and heart-rending injustices. Involvement in such situations is bound to affect the personal lives of helpers in some fashion.

The Effect of Helping Roles on Personal Life

Leaving behind your feelings and thoughts from a stressful day may prove to be very difficult. A certain client may have profoundly affected you; your work may have triggered an unresolved issue of yours; or you may have felt overwhelmed, ineffective, and powerless to help. Often, you may not even be conscious of the impact until you are on your way home or later when you are alone or interacting with family or friends.

SCENARIO

When Lupe was working with spinal cord injury patients, she would often find herself upset when driving home. She felt sadness for the innocent victims of accidental injuries and random crimes. She was often unable to let go of her feelings once she arrived home and finally had to grapple with the possibility that she could not work with this disabled population without allowing it to affect her personal life. After several years of working in this rehabilitation setting, she chose to change to mental health. There, she felt more able to empower clients and engage in interventions that lead to positive change.

Many helpers find that they are not suited to working with certain populations. Others can adapt to their initial difficulties, learning to work with clients effectively and successfully managing their emotions as well. Fieldwork can allow you to determine which client populations set off strong feelings and which you can more comfortably work with. Fieldwork also permits you to experiment with ways to manage your feelings in relation to your work with certain client populations.

THINKING THINGS THROUGH

In your work as a helper, have you experienced a situation that has left you upset or overwhelmed? Have you been able to manage your emotions effectively in your fieldwork experience? If so, how did you accomplish this? If not, how did your emotions about your work affect your personal life?

In contrast, the helper's role can be most rewarding. The prospect of helping another can strengthen one's self-esteem, which can contribute to improved interpersonal and professional relationships. Many helpers also report that working with disadvantaged people made them appreciate the positives in their own lives. Additionally, working with clients can temporarily take the focus off the helper's own emotional struggles.

SCENARIO

A helper comes to her fieldwork placement after a fight with her boyfriend. She had given in to her anger and engaged in some yelling. In her placement, she co-leads a women's assertiveness training group at a battered women's shelter. This particular session was especially moving as many of the women were able to express their hurt feelings and practice assertiveness. The intern realizes that she did not express her own feelings assertively with her boyfriend, but rather attacked him verbally for his negative behavior. She understands her own dynamics more clearly and decides to express herself assertively to her boyfriend the next opportunity she gets.

THINKING THINGS THROUGH

Discuss a situation where your work with clients has enhanced your view of yourself and your interpersonal relationships with family, friends, and colleagues.

Most of us find that our work as helpers affects us in many ways. With experience and practice, it becomes easier to separate the emotions involved in our work from those we experience in our personal lives.

Impact on the Family

One's family can be greatly influenced, negatively and positively, by one's work in human services. Let us look at how several familial relationships are affected.

Intimate Relationships

The relationship with a significant other—a spouse, a boyfriend or girlfriend, or a life mate—are often most affected. This is due in part to the deep level of trust that exists in such relationships, fostering an open environment where sharing can readily occur.

Relationships where mates work in different professions There are as many helpers who prefer to have mates in a different career as there are who wish that their mates could share in their work. Advantages and disadvantages can be identified for either. Among the benefits of having a mate who works in a different field are:

- Sharing different aspects of their work
- Ease of putting work aside in favor of the personal relationship
- Supporting one another in career ventures
- Financial stability—if one partner's work is facing budget cuts, the other's may be unaffected
- Teaching the significant other about human behavior and the helping process
- Helping the significant other to become more self-aware and introspective

SCENARIO

Your wife is employed as an accountant in a small family business, while you are a human service student at a major university. You are feeling somewhat down one evening after your fieldwork because your clients' histories reflect so much abuse. You are open to expressing your feelings, but you would rather go out and have some fun this evening as a way of coping. Although your wife is open to hearing your feelings, she accommodates your desire, and the two of you go out and have fun. Because she has not been working with such heavy issues, she is able to help you refocus and enjoy some of life's pleasures.

In contrast to these benefits, there are also drawbacks to having a mate in a different profession:

- The significant other may be unwilling or unable to relate to issues about clients.
- He or she may find it extremely difficult to deal with the emotional issues presented by the helper.
- Because the mate is not involved in the profession, he or she may be uncomfortable discussing issues he or she does not understand.
- The helper may inadvertently displace work-related feelings onto the mate.
- The helper may talk too much about work and clients, leaving little time for communication on other issues.
- Helpers approaching the end stage of burnout may completely withdraw from everyone, including their mates.

In any of these instances, stress can be high. In extreme cases, such stress can contribute to erosion of the relationship. Helpers may reach the point where they are more involved in work issues than in the personal relationship, or communication may break down. Each partner may then live a separate life, oblivious of the other person.

SCENARIO

You are married to a partner who works as a carpenter and cabinetmaker. You are a social work intern at a hospital, working with drug and alcohol clients. Today you were working with a client who was extremely depressed and suicidal. You come home stressed, worried, and sad about the client. When you try to share your feelings with your partner, he tells you to cheer up. "The client would have to be stupid to kill himself." You feel hurt by his remark and his attitude. Instead of talking about it with him, you become angry and withdraw from any further contact during the evening.

THINKING THINGS THROUGH

What effects could this type of interaction have on the relationship over time? What could you do in the preceding scenario to deal with your feelings more effectively? Discuss how your interpersonal relationships have been affected thus far by your involvement in helping.

Relationships where mates are both in the helping profession As a married couple working in the same field, we have found that our relationship can strengthen our bond but also poses additional issues and dilemmas. Because our work is a common bond between us, we automatically share, consult, and support one another in our respective positions. We even work together sometimes—for example, in writing this text, teaching together, and co-leading groups. This is most often an enjoyable and productive experience, in which we complement one another and learn about each other's abilities and styles. On some occasions, however, working together has created additional stress in our relationship. We have different styles and, in some instances, markedly different viewpoints. Furthermore, although our similar careers have created healthy competition, this has sometimes led to feelings of jealousy that have been challenging to cope with. In addition to our joint ventures at work, we have developed joint relationships with each other's colleagues. What has perhaps helped us the most has been open communication, a willingness to continue working on issues as they arise, and the ability to maintain our separate identities. In this regard, we have endeavored to cultivate separate friends, projects, and interests.

SCENARIO

Randy and Lupe, who are both licensed clinical social workers, are supportive and helpful to each other when discussing their work at the end of the day. On several occasions when one has been upset or perplexed by a situation, the other has provided constructive professional feedback, along with support and validation. The fact that they both understand the intricacies of the field has enhanced their ability to help one another.

Unlike us, some couples who both work in human services find it to be problematic. They may maintain a healthy relationship by intentionally separating their careers from their relationship. Once again, it is all-important to keep open channels of communication and continually strive to maintain a balance between work and personal life.

SCENARIO

Two partners are both human service workers in an adolescent inpatient unit at a psychiatric hospital. The work is sometimes so exhausting and intense that they would rather not discuss their work at home. They have made a conscious decision to keep their work at the office, not to bring the day home with them. This appears to work well for them. They say that they are satisfied with keeping the focus on their home life more than on their work.

THINKING THINGS THROUGH

Can you imagine how such an arrangement might have adverse effects on their relationship at some point? Can you give examples where two individuals from the same profession might have difficulties related to their work? How might their working together enhance the relationship?

Impact on Family Members

We define *family* here to mean the family of origin or a chosen family, in which a couple forms a family of their own. In both instances, working in human services can certainly affect family relationships. Family members can be sensitized to the plight of society's most vulnerable members and can vicariously enjoy the satisfaction of helping. On the other hand, the stress on the helper can lead to misunderstandings, disagreements, and heated conflicts with family members. Some instances are listed here:

The helper is unable to separate work from home life.

The helper and family members hold vastly different viewpoints regarding social services and the purpose of helping.

The helper has high expectations for family members' understanding of human nature and empathy.

Cultural values may separate the helper from the family. For example, a helper's parents may value medicine or law more highly than the helping professions. Or they may not be comfortable with the view that openly expressing your feelings is best. Or a female helper may be criticized because her proper role is thought to be staying at home and taking care of a family.

The communication within a family system may be closed.

When children have a parent who works in the helping profession, they may learn about the value of open and direct communication, caring, and empathy, and feel proud of their parent's role in helping others. However, some children may feel abandoned or resentful because the parent dedicates so much time to helping others at the expense of the children's needs. Helpers must be cognizant of the total impact of their work on their families. By engaging in attentive listening and careful observation, we can begin to notice if our role as helpers is having a positive or negative influence on the significant people in our lives. If the signs are negative, we must be willing to make necessary adjustments or changes so that our work will be less of a liability and more of an asset.

SCENARIO

Having a deeper understanding of child development and the importance of separating the personhood of children from their behavior, a human service worker may be able to apply discipline with good results. Applying consequences, punishments, and reinforcements in a way that does not attack the child's self-esteem allows the child to grow up with a stronger sense of self, a deeper knowledge of personal feelings, and the ability to express himself or herself assertively. (This is not to say that only human service workers can be effective parents, or that all human service workers are able to apply their knowledge.)

SCENARIO

You are an intern working in the children's unit at a hospital. You often engage in play therapy; children seem responsive to this approach. Reaching home, your 8-year-old daughter begs you to play Barbie with her. "Please, Mom, you promised me." You hesitate. You have engaged in play with dolls all day and are really not in the mood. Nonetheless, you feel bad. Now that your own daughter wants a little of your time, you don't have any energy left. You make an effort and actually find the experience rewarding.

What if you had resisted the play? How might your daughter have felt? How about you?

THINKING THINGS THROUGH

What positive influences has working in the helping professions had on you and your family? Specifically, describe an incident that highlights some insight or knowledge that you were able to put to use in your own relationships.

■ RELATIONSHIPS WITH COLLEAGUES

Working with colleagues in a professional capacity requires sensitivity, open communication, and conflict resolution skills. We may well spend more waking hours at work than home, so our colleagues are often like an extended family. Therefore, we should strive to maintain positive relationships with our colleagues, who can offer support and professional consultation.

Benefits

Ideally our colleagues would be close friends who knew us well, enjoyed quality time with us, and shared interests with us. However, only some colleagues are friends; others are primarily coworkers; and with others you would prefer to have no involvement at all. McMahon (1996, p. 181) makes an important distinction: We choose to spend time with friends and usually find these relationships fulfilling, but at work we generally have no choice who our colleagues are.

Regardless of what we might prefer, it is important to take time to develop quality relationships with coworkers. This enhances our work environment and allows for increased cohesion and collaboration. The agency can function more smoothly in its mission to help clients, and workers feel better overall about their roles.

SCENARIO

Randy has always endeavored to create close relationships with others at work. For example, he is friends with a fellow clinician who also has a family and similar interests. They can discuss feelings in both their personal and their professional lives. Often, they consult on cases, discuss relationships with coworkers, and give one another support and feedback. Relationships like this add immensely to enjoyment and energy at work.

THINKING THINGS THROUGH

In your current agency, which individuals do you feel closest to? Is there a potential for a closer working relationship? Write down at least two risks you may be willing to take to help these relationships become sources of support for you. Attempt to follow through, keeping in mind that the individual may not respond favorably or become a source of support for you.

Creating a Supportive and Cohesive Environment

Although you may not choose those with whom you work, you do have control over how you choose to interact with them. Encouraging an environment of mutual support can reduce stress, help prevent burnout, and build a cohesive atmosphere.

SCENARIO

As an intern working in a child guidance clinic, you ask to be included in meetings where the professionals discuss cases. An undergraduate intern has not made such a request before. You also have asked one coworker, whom you particularly admire, to give you consultation on the children you will be working with. Although he is not your supervisor, you feel an affinity for this worker and have enjoyed your interactions with him. You have discussed this with your supervisor, who agrees that this would be a great learning experience. Because of your motivation, dedication, and obvious interest in learning, the agency agrees for the first time to allow an undergraduate intern into the case consultation meetings. Later, when you are about to leave your internship, the staff takes you to lunch, where they share how your presence will be missed. They have valued your ability to work as a team member.

THINKING THINGS THROUGH

Discuss a situation where teamwork and cohesion in your agency contributed to a more positive outcome for the clients that the agency serves. Then tell about another situation where a lack of teamwork and cohesion played a part in a negative client outcome.

Creating Alliances and Friendships

This may not always be desirable, but it can provide you with added support and make your workday more enjoyable. Humor and playfulness in the workplace have been shown to release tension.

Nonetheless, in some instances, friendships in the workplace can lead to conflict. Differences in values and preferences involving work-related or personal issues may cloud the relationship. Healthy competition can result in jealousy. It is therefore essential to maintain healthy boundaries with friends in the workplace. It is also important to be open to engaging in conflict resolution early on in such relationships to help prevent future problems.

SCENARIO

One of your friends at the agency is suddenly promoted to supervisor of your unit. Your relationship has been close, and you have shared a lot with her, but this seems to change when she assumes her new position. When you talk with her, she becomes "official" and seems to discourage your friendship. Your feelings are hurt; you are disappointed and frustrated.

THINKING THINGS THROUGH

How would you handle the preceding scenario? Can you think of any negative experiences you have had with friends, either at work or personally? Describe the situation, your feelings, and ways you could have handled the situation differently.

Problems with Colleagues

Difficulties with colleagues create added strain in the workplace and can be difficult to cope with. Problems may result from power issues, personality conflicts, divergent views of the helping process, or agency policies and practices. When a cohesive and collaborative tone has been set, such problems are more easily resolved. When the staff is fragmented or split, however, and strong leadership is absent, problems are likely to mount. Working can then be emotionally draining, and helpers may feel disillusioned and apathetic about their work. Clients may also be affected indirectly: Helpers become preoccupied by their interpersonal conflicts with colleagues and neglect to focus on their clients.

Understanding, respecting, and learning to work with colleagues' different styles and personalities is certainly a challenge. Although you may not agree with them, it is important to respect them and accept their differences. An accepting atmosphere makes conflict less likely—and more easily resolved when it does occur. The principles of self-awareness, knowledge, and skill (discussed in connection with diversity in Chapter 5) can fruitfully be applied to our interactions with colleagues.

When colleagues are not amenable to our efforts to engage in appropriate conflict resolution, other options must be explored. The particular individuals may be difficult to deal with or may have personality problems that make them unresponsive to problem-solving techniques, compromise, or attempts at negotiation. In such cases, it may be more important to focus on your own needs and maintain effective involvement with your clients. McMahon (1996) suggests focusing on taking care of ourselves and owning responsibility for what we do and who we are (p. 188). We must not allow such negative relationships to discourage us; rather we can learn from them. Strive to maintain a positive attitude. If assertiveness skills do not work, explore ways to keep working with the difficult individuals without depending on any support, encouragement, or feedback from them. Scott (1990) advocates protecting yourself emotionally by learning to let go and not personalizing such conflicts. Others must not be allowed to hold you back from becoming the best helper you can be. Avoid negative people who won't or can't change, do not waste energy on skeptics who will not listen, and do not allow others to manipulate you. Remain aware of your skills and capabilities, and keep in mind your priorities. Seek support and feedback from those you trust and feel comfortable with.

SCENARIO

At the teen shelter where you are an intern, there is a coworker who is very critical of others. She appears to be burned out, but she continues to work in the agency despite her apathy and negativity. Other workers ignore her or distance themselves from her when she becomes particularly critical. You are finding this to be a challenge. You have never encountered an individual like her before and you want her to like you.

THINKING THINGS THROUGH

How would you handle this situation? Would you use the same strategy as the other workers and withdraw from her? In what other ways might you deal with this situation? How might you deal with your desire to be liked by her?

Assertive Communication and Conflict Resolution with Colleagues

Applying these skills, helpers can continue to feel positive about their position and the agency. Workers who hesitate to speak up or are reluctant to resolve conflicts often lose enjoyment in their work, become less effective, or ultimately withdraw (either emotionally or literally) from their positions as helpers.

Assertiveness with colleagues is not always easy. You may fear reprisals or increased conflict, and some colleagues would rather that you remain passive. Assertiveness is nonetheless likely to produce the best overall outcome. Listed here are several suggestions for practicing assertiveness skills at work (Alberti & Emmons, 1995):

Improve your decision-making skills by practicing assertive action.
Negotiate more effectively by being assertive.
Deal with angry coworkers assertively.
Learn to say no, so that you do not lose yourself. Most organizations will take all you'll give, and expect more. If coworkers know you can't say no, they won't be reluctant to make unreasonable requests.
Be persistent. Plant a seed and nurture it.
Be assertive about goal setting. Ignore manipulation. If requests are unreasonable, say so. In response to reasonable requests you may not be able to handle, explain why.

Conflict resolution, problem solving, and negotiation with colleagues can make your workplace more enjoyable. You can also direct more of your energies to working directly with clients and their difficulties. Alberti and Emmons (1995, p. 180) suggest that conflict resolution "can be aided through creating an atmosphere which allows—even encourages—constructive disagreement. As different ideas are expressed, it becomes possible to work out compromises which build on the strengths of everyone's contributions."

SCENARIO

As a Spanish-speaking intern in a school you are being asked to do more and more: home visits, interpreting, translating, and meeting with students who speak mostly Spanish. You are being asked to put in more and more hours without pay. At first, you are reluctant to say anything, but it becomes increasingly overwhelming. You decide to use assertiveness skills to take care of yourself and to avoid becoming burned out. You tell your supervisor that although you have learned a lot from the experiences, you are frustrated by the large time commitment. You are finding little time for your personal life and schoolwork. You let your supervisor know that you will have to cut back to working only the ten hours that are required in your internship agreement. After you have finished, you notice your hands are shaking, and you have mixed feelings.

THINKING THINGS THROUGH

How would you have handled the above situation? How else could this situation have been resolved in an assertive manner? How do you think the supervisor will respond? Has something like this ever happened to you?

RELATIONSHIPS WITH ADMINISTRATORS

In some agencies, contact with administration is minimal. Administrators may be mostly absent and delegate many of their responsibilities to assistants, department chiefs, or supervisory staff. This is common in large organizations, where administrators may not be integrally involved with line staff. In smaller agencies, administrators may be closely involved with staff. In either case, concentrate on developing positive relationships with administrators.

Administrator's styles can be passive, distant, and withdrawn; assertive, communicative, and open; or aggressive, hostile, and noncaring. Try to understanding the personality traits and management styles of the administrators of your agency. You do not have to agree with or even like your administrators, but you can learn to accept and respect them. Respond empathetically; try to put yourself in their shoes. Be sensitive to the arduous tasks that management often faces.

Conflict and the Role of Administrators

Conflict is inevitable in any setting, including human service agencies. What is most important is how this conflict is managed. Sources of conflict within organizations are outlined in the list that follows (adapted from Alberti & Emmons, 1995; Bolton, 1979; Scott, 1990; Weiner, 1990):

- The manner in which an organization is structured may partly determine conflict. The more centralized and bureaucratic an agency is, the more likely it is that there will be conflict. Conversely, when agencies are flexible and encourage open communication, conflict is less likely, because of the prevalence of collaboration and teamwork.
- The personal style of the administrator is also influential. This often dictates the style of management and the overall effectiveness of the manager. A supportive,

nondefensive leader is more effective than one who is distant, unsupportive, and overly defensive.

- The administrator's personality and style can also contribute to the success of conflict resolution within the agency. Some administrators encourage a certain amount of conflict, which can be healthy and beneficial within a spirit of collaboration. When this is the case, there is a greater chance for sound conflict resolution.
- Agency policies and practices, set or reinforced by administrators, can have an impact on the amount of conflict generated within the agency. Policies and procedures that are overly structured, as in many rigid bureaucratic systems, tend to engender conflict. On the other hand, if agency rules and regulations are realistic and upheld by all, the risk of divisive conflict is significantly lessened.
- Policies and practices sometimes change rapidly, or new policies are introduced abruptly. The chance of conflict is then heightened. In contrast, success is more likely if such changes are presented in a way that allows staff involvement and negotiation.
- Having established methods for grievances, applied equally to all, is also important. When management shows favoritism and interferes with the process, conflict is certain to arise.
- Finally, effective administrators are usually well trained in assertiveness, conflict resolution, and management skills. They are also open to learning from their mistakes and receptive to accepting constructive feedback. When such skills are undeveloped or used ineffectively, there is a greater likelihood of breakdown in communication, trust, cohesion, and follow-through.

SCENARIO

Imagine that you are working in a large human service organization that is threatened with layoffs and budget cuts. The administrator is rarely available; when he is, he usually seems annoyed if you approach him with any questions about the budget and your concerns for the future. At meetings, he is vague. He does not seem to have a particular style of management and is mostly erratic and unpredictable. Staff morale is deteriorating, and you are also beginning to feel apathetic. You attempt to understand the stress he must be under, yet you feel that he could be more communicative and respectful of the staff members and their concerns.

THINKING THINGS THROUGH

In the scenario above, identify possible sources of conflict. How do you feel about the administration? What might you do to help yourself feel more grounded? Do you have any ideas about how you might approach management?

Assertiveness and Conflict Resolution in Relationships with Administrators

This can be difficult, given the power differential, the threat of being viewed negatively, and the risk of losing one's job. Nonetheless, it is sometimes a necessity. Assertiveness skills may reduce some of the anxiety you are likely to feel. You should stand up for your rights in an unaggressive manner that respects the other person. If your feelings, thoughts, or requests are expressed assertively to administrators, there are several advantages:

There is a greater likelihood that you will be heard and possibly have some or all of your needs met.

Administrators may develop a new respect for you because of your assertiveness skills.

You will feel less frustrated because you will not have to keep all your feelings bottled up inside.

You are likely to feel better about yourself and more confident in continuing to be assertive.

Conflict resolution skills may also prove helpful when you experience conflict with an administrator. Approaching the administrator directly and attempting to resolve a conflict in an open, direct, and mature manner is far better than allowing the matter to simmer to an explosive state.

SCENARIO

Your work in a small agency has become increasingly stressful. You are the only full-time worker besides the agency manager. All the other staff are part-time caseworkers, undergraduate and graduate interns, and volunteers. When you first accepted the position, you were thrilled. No other intern had been hired there full time after graduation. You were promised supervision for clinical hours and the opportunity to engage in some special outreach projects. It has been a year and neither of these promises has been fulfilled. In fact, it seems as though they never will be. You have become the informal supervisor for all staff, due to the manager's frequent business trips, meetings, and absences. You have hesitated to talk directly to her because she appears overwhelmed herself. You planned to take her out to lunch to discuss your concerns. Unfortunately, your plans aren't realized. You finally lose it when you learn that your long-awaited vacation has been canceled because there is no available back-up. You march into your supervisor's office, where you blow up. Both she and you are astonished at all the anger you have been storing inside.

THINKING THINGS THROUGH

In the scenario, what happened to your plans to assertively discuss your feelings with the manager? What are some of the issues? How might you have responded more assertively? How could you have dealt more effectively with your vacation being canceled?

■ RELATIONSHIPS WITH AGENCY STAFF

These include relationships with secretaries, receptionists, transcribers, housekeepers, and security personnel. Their work makes your workday go as smoothly as possible. For instance, a secretary may help in scheduling clients, making necessary calls to clients or other agencies, and informing you of deadlines and policy changes. Receptionists sometimes relate directly to clients and can therefore provide you with invaluable information about their behavior in the waiting area and can alert you to possible problems. Housekeeping staff keep your work environment clean, allowing you to feel comfortable and present a positive image to clients and professionals alike. Lastly, security personnel guarantee safety and give you protection and reassurance in potentially dangerous situations.

Because ancillary staff members have a direct impact on your daily life as a helper, it is advantageous to cultivate positive relationships with them. Accept them, and be respectful of them and their role within the agency. Their jobs are often just as stressful as ours, if not more so. Thus, we must be sensitive to the stressors and limitations of their jobs. Take time to work with agency staff, not against them. This has clear benefits and is in alignment with the human service principle of working with others.

SCENARIO

Imagine that you exchange jobs with the department secretary for one day. Your day would be filled with phone calls, competition for your attention by social workers wanting you to schedule clients, requests by physicians to order charts and blood tests, complaints by clients who are put on hold or who cannot get answers to their questions, and demands by your supervisor not to fall behind. To top it off, when you finally leave that evening—your desk a mess and your head throbbing—you open your paycheck to find only a third of the pay you ordinarily receive. Are you ready to return for another day?

THINKING THINGS THROUGH

Consider all the ancillary staff with whom you work. List the stressors they typically face in a day. Compare those you experience. Can you recognize that their jobs hold their own special kinds of stress?

Take these relationships seriously and respect all staff members, regardless of their status in the agency hierarchy. This sets a solid foundation from which assertiveness and conflict resolution skills can be effectively implemented. Deal with problems or issues early on, when they are small and resolvable. Also, regard conflict in a positive light—an opportunity for growth or change. We have found this attitude helpful when working with staff.

PROFESSIONAL RELATIONSHIPS IN THE COMMUNITY

The role of professional helpers may involve helping more than clients, families, or small groups. Working within the community is another area where effective helping can take place. We can work on a "macro" level to alleviate broad social problems in the community. In our work within the community, we can adopt many roles. Some apply directly to work within an agency setting, and others relate to community change and mobilization. Homan (2004, p. 24) views a community as a client: "Just like an individual or a family, a community has resources and limitations. Communities have established coping mechanisms to deal with problems." Netting, Kettner, and McMurty (1993, p. 4), who emphasize "macro" practice, enumerate community activities that human service practitioners can perform:

Negotiating and bargaining with diverse groups
Encouraging consumer participation in decision making
Establishing and carrying out interagency agreements
Conducting needs assessments
Acting as advocates for client needs in various community systems
Participating in policy-related activities: working as coalition builders, lobbyists, and overseers of legislative developments

THINKING THINGS THROUGH

Are you already fulfilling or partaking in any of these roles or activities? Which others might you consider, in your current position? How might you proceed?

Collaborating with Professionals from Other Disciplines

There will be many occasions when you will work with specialists from other disciplines. This may include relationships with physicians, nurses, occupational or physical therapists, speech pathologists, attorneys, court representatives, law enforcement officers, educators, and school administrators. Developing a positive working relationship with such individuals is an asset. Working together as a team can help clients receive more comprehensive treatment. Important information and referrals can result from such interactions, as well as additional support. Once a solid foundation is set, such relationships should be maintained and reinforced. To be effective in our work with clients, we must strive to build reliable, open, and mutually respectful relationships with professionals from other disciplines.

SCENARIO

You are a new intern working in a family service agency where there are two main departments: crisis relief, and child and family counseling. You are assigned to work in both. Newly arrived in the community, you are overwhelmed by the community resources you need to know about to be effective in either department. In the crisis unit, you must know about shelters, food drives, charities, bus passes, application procedures for aid and disability, and a myriad of other details. In working in the counseling section, you need to know about parenting classes, senior service resources, local schools, health centers, and legal aid. In making these community contacts for your clients, you begin to develop a relationship with key people. As time passes, you realize how much easier it is to seek resources once you have established relationships with other professionals in the community.

THINKING THINGS THROUGH

Describe how working with other professionals in the community has facilitated your job. What would happen if these contacts did not exist, or if they were negative in nature?

Representing the Agency and the Profession

As a representative of both the agency and of the helping professions in general, we must present a professional image. This facilitates negotiation and collaboration with others, and makes us proud of ourselves. There are three major goals:

Increasing community awareness and sensitivity to social services issues
Increasing community awareness of available human service resources
Acting as an advocate for clients by connecting them with community resources and by promoting their involvement in community activities that can enhance their functioning

You can develop goodwill for your agency when you strive to build strong and positive community contacts for yourself. Similarly, strong community relations can assist you in developing a positive and sound reputation that can serve to enhance your career for years to come.

SCENARIO

When Lupe worked in a community mental health center, she presented in-services to various community agencies. She conducted workshops on behavioral problems of children for staff at a Head Start program, led training on crisis intervention for career counselors at a community college, presented an overview of mental health services to the Latino community on a Spanish TV station, and made a proposal for a senior home-bound program to community leaders. Lupe always prepared carefully for these talks, so she presented a professional image and gave community groups the information they needed.

THINKING THINGS THROUGH

Consider ways you might work within the community as an agency representative. What might your role be? What are some of your specific job duties? Which would you consider to be the most important?

Community Change

Our work with clients often sensitizes us to the problems that exist within the community. Because of our commitment to helping, we may want to expand our role to become an agent of community change. A community change agent must be open to community involvement. He or she will come to understand the community, its pressing problems, and the resources available to support change. Homan (2004, pp. 26–27) describes five targets for community change:

Neighborhood empowerment
Community problem-solving systems
Developing community support systems
Community education
Developing a broad-based community organization

These targets highlight the fact that community work may help clients on a larger scale. For community change to be effective, both the community change agent and the community "must believe in its own ability to change and must take responsibility for its actions or inactions" (Homan, 2004, p. 25).

SCENARIO

You are interning at a homeless shelter that is minimally funded and can provide food and shelter for women and children only for a limited time. No social programs have ever been established at the shelter, and the skeleton staff is overwhelmed with keeping up with the cooking and feeding. You realize that many needs go unmet in this shelter. You embark on a project to involve the community in the work of the shelter and to educate people about the diversity of the homeless population. As a result, a public health nurse begins to come weekly to examine ill children and pregnant mothers and to give immunizations. Volunteers (often retired teachers) come regularly to read to the children and teach them basic skills. A representative from a county drug and alcohol program holds weekly AA/NA meetings there. A local church plans to hold a fund-raising fair for the homeless. You contact local colleges and universities to enlist their help in providing interns who can begin to develop a program for the women and children.

THINKING THINGS THROUGH

Based on the preceding scenario, what changes might result from soliciting involvement from various community agencies? How else can community change be effected to provide the homeless with more of the services they need to maintain an independent living situation? How can the community be educated about this population and its needs?

Consultation in the Community

Various definitions of consultation exist, with no universal agreement yet in existence. In 1970, Caplan (1970, p. 19) defined consultation as "a process between two professional persons, the consultant, who is a specialist, and the consultee, who invokes the consultant's help in regard to current work problems." According to Gladding (2004), one of the best definitions of consultation was given by Kurpius in 1978; he views consultation as "a voluntary relationship between a professional helper and help-needing, individual, group, or social unit in which the consultant is providing help to the clients in defining and solving a work-related problem or potential problem with a client or client system" (p. 335). For our purposes, *consultation* means connecting with others in the community to share information about client needs or general issues. For effective consultation, one must have positive interpersonal skills, a genuine caring for clients, and respect for the knowledge and resources available in the community. Persistence, patience, and planning are all crucial if consulting is to be productive and mutually beneficial.

SCENARIO

You are working with children who have various behavioral problems. Many of the referrals are from schoolteachers and principals and involve children suspended for behavioral problems; some have also been diagnosed with attention deficit hyperactivity disorder. You recognize a need to consult with school personnel about their perceptions. You are careful to have clients sign a release form that allows you talk to others in the community about them. You are able to get an overall picture of your clients outside of the office setting, which helps greatly in your assessment and treatment plan.

THINKING THINGS THROUGH

List some types of consultation you have been involved in. Describe a specific client about whom you consulted with an outside resource. Was it helpful? Can you think of any other areas you may need to consult in?

Cultivating Positive Relationships in the Community

To build relationships in the community, you must seek opportunities to take an active part in the affairs of the community. Participate in community activities, present a genuine, professional, and respectful image, and represent your agency with integrity and sincerity. We support Homan's (2004, p. 51) contention that "working relationships require communication and trust." Without the support of the community, effective collaboration or community change will not be accomplished. Successful outcomes are more likely when everyone works together. "An organization characterized by the ability to develop and maintain healthy relationships will have more available resources, an increased ability to weather unexpected storms and challenge setbacks, and will be able to use more of its energy for focused action. It will be more powerful as a result" (Homan, 2004, pp. 51–52).

THINKING THINGS THROUGH

Consider ways in which you might be able to cultivate positive relationships within your community. Who would you contact? What might you do? What might the outcome be?

Assertiveness and conflict resolution skills are necessary here also. Conflicts with the community often result from community perceptions of a client's behavior, needs, or motivational level, or from the scarcity of a needed resource. It is critical to act assertively and be an advocate for the client whenever possible. Some conflicts may have no apparent solution; conflict resolution skills may be needed to reach a compromise or resolution.

CONSIDERING CULTURE

Communication and trust are two of the most fundamental elements in any relationship. In working within communities, the challenge is to maintain open and trusting communication between all its members. As communities become more diverse, this challenge becomes increasingly that much greater. Homan (2004, pp. 13–14) highlights this point in the following statement:

> Cultural differences provide a unique broadening of perspectives and a variety of skills. They also provide challenges to communication and trust. Ignorance of different ways of experiencing the world and expressing that experience can lead to mistakes in understanding and limiting access ability. . . . Effective change agents understand that issues of trust and possibilities for miscommunication are present in any relationship, and that these are more likely when people from various backgrounds are working together. The degree of difficulty is heightened when conflict and power imbalances have characterized the history of the groups' interactions. Thus, being culturally competent is imperative for change agents.

Clearly, being culturally aware and competent remain essential skills for all practitioners, whether they work with individuals, families, groups, or communities. It is important to respond appropriately and with sensitivity to the various cultural environments we live in. Working in a collaborative fashion helps strengthen our relationships and ensures the realization of our mutual goals.

In closing, we hope that you will come to value the many relationships in your life. It is especially important for helpers to develop and maintain positive relationships with everyone they interact with, both personally and professionally.

■ CONCLUDING EXERCISES

1. It is difficult to take the time, in our busy lives, to create meaningful friendships. Or is this simply an excuse to remain isolated and withdrawn? Imagine taking a risk to be more open with one person in your fieldwork placement. Who would you choose if you wanted to create one more supportive relationship in your life?

2. How do you see your personal life affecting your professional roles at the moment you read this? Would you consider your personal relationships a liability or asset at this time? How so? What feelings are you aware of as you consider this question? What changes do you think you might need to make for your personal relationships to become a more positive influence in your life?

3. You work in a large human service agency that has recently experienced personal and professional conflicts among staff. As a result, a new staff development coordinator has been hired. She gets to know you and likes your style, people skills, and approach to conflicts. You admire her as well, and you are honored when she asks you to participate in a staff development seminar entitled "How to Build Positive Relationships at Home and at Work." She asks that you prepare a handout listing potential problems in personal and professional relationships, followed by constructive methods for preventing, minimizing, or eliminating them. The ultimate goal is to help staff become more aware of their relationships and work toward improving their skills in maintaining or developing healthy relationships. Please address these issues in a simulated handout.

CHAPTER NINE

Keeping Alive in Agency Settings

▓ AIM OF THE CHAPTER

In this chapter, we will discuss some factors that human service workers should bear in mind when working in agencies. The advantages of agency work are covered, along with major sources of stress. Many of you already appreciate the beauty of working with people and the introspective aspects of the helping professions, but you must also cope with overwhelming feelings, the demands of paperwork and documentation, and self-imposed limitations.

The last section focuses on stress and burnout. These are endemic in human services—major pitfalls threatening every helping professional. Examining common causes and useful preventive measures can help you learn how to manage stress to avoid burnout.

▓ ADVANTAGES OF WORKING IN AGENCIES

Working in an agency, you are at the heart of helping. The satisfaction of knowing that you are instrumental in assisting others can make up for all the difficulties inherent in working in this field. Knowing that you have made a positive impact on another person's life is often a reward in and of itself. Most of us entered this profession because we wanted to help people in need; seeing our efforts bear fruit reinforces our desire to continue working with others.

Interaction with Others and Opportunity for Teamwork

Another positive aspect of working in agencies is the constant interaction with clients and coworkers. If you are a "people person" and enjoy being at the center of activity, then agency work is for you. The collaboration involved in working as a team can be beneficial to you as well as the client. You can share your fears and frustrations; gain validation, support, and feedback; and learn from other team members. Working in a supportive environment can also create cohesion and a sense of unity. For clients, teamwork leads to the delivery of more complete and higher-quality services. By comparison, working alone can be isolating and overwhelming, because the helper lacks support or the opportunity to consult with other professionals.

Sharing the helping role with others can also involve creative ways of working together; this can keep you motivated, inquisitive, and emotionally and intellectually keen. Just realizing that you are not alone can be reassuring. In many agencies, you may be able to choose, to some extent, how much or how little involvement you wish to have with others. In some agencies, staff tend to keep to themselves, while in others they constantly work together. The nature of your job, your personal style, and those of your coworkers all help determine how much teamwork there will be. If you take an active stance, as we have advocated throughout, you are likely to create a working environment that suits your needs well.

SCENARIO

When Lupe worked as part of a multidisciplinary team in a rehabilitation facility, she benefited from the teamwork there. She was able to share the overwhelming feelings of loss engendered by the patients' spinal cord injuries. Also, she felt valued because each member of the team had an important role in providing patient care.

SCENARIO

At a mental health center where Lupe once worked, the staff all shared responsibility for the day treatment clients. One day, the staff had to deal with crisis after crisis. At lunchtime, they ordered pizza and, for dessert, German chocolate cake. All was devoured in no time. They decided that on stressful days from that point on, that would be their lunch menu. Sharing cases, frustrations, and lunch seemed to help them survive the difficult aspects of working in mental health.

SCENARIO

Randy feels good about the teamwork he and his graduate assistants engage in at the university. They meet regularly to discuss ways to improve programs and to give one another support and validation. Their collaborative efforts enhance their work and benefit the program as well. When someone on the team is missing it is noticeable. Each member adds to the overall functioning of the unit. . . . Randy remembers being the only one on the unit at work on spring break; this is when he became aware of how important teamwork was. Although he was able to do his work well, he missed the feedback and support that his fellow coworkers usually contribute.

Agency work is seldom boring or repetitive. Each new day presents a new situation, helping to keep workers stimulated and eager to learn. The vast variety involved in working with others is intriguing, although sometimes also overwhelming. Learn to appreciate the uniqueness of your work, and you will cope better with the difficult aspects of agency life.

SCENARIO

Randy's work as a community outreach worker was intriguing and certainly never dull. His job entailed visiting community-based agencies, where he met interesting people and worked with many different client populations. No day was ever the same; each was an adventure.

A great advantage of agency life is the opportunity to develop strong relationships with coworkers. Many will share some of your values and interests.

SCENARIO

Both Lupe and Randy maintain special bonds with coworkers from years past. Having shared a special time together in a helping capacity makes for pleasant memories. Some of their closest friends are those they met at work. They often engage in reunion dinners and frequently communicate with past and current coworkers via phone, email, voice mail, etc. The support provided by these colleagues along with their shared beliefs and values is a primary reason why their relationships often extend outside of work.

An Introspective Profession

In the helping professions, we can learn about ourselves and develop a greater capacity for growth and self-awareness. Of course, for some people, this introspection can be a stressful aspect of helping.

Dealing with clients, coworkers, and administrative personnel will no doubt challenge you to look at yourself. Such self-analysis is often difficult. Helping others involves close engagement with other people. If you enter this field without a willingness to examine your own feelings and beliefs, it will ultimately hinder your ability to help others. Introspection can encourage growth in both you and your clients.

SCENARIO

You are newly employed at a children's hospital. In the pediatric oncology unit, you will work with children who have been diagnosed with a terminal illness. Early in your orientation, you become exceptionally sad. Your supervisor helps you realize that repressed feelings about the loss of your own brother, who died of cystic fibrosis years ago, have been awakened by your work with terminally ill children. Your supervisor is supportive. She suggests that you might seek counseling, to work on the unresolved issues that are beginning to surface. Initially reluctant, you ultimately agree that this is a good idea.

SCENARIO

Working in a senior day-care center, you encounter clients who are vulnerable and in need of human services. As a new staff member, you are eager to work with this often neglected population. During the first few weeks at the center, you become aware that working with some clients brings up your own issues of loss and fear of your own mortality. In addition, you wonder if you can cope with all the emotional and physical pain you witness.

THINKING THINGS THROUGH

During your fieldwork training, have you been aware of the importance of introspection in working as a helper? Explain how you first became aware of this, and how you initially reacted. Consider how you might react now. Share what you feel is advantageous about being open to examining your deeper feelings and thoughts.

We believe that the benefits of introspection far outweigh the negatives. You can learn about yourself, recognize your strengths and limitations, and develop a greater capacity for growth and self-actualization. Next we present some examples of how being open to self-exploration can be beneficial to you, both personally and professionally.

SCENARIO: Learning About Yourself

As they become involved in their fieldwork experience, many students are surprised to learn more about themselves than they had anticipated. Some choose this field to help others but learn that they can also work through some of their own issues in the process.

SCENARIO: Recognizing Your Strengths and Limitations

Many of our students are pleased to learn that some of their major strengths involve being empathetic, sensitive, and caring. The limitations they may discover include being impatient, having a tendency to rescue others, and feeling overwhelmed. By learning about their limitations, they can set appropriate goals for self-improvement.

SCENARIO: Greater Potential for Growth

In placements where students are encouraged to look at themselves, they are often amazed by the learning and growth they experience by being assertive and responsible for their learning.

■ MAJOR SOURCES OF STRESS

Encountering stress is inevitable, especially when working within the human service field. Corey and Corey (2003) categorize sources of work-related stress for helping professionals into individual and environmental. Individual or internal stressors refer to the attitudes and personal characteristics of the helper, while environmental or external stressors include "physical aspects of the work setting or the structure of the position" as well as "organizational politics, restrictions imposed by insurance companies, an abundance of paperwork, and cutbacks in programs, . . . quality of working relationships with colleagues and stressful client behaviors." Such behaviors may involve suicidal threats or attempts, severe depression, anger and hostility, apathy, and lack of motivation or cooperation. Here we will address some of these issues in more detail.

Coping with Overwhelming Feelings

Because of the suffering of many clients, working with people in a helping capacity can be an overwhelming experience. The intensity of human emotion and dysfunction encountered in working with abuse, suicide, homicide, depression, domestic violence, pain, and loss arouse emotional reactions you may not have anticipated. You may not feel prepared to deal with the impact of such pain and trauma. Having such feelings is quite normal and appropriate. If you were so out of touch with your own feelings that you were not affected by clients' suffering, you would most likely not be an effective helper.

Maintaining a balance between your emotions and your focus on the necessary interventions is the key to helping others without becoming too overwhelmed. Learn to recognize when you are becoming too emotionally involved. Talk openly about your feelings with your supervisor, colleagues, field instructors, and fellow students. You will probably gain support, validation, and a new perspective. Perhaps you will discover that your feelings are connected to some unfinished business or unresolved issues in your own life. Such an awareness could be the impetus for you to work through these issues and enhance your own coping skills. However, if you discover that the feelings are so overwhelming to you that discussing them with others is not sufficient, we encourage you to seek counseling.

Some beginning helpers may regard seeking counseling as a sign of weakness, but we prefer to view it as an opportunity for growth. Indeed, willingness to seek help can be considered a strength. If you are able to recognize and express your deeper emotions, you will more likely succeed in eliciting the same from your clients.

SCENARIO

One of our students in a fieldwork class took an internship working with abused children because this had long been an interest of hers. She knew that she had been abused as a child, but she was only partially aware of the aftereffects of the abuse. As she encountered the feelings of the children, she began to feel overwhelmed by memories of her own abuse. In class, she courageously talked about her countertransference. This process of venting helped her realize that she needed counseling. During the course of the semester, she did pursue counseling. It was instrumental in helping her be more effective in working with abused children at the agency. At the end of the semester, she was unsure whether she wanted to continue working with this population. Nonetheless, she felt positive about her work at the agency and appreciative that she had accepted the opportunity to work on her own unresolved issues.

What unfinished issues can you identify for yourself at this point in your training? Have you learned anything new about yourself through your field placement?

Working with Difficult Clients

Clients can be resistant, demanding, and potentially abusive. This can also take a toll on you emotionally, physically, and intellectually. The demand for increased productivity can compound such exasperation, particularly for the beginning helper. Because of continued reductions in funding, many agencies must provide the same services (or more) with fewer staff. This creates stress, above all for those who are not accustomed to the high level of stress inherent in agency work. Learning to set limits, prioritize, reframe, and seek support are crucial to your emotional survival. Later in this chapter, we will elaborate on preventive measures you can take to remain energized.

Hostile Coworkers

Having to work with hostile or difficult staff members can also be quite stressful and frustrating. Because you may have no choice about who your coworkers will be, you may have to work with people you do not particularly like. Learning to accept others while maintaining your own identity is important. Building and maintaining support systems within and outside of one's work environment can often be the most helpful response.

Self-Imposed Limitations

In the helping professions, you constantly face self-imposed limitations. Being afraid to take risks is a common human experience; sometimes safety and comfort win out over trying something new. However, such a style may lead to stagnation, burnout, and ineffective helping. Taking the risk of stretching yourself professionally requires courage and willingness to endure the resulting anxieties. Allow yourself to make mistakes, and accept your limitations. This is necessary if you are to be successful in any new venture.

SCENARIO

Joel, a second-year graduate student, was unsure what area of specialization to choose for his last year of school. He had worked with families in a drug rehabilitation center during his first internship and enjoyed it. What interested him most was working with the few AIDS patients there. He was struck by the added pressures these families face. Joel contemplated working in an agency dedicated to working with AIDS patients and their families. However, whenever he thought of calling the director to ask for an interview, he would become anxious and fearful. "What if I can't help them?" "What if I can't handle the constant pain and suffering and death?" "What if people assume I am gay and also infected with HIV?" "What if I'm exposed in some way to the AIDS virus?" Finally, when the deadline for selecting his internship was near, he decided to consult with his previous field instructor, who helped him recognize that his aspirations to work with this population far outweighed his fear. He was encouraged to confront his fear and challenge himself to develop his expertise. Once he made his decision, he felt relieved and excited by the new opportunities to help in such a critical area of human services.

Can you think of a professional area you may wish to explore further?

Documentation and Paperwork

Proper documentation and paperwork are not self-imposed, but rather dictated by the agency. Helpers often detest these aspects of agency life the most. If not managed properly, the time spent documenting and keeping records can certainly detract from providing direct services. But understanding the value of documentation may put this seemingly unnecessary work in perspective.

Besides being required by the agency, documentation is also done for ethical and legal reasons. Many people who enter human services have as their primary motivation to work with people. Such eager, enthusiastic beginning professionals are sometimes disappointed to learn that a portion of their time will be taken up with tasks such as writing evaluations, assessments, progress notes, and incidence reports; making and recording telephone contacts; and filling out forms. According to Moore and Hiebert-Murphy (1994, p. 433), "[p]aperwork and writing reports is an integral part of your work as a helping professional. Although we all, from time to time, feel as though we are about to be drowned in a sea of paper, keeping accurate records and developing your skills as a writer are important."

Others view paperwork as a matter of accountability. Filip, Schene, and McDaniel (1991, p. 81) state, "It is important to recognize that the problems of your work are not tangible except as they are documented." It may be difficult to prove your high level of responsibility and commitment to your clients without paperwork to verify your work and interventions.

Also, in today's volatile economic and political times, many agencies are forced to justify their existence. This can generate even more paperwork; workers are often required to document their clients' need for continued services. For many helpers, this is a time-consuming and stressful endeavor.

Certainly, not every ounce of paperwork is a necessity. However, documentation is a necessary component of professional helping in any agency. The subsections that follow discuss various reasons why documentation is important.

Record Keeping

Just for the sake of keeping a record of your client interactions, it is important to document some aspects of your work. Depending on the nature of the funding and the particular agency, the type of documentation will vary. Some agencies may require a brief note about the service provided; others may want a detailed account of the treatment rendered.

SCENARIO

In many agencies, human service workers are asked to see many clients daily. Beginning professionals may be overwhelmed and unable to recall the people they helped and how they intervened. By documenting their interaction with clients in a timely way, this problem is lessened. Workers can turn to their records for reference points to jog their memory and help them plan appropriately for future interventions.

For Statistical Purposes

Many agencies' outside funding depends on the number of clients served and the outcome of the service provided. Statistics on both quantity and quality are therefore needed, which requires documentation.

SCENARIO

While you are working at a community center where programs for seniors are being developed, you are asked to keep track of the age groups seen (55+, 65+, and 75+). The results may have a direct impact on program development and funding. Also, the clients who speak a different language need to be counted, to ascertain whether hiring bilingual staff is warranted. The clients' presenting problems must also be documented because this will affect the focus of specific programs.

For Ethical Purposes

Ethically, documentation is necessary to protect both the helper and the client. Keeping records helps ensure continuity of care.

Strive to learn more about clients' past treatment history and involvement with other human service agencies. Requesting previous records is advantageous in understanding the client fully, just as in the case of a person seeking medical treatment, where the physician inquires about past treatment and medications. Documentation is important. If no accurate records were kept, the helper would be at a disadvantage in the current helping relationship.

SCENARIO

You are newly employed as a crisis worker at a family agency. Part of your job will be to assess suicide risk in clients and determine whether a referral for ongoing counseling or a psychiatric evaluation is warranted. One of your first clients is well known to the agency; she is frequently in crisis. She is now depressed about side effects of her new medication. She doesn't know the name or dosage, however. She wants to meet with her previous crisis worker but does not recall his name. You look into her chart for specifics, and no pertinent data is there. The last chart note is vague. There is no mention of a medication or information on interventions used. You feel overwhelmed anyway, and even more frustrated when necessary information is not well documented. As a result of this case, you learn to be more thorough in your own documentation.

How would you feel if you were in this situation? How might you handle the situation?

Documentation for Legal Reasons

Helping professionals must be aware of the legal implications of their work. In this context, documentation is extremely important. Accurate records validating your actions can be helpful if there are ever any questions regarding your interventions.

SCENARIO

If a helper's conduct were ever questioned, accurate records would provide validation for his or her actions. Similarly, a client's claims about his situation could be verified by a review of the documentation in the records kept by the helper. A colleague of ours was called into court to testify in a case where a client had sought treatment in another agency. He was asked to show that he had followed up on the medications given to the client by a staff psychiatrist. Because he had kept ongoing records, he could refer to them to show that he had asked the client about his medication.

Moore and Hiebert-Murphy (1995) cite the following reasons why paperwork and report writing are important:

Records and reports are your first line of defense (or vulnerability).

Your notes, summaries, or reports are likely to be read by others, so they should be concise and accurate and reflect your professionalism.

Paperwork is often a way for professionals to communicate. This is true within and outside of the agency.

By regularly documenting your work, you will increase your understanding of the clients you see. Charting behavioral changes will allow you to tailor the treatment appropriately. Often, changes in the client situation necessitate re-evaluation of your initial plan, and keeping up with paperwork will facilitate this process.

As a competent, responsible, and ethical professional, you must maintain up-to-date, comprehensive records that document your work with your clients.

Maintaining a Positive Attitude Toward Paperwork

In agency settings, we have found that it is most helpful to keep a positive attitude when it comes to doing paperwork. If you accept this requirement as a part of your work and recognize its value, rather than resisting and complaining about it, life will be much easier. Learn to make paperwork your own—tailor it to meet your own needs when possible. This will help you develop a positive attitude.

SCENARIO

Randy used to spend hours irritated about the amount of paperwork he was expected to do. He decided to try cognitive restructuring with regard to his thoughts about the paperwork. "I will get to the work when I get to it" replaced his former thoughts, "When will I ever get all that work done" and "It's terrible that I have so much paperwork." This alone was enough to help him relax. He also found that, relaxing more, he was actually able to complete the paperwork faster and more efficiently.

Have you experienced any of these frustrations when having to do paperwork?

THINKING THINGS THROUGH

Give an example about your experience with documentation. Describe how you felt and how you responded to the demands of paperwork. Can you see any ways in which you could use cognitive restructuring to decrease stress in your agency work?

Helpful Tips on Keeping Up with Paperwork

Keep a positive mindset; minimize resistance.

Prioritize your work daily; make an effort to eliminate unnecessary paperwork.

Keep an organized system to maximize efficiency and accuracy.

Learn from previous efforts; use previous reports, evaluations, and memos as guides for current work.

Schedule a block of time, even a half-hour at a time, to complete paperwork.

Develop a routine that is appropriate to your personal style and work demands.

Consider giving the agency advice regarding streamlining certain aspects of paperwork, such as with repetitive or long-winded forms.

Recall that paperwork has many positive aspects. It does not have to be "just a waste of time." Recognize its value and tailor it in such a way that it will be useful and beneficial in your work with clients.

Budget Cuts, Realignment, and Layoffs

Today, society's commitment to human services is in question. In the present turbulent economy and the ongoing political and philosophical debates regarding the need for social programs, human service programs are being cut or eliminated. Working in this climate of uncertainty can be very stressful for all concerned. Not knowing whether you will have a position next week affects you emotionally as well as professionally. Your mind may be less on your clients' needs than on your own worries about the future.

SCENARIO

You were hired by a large human services organization whose funding was strong. Within several years, however, budget cuts began to loom over the agency. As the last one hired, you would be the first one to go. You really enjoyed your work and the contact with colleagues, but you became demoralized at the constant threat of layoffs. No one could give you any clear idea about the future of your job. You began to look for other opportunities and finally left this position due to the uncertainty of your future.

How would you feel if this were to happen to you? Would you leave? Or would you stay and weather the storm?

Morale suffers in agencies when there are cuts, realignment of staff, layoffs, or the threat of any of these. Frustration, resentment, and anger mount. Staff may feel caught in a trap: "We need our jobs, so we will have to tolerate these changes." The message they receive loud and clear from their employers is "You should be glad that you have a job."

SCENARIO

Morale is especially low at a human service agency where there have been years of budget cuts, staff realignments, and threats of further cuts and layoffs. This has taken a toll on the morale of the entire organization. Workers resent the system, and coworkers and clients are caught in the middle. The staff members seem irritable, and the cohesion that once existed is now shattered.

THINKING THINGS THROUGH

Give an example of an agency or individual you know that has been directly affected by budget cuts, reorganization, downsizing, or layoffs. Explain how this has affected those involved on an emotional level. How would you cope in such a situation?

To summarize, stress can be experienced on various levels. Sweitzer and King (2004) identify these stressors as issues tied to the type of work involved; issues with the various people one interacts with, including clients, supervisors, coworkers, and peers; issues with oneself and the impact of change, plus expanding one's knowledge and one's skills; and, issues tied to the organizational system.

Working under these conditions can "turn you off" to working in agencies. We hope you can find a way to balance the negatives with the positives and strive to maintain a positive outlook toward helping.

▪ BURNOUT

Preventing burnout and remaining professionally vital are among the greatest challenges of all for the human service worker. No one is immune to burnout. If you believe you are impervious to stress, you are among those who are most susceptible to the phenomenon of burnout. Gilliland and James (1990, p. 572) highlight the nature of burnout and the importance of early recognition:

> Burnout is not simply a sympathy-eliciting term to use when one has had a hard day at the office. It is a very real malady. . . . It is prevalent in human services professions because of the kinds of clients, environments, working conditions, and resultant stresses that are operational there. No one particular individual is more prone to experience burnout than another. However, by their very nature, most human service workers tend to be highly committed to their profession, and such commitment is a necessary precursor to burnout. . . . All human service workers, public or private, tend to be unable to identify the problem when it is their own. No one is immune to its effects.

Burnout is increasingly recognized as a serious problem for professionals in the human services (Zastrow, 1995). Human service professionals are considered especially vulnerable: "The kinds of people who select human services as a career are typically concerned with individuals and their problems, attuned to human suffering, and anxious to make a difference," which makes them vulnerable to "working too many extra hours or to putting their clients' needs ahead of their own" (Sweitser & King, 2004, p. 127). Farber (1983, p. 1) states that burnout "is a complex individual-societal phenomenon that affects the welfare of not only millions of human service workers but also tens of millions of those workers' clients." Paine (1982) and Maslack (1982) emphasize the tremendous impact stress and burnout have on individuals, society, and organizations. Impairments in productivity, interpersonal relationships, and emotional and physical health are likely outcomes of job stress. The costs are certain to be high and cannot be ignored. In sum, burnout must be taken seriously by all human service workers; it has "major ramifications for both individuals and institutions" (Maslack, 1982, p. 39).

Ultimately, the key to minimizing burnout is learning to deal with stress more effectively. Being able to recognize the signs of stress early can help you take appropriate steps to reduce stress.

Burnout Defined

A myriad of definitions of burnout exist. For example, burnout has been described as a state of physical, emotional, and mental exhaustion that results from constant or repeated emotional pressure associated with an intense, long-term involvement with people. It is characterized by feelings of helplessness and hopelessness and by a negative view of self and negative attitude toward work, life, and people (Corey & Corey, 2003).

Burnout is the state of becoming emotionally or physically drained to the point that one cannot perform functions meaningfully. It is the single most common personal consequence of working as a helper (Gladding, 2004).

Many of the symptoms of depression are also considered signs of burnout. The *Diagnostic and Statistical Manual of Mental Disorders* lists depressed mood, hopelessness, irritability, markedly diminished interest of pleasure in some or all activities, apathy, amotivation, fatigue or loss of energy, and feelings of worthlessness as symptoms of clinical depression (American Psychiatric Association, 2000). Burnout is also similar to chronic depression in that it is typically a gradual process. An individual's feelings, attitudes, and behaviors slowly deteriorate to a state where fatigue, negativity, and pessimism prevail.

Compassion Fatigue and Vicarious Traumatization

Slater and Spetalnick (2001) have found that many human service professionals who work with the most difficult and sad sides of humanity are apt to develop what is called "compassion fatigue/vicarious traumatization." These workers need to be aware early on that their work can lead to serious repercussions if not managed appropriately. Sometimes simply hearing clients share their life stories or watching their painful reactions can be enough to greatly impact a helper who must deal with these situations daily. Some become forever traumatized, others may react by becoming detached and appear "uncaring," while still others are affected but are able to stay focused.

Symptoms of Burnout

Numerous authors have identified the symptoms of this phenomenon. The following are core symptoms.

Behavioral

"A marked departure from the worker's former behavioral norm" is typical (Forney, Wallace-Schutzman, & Wiggers, 1982). An individual may become easily irritated, act irresponsibly, show little empathy, or simply withdraw and remain in isolation.

SCENARIO

Al is a well-respected clinic manager who suddenly becomes abrupt in his manner and increasingly demanding of support staff. When confronted with his change in behavior, he acknowledges feeling overwhelmed by the upcoming audit and budget renewals.

Physical

Loss of energy, depletion, debilitation, and fatigue are symptoms of burnout. Poor eating habits, disturbed sleep, little exercise, and a range of chronic health problems may accompany this fatigued state (Maslack, 1982; Pines & Aronson, 1988).

SCENARIO

Jane, an outreach worker for a community services center, is usually energetic and enthusiastic. Lately, she has felt tired even upon awakening in the morning. She drags herself to work and finds it difficult sometimes even to complete the day. She has been leaving work early, complaining of stomach problems. Jane's supervisor is concerned: she has previously been the most productive of the entire staff. Her recent physical complaints are indeed worrisome.

Interpersonal/Relational/Social

Relationships at all levels are affected. An individual who is burned out may withdraw or relate to others in an apathetic or negative manner (Corey & Corey, 2003; Watkins, 1985).

SCENARIO

Ruben is so overwhelmed by increased work pressures that he rarely attends staff meetings or goes out with coworkers, as he used to do. At home, his family has noticed a similar pattern. Ruben tends to spend more time alone in his study and is less receptive to taking part in family gatherings. His apathetic attitude is now interfering with both work and family, and his increased irritability is beginning to spill out inappropriately in all areas of his life.

Emotional/Attitudinal

Burnout is characterized by a significant loss of commitment and moral purpose in one's work (Cherniss & Krantz, 1983; Pines & Aronson, 1988). Symptoms of depression may appear: hopelessness, apathy, lack of motivation, and low self-esteem. Anxiety may also be manifested as excessive worry, preoccupation, fear, and irritability.

SCENARIO

Tina has worked at the crisis hotline for five years now. She has been promoted several times because of her excellent clinical and managerial skills. However, recently she has seemed less interested in her clients—more skeptical, sarcastic, and even cynical. On occasion, she has talked about her identification with suicidal callers: "I can sure relate. Life is not worth it sometimes!"

Intellectual

Burnout involves a gradual loss of curiosity and desire to learn. In extreme cases, this can be seen as a loss in mental abilities (Corey & Corey, 2003). A negative attitude can also accompany such a decrease in intellectual functioning.

SCENARIO

Brad is well known for his enthusiasm and dedication to program development at his agency. He is responsible for several new programs being funded at the community center where he is assistant director. However, lately he seems uninterested and lacks the initiative to pursue any new ventures. A sure sign of change appeared when he put together a grant proposal haphazardly. When he was told that the grant did not get approved, he seemed not to care.

Spiritual

Burned-out individuals may suffer a decline in their spiritual beliefs. The meaning of life is questioned, and cynicism may develop.

SCENARIO

Luanne is a well-respected Christian counselor. Recently, she has become less involved in the church functions she used to thrive on. She has even become skeptical about her spiritual beliefs and critical of people whom she used to stand up for. Family, friends, and coworkers are stunned by her character change and worried about how this will affect her work.

THINKING THINGS THROUGH

Describe how the symptoms of burnout have applied to you. Give an example of each symptom you have experienced, and how you dealt with it. Develop a plan of action to deal with your list of symptoms, should they become more serious.

Kottler (1986, pp. 92–95) offers examples of burnout symptoms:

Celebration when clients cancel
Daydreaming and escapist fantasies
Tendency to abuse alcohol or drugs or to indulge in any type of addictive behavior
Loss of excitement and spontaneity in work with clients
Getting behind in paperwork
Deteriorating social life
Reluctance to explore the causes and cures of the condition

Stages of Burnout

We are all susceptible to some degree of burnout. Burnout is most often characterized by certain developmental stages; knowing them can help in early recognition. As cited by Edelwich and Brodsky (1982, pp. 133–154) the stages are as follows:

Enthusiasm. Early in their careers, helpers are extremely motivated to set the world on fire. This stage is filled with high hopes, unrealistic expectations, and a zealous desire to make a difference.

Stagnation. When personal, financial, and career needs are not being met, stagnation may begin to set in. Opportunities for promotion may be absent or seem impossible to attain. The person may feel stuck, trapped, or stagnant and therefore begin to lose the will to work in a helping capacity.

Frustration. This stage is reached when helpers begin to question their effectiveness, value, and impact. They may feel so overwhelmed by obstacles that they lose sight of their original motivations for becoming helpers. A domino effect occurs in some organizations: burnout can be contagious.

Apathy. When a person feels indifferent, unresponsive, and listless, he or she is suffering from the most serious stage of burnout: apathy. This is truly a crisis situation. There is disequilibrium, immobility, denial, and loss of objectivity, which make the condition that much more difficult to treat.

THINKING THINGS THROUGH

Consider your own experience with burnout or that of someone you have known. Use the preceding list to determine at what level the burnout seemed to be. What recommendations do you have to treat this problem? If you were an administrator responsible for designing a program to prevent burnout at your agency, what would you recommend?

Causes of Burnout

Burnout is a reaction to high levels of stress (Zastrow, 1985). Stress contributes to many physical illnesses, as well as to emotional and behavioral difficulties (McQuade &

BOX 9.1 Stress Questionnaire—How do you define stress?

In Chapter 7, challenges and stressors were divided into three categories: personal, organizational, and environmental. The questions here will follow the same general format. We recognize that there will be overlap. Make your best attempt at dividing your responses into these categories, but do not stress yourself out by trying to make your responses fit exactly.

What intrapersonal and interpersonal stressors affect you in your everyday work setting?

What organizational stressors affect you in your agency?

What environmental stressors affect you in your agency?

How do you cope with stress? Name the positive and negative coping skills you use.

What do you feel is the impact of continued stress on your life, both personally and professionally?

Aikman, 1974; Greenberg, 1980). The Stress Questionnaire presented in Box 9.1 may give you added insight and allow you to personalize this issue.

Zastrow (1985, p. 281) defines stress as "the emotional and physiological reactions to stressors. A stressor is a demand, situation, or circumstance which disrupts a person's equilibrium and initiates the stress response." Corey and Corey (2003) separate stressors into the categories *individual* (illness, allergies, obsessiveness), *interpersonal* (death of a close friend, relationship conflicts), and *organizational* (deadlines, pressures to produce, bureaucratic constraints). Others label these categories slightly differently—for example, *biogenic* (coffee, amphetamines, exercise, and electrical shock), *psychosocial* (relationship based), and *environmental* (organizational, societal, or man-made stressors).

THINKING THINGS THROUGH

Based on your responses to the stress questionnaire above, can you identify any additional stressors?

Farber (1983, p. 6) identifies cornerstones of burnout, which are also helpful to know if burnout is to be understood and prevented.

Role ambiguity Responsibilities, methods, goals, status, and accountability are unclear or not at all specified.

SCENARIO

You begin a new job as a supervisor at a drug rehabilitation agency. You were told that you would supervise graduate students who are working on obtaining their clinical hours. Several weeks into the job, you inadvertently see your name on a seminar schedule and as the designated supervisor of several staff. When you inquire about these additions to your job description, you are told, "Oh yes, didn't we tell you about these? They are also a part of your job."

How would you feel? How would you react?

Role conflict The demands on the employee are incompatible, inappropriate, and inconsistent with his or her values and ethics.

SCENARIO

You work as mental health counselor for a community mental health center. You have a fairly large caseload. Your clients are all chronically mentally ill and need ongoing management. One day at a staff meeting, you are told to terminate your clients as soon as possible because there will be a reorganization. Clients will have to reapply, at which point their eligibility will be determined. You are in shock. You find this completely unethical and not in line with your values of helping people in need.

How might you cope with this role conflict?

Role overload The demands placed on a worker have become too great.

SCENARIO

Your work role has gradually been expanded to the point where you now feel totally overwhelmed by your job. You attempt to discuss this with your supervisor, but she is also overwhelmed and feels at a loss to change the current situation. Even more demoralizing is the indirect message you get: Your job is at stake. There are many people who would love to have your job, so you feel forced to comply, despite the unfairness of your caseload.

How would you handle this situation?

Inconsequentiality Helpers have a feeling that no matter how hard they work, the outcome means little in terms of recognition, accomplishment, appreciation, or success.

SCENARIO

You work at a juvenile detention center, where more and more teens are admitted each day. The staff/client ratio is so high that you feel you are ineffective in your work. Instead of helping to rehabilitate the young people, you find it possible only to police them and keep up with the required record keeping. Also, you are criticized when one little error is found and rarely commended for diligence.

How would you feel?

THINKING THINGS THROUGH

Thinking about the four cornerstones of burnout, can you cite any examples that apply to you in your agency? Elaborate and offer intervention strategies.

Recognizing the causes of stress can help you to determine what influences are at work in your own life. You can plan strategies to cope more effectively, thereby reducing the risk of burnout. The following factors are likely to increase a person's stress level:

Negative perception habits
Family pressures
Environmental demands
Work problems
Rescuing mentality
Responsibility without authority, resources, or gratitude
Negative coping skills
Undeveloped stress management skills
Personal tragedy

THINKING THINGS THROUGH

Can you identify any of these factors in your own life? Tell what impact they have had on your stress level with clients, coworkers, friends, and family.

The Impact of Stress and Burnout

The impact of chronic or excessive stress can be great. When an individual reaches the last stages of burnout, fragmentation, impaired professionalism, and erosion of the helper–client relationship are likely to occur. Agencies can also be affected and become increasingly dysfunctional as a result. Most critical, however, is the damage such stress can do to the helper and client. When helpers become incapable of fulfilling their roles adequately, services to their clients are sure to suffer. Regardless of your professional position, prevention and early intervention must remain a high priority.

Prevention and Intervention

Stress and burnout are natural risks faced by human service workers employed in agencies. Awareness of the symptoms and causes can help you develop appropriate interventions to prevent or cope with stressors. There are as many suggestions on how to deal with burnout as there are definitions.

Myers, Sweeney, and Witmer (2000) are credited with developing a model called the "Wheel of Wellness," which helps human service professionals to assess their level of health in the areas of spirituality, work and leisure, friendship, love, and self-direction; these areas are further divided into 12 subtasks in Box 9.2. Each task is ranked between 1 to 5 (5 being the area needing more focus) to serves as a guide for helpers to use in developing an appropriate plan to combat burnout.

From a review of the literature, and from our professional experience, we now list some interventions that we believe are appropriate and practical. Our list is by no means exhaustive; be on the lookout for new and improved ways to reduce your stress and avoid the end stages of burnout. Remember that a person's reaction to stress is an individual matter and that stress-reduction techniques alone are hardly enough. Work concerns, unresolved personal issues, organizational problems, and conditions outside of work must all be taken into account in preventive and interventive efforts.

Personal awareness As discussed in Chapter 1, personal awareness is a must for the human service professional. Be aware of yourself and both the joys and frustrations of working in this field. By modifying an unrealistic attitude, you can avoid disappointment and eventual apathy, thus helping to prevent burnout.

Realistic expectations and limit setting Be realistic, not idealistic. Set limits; beware of over-involvement; keep an eye on your motives (watch for countertransference) and the limitations of your professional role.

Professional support Seek peer, supervisory, administrative, community, and faculty support. You can gain validation, consultation, and education.

Social support systems Social support can buffer our intense involvement with our clients and colleagues. If clients are depressed and burned out, don't you encourage them to seek outside support through social channels?

Goal setting Setting appropriate goals can enhance self-esteem, increase efficiency, and help you deal with organizational issues. Also, self-esteem and confidence are elevated when goals are successfully met, so goals must be attainable.

Time management Time management helps people set short- and long-term goals and use their time effectively in reaching these goals. It limits feelings of being overwhelmed. We recommend "to-do" lists and prioritizing as a part of time management. Be flexible enough to reevaluate your goals, plans, and priorities.

Corey and Corey (2002) state that "using your time wisely is related to living a balanced life and is thus a part of wellness" and "time management is a key strategy in

---

BOX 9.2 Wheel of Wellness

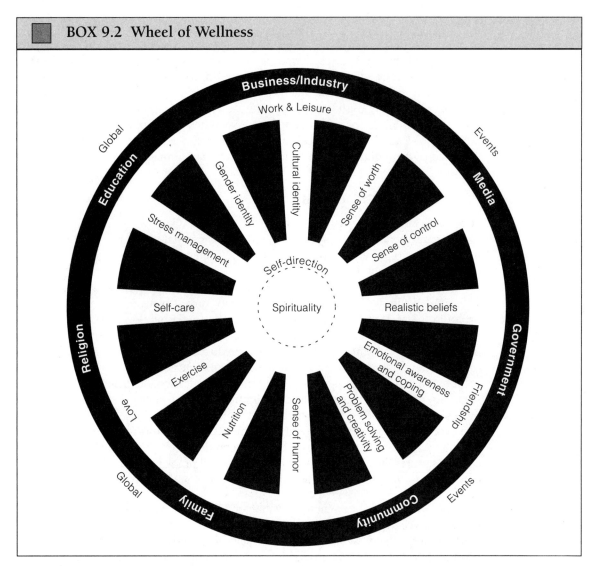

Source: Adapted from Myers, J. E., Sweeney, T. J. & Witmer, J. M. (2000). The Wheel of Wellness Counseling for wellness: A Holistic Model for Treatment Planning. *Journal of Counseling & Development,* 78, 251-266. Reprinted wih permission. No further reproduction authorized without written permission from the American Counseling Association.

---

managing stress" (p. 169). Overloading ourselves with too many projects, using our time ineffectively, and procrastination are areas that can be targeted via time management strategies.

Positive thinking The more negative one's attitude is, the higher the risk of burnout. Positive thinking is essential in this field, where so much pain, sorrow, and tragedy are encountered. If you have a positive attitude, others will appreciate and like to be around you, which is likely to help you feel good about yourself as well.

Cognitive restructuring Challenge yourself to change any negative and irrational thoughts, such as "I'm only going to fail," or "It's not worth it, I'm giving up," or "I always have to be perfect!" Corey and Corey (2003, p. 367) believe that because we create much of our stress by the beliefs we hold and the statements we make, we also have the capacity to lessen its impact by using cognitive strategies that can help "retain our vitality on both the personal and professional levels."

Relaxation techniques The most common techniques are related to relaxation response: deep breathing exercises, progressive muscle relaxation, meditation, guided imagery exercises, self-hypnosis, and biofeedback. Relaxation training through deep breathing and muscle-relaxing exercises has been shown to "help treat individuals with various stress-related physical symptoms, to reduce extreme emotional reactions to crisis situations, to reduce tension associated with performance anxiety, and to reduce anger in provocative situations" (Hepworth, Rooney, & Larsen, 2002, p. 424).

Maintaining a balance Being devoted solely to your work is admirable, but not very wise. To give to others, you must be able to give to yourself as well. Get involved in outside activities that can help you relax, take your mind off your work, and rejuvenate you so that you can be more effective in your work.

Take care of the basics Eat sensibly, get enough sleep, and exercise. Proper nutrition is proven to help you think more clearly and feel energized. Regular, restful sleep is also a must; sleep deprivation can cause fatigue, depression, and an unpleasant disposition. Regular exercise, be it walking, working out, dancing, or taking part in a sport, has also been shown to reduce stress.

Personal goodies Reinforce yourself for your diligence and hard work. Providing yourself with "goodies" is a way to reward yourself for your efforts and commitment to helping. Personal treats might include traveling, listening to music, laughing, being hugged, window shopping, or sitting in a tub of warm water.

Humor and play Laughter is often the best medicine for stress. Humor helps us relax and enjoy our work more fully, and it also takes the edge off intense emotional situations. "Humor not only acts as a buffer against stress but provides an outlet for frustration and anger" (Corey & Corey, 2002, p. 166). Some consider laughter a form of "internal jogging," saying that it serves a physiological purpose as well. Vergeer (1995) has studied the physiological changes caused by laughter that have been shown to release endorphins, lower heart rate and blood pressure, stimulate respiratory activity and oxygen exchange, and enhance immune and endocrine functions. Humor also allows us to perceive the paradoxes of life from an emotional distance. We can separate ourselves from worrisome incidents and perhaps gain a more realistic perspective.

Acceptance, adaptation, or change Assess the nature and scope of the stressors in your life. Determine whether they can be reduced, changed, or eliminated. This may involve healthy risk taking and assertiveness. At times, stressors cannot be changed, so acceptance and adaptation are then necessary. Sometimes energy spent fighting an inevitable situation only compounds the stress already being experienced.

Active participation in workshops and continuing education Strive to keep learning new insights and remain intellectually stimulated. This adds a change of pace to our daily routines and allows us to interact with others in the professional community.

Personal stress management plan Develop a personalized plan for how to deal with the stress in your life. As with a diet, results will only occur if the plan is followed properly.

Weiten and Lloyd (2000) have identified various constructive coping strategies to combat stress:

- Confronting a problem directly
- Staying in tune with reality
- Accurately and realistically appraising a stressful situation rather than distorting reality
- Learning to recognize and manage harmful emotional reactions to stress

- Consciously and rationally evaluating alternative courses of action
- Learning to exert behavioral and self-control

THINKING THINGS THROUGH

Using the scale in Table 9.1, rate your level of effectiveness in managing each of the areas discussed, in your fieldwork and in your personal life. In what areas are you strongest? In what areas do you need to make some improvements? Using your ratings, plan to take care of yourself better. Make one concrete suggestion for change in each area where you gave yourself a low rating. Discuss one such goal with your class, seminar instructor, supervisor, friend, or partner, and then follow through with your plan.

THINKING THINGS THROUGH

Based on your responses in Table 9.1, develop a personal stress management plan to meet your needs. Make a realistic plan, one that you can live by.

To counter the stress you may be feeling just thinking about the dangers of burnout, we would like to close this chapter on a happy note. We leave you with the following words of Maya Angelou (1993): "A Day Away."

We often think that our affairs, great or small, must be tended continuously and in detail, or our world will disintegrate, and we will lose our places in the universe. This is not true, or if it is true, then our situations were so temporary that they would have collapsed anyway.

Once a year or so I give myself a day away. On the eve of my day of absence, I begin to unwrap the bonds which hold me in harness. I inform house mates, my family and close friends that I will not be reachable for twenty-four hours; then I disengage the telephone. I turn the radio dial to an all-music station, preferably one which plays the soothing golden oldies. I sit for at least an hour in a very hot tub; then I lay out my clothes in preparation for my morning escape, and knowing that nothing will disturb me, I sleep the sleep of the just.

On the morning I wake naturally, for I will have set no clock, nor informed my body time-piece when it should alarm. I dress in comfortable shoes and casual clothes and leave my house, going no place. If I am living in a city, I wander streets, window-shop, or gaze at buildings. I enter and leave public parks, libraries, the lobbies of skyscrapers, and movie houses. I stay in no place for very long.

On the getaway day I try for amnesia. I do not want to know my name, where I live, or how many dire responsibilities rest on my shoulders. I detest encountering even the closest friend, for then I am reminded of who I am, and the circumstances of my life, which I want to forget for a while.

Every person needs to take one day away. A day in which one consciously separates the past from the future. Jobs, lovers, family, employers, and friends can exist one day without any one of us, and if our egos permit us to confess, they could exist eternally in our absence.

Each person deserves a day away in which no problems are confronted, no solutions searched for. Each of us needs to withdraw from the cares which will not withdraw from us. We need hours of aimless wandering or spates of time sitting on park benches, observing the mysterious world of ants and the canopy of treetops.

If we step away for a time, we are not, as many may think and some will accuse, being irresponsible, but rather preparing ourselves to more ably perform our duties and discharge our obligations.

When I return home, I am always surprised to find some questions I sought to evade had been answered and some entanglements I had hoped to flee had become unraveled in my absence

A day away acts as a spring tonic. It can dispel rancor, transform indecision, and renew the spirit

▨ **TABLE 9.1**

	1	2	3	4	5
	Least Effective	Less Effective	Moderately Effective	More Effective	Most Effective
	Preventive Interventions	Fieldwork		Placement Rating	Personal Life Rating
Personal awareness					
Realistic expectations and limit setting					
Professional support					
Social support systems					
Goal setting					
Time management					
Positive thinking					
Cognitive restructuring					
Relaxation techniques					
Maintaining a balance					
Taking care of the basics (sleep, eating, exercise)					
Personal goodies					
Humor and play					
Acceptance, adaptation, or change					
Workshops, training, and continuing education					
Personal stress management plan					

We conclude with the following "Tips for Managing Stress" that Corey and Corey (2002, p. 167) offer in their text *I Never Knew I Had a Choice*:

- Find ways to simplify your life.
- Become more aware of the demands you place on yourself and on others.
- Make the time each day to do something that you enjoy.
- Practice consciously doing one thing at a time.
- When you feel stressed, pause to take a few deep breaths.
- Regularly practice one or more relaxation techniques.
- Strive to live in the present. If you become aware of regretting past actions or worrying about the future, tell yourself that this moment is what counts.
- Practice being kind to yourself and to others.
- Ask others to help you learn to deal effectively with stress.

■ CONCLUDING EXERCISES

1. Interview a professional who could be a role model or mentor with regard to continuing to find enjoyment in work. Ask how this person manages to cope and maintain a constructive perspective on his or her work. Interview several individuals if possible; see if there is a common theme to the methods they employ.

2. Take "a day away." Pick a day to treat yourself to something enjoyable that you can do on your own: taking in a movie, getting your nails or hair done, taking a walk or hike in the forest, listening to beautiful music, or lying in the sun at a park. Whatever it is, recognize that you deserve to take good care of yourself.

3. You have been working in a small clinic for five years now. You enjoy both your work with clients and your relationship with staff. You feel especially fortunate that you work closely with a staff that is very cohesive and committed to teamwork and collaboration. In the last few months, however, the clinic manager has become irritable, defensive, and difficult to approach. It hurts you to see your supervisor (who is also a colleague and friend) struggling, but it has begun to affect your work and the overall functioning of the agency. You and a coworker agree that burnout seems to be the culprit. You decide to share your concerns with your supervisor. She becomes defensive. "Burnout! What do you mean? Why do you say that? You're totally off base." Consider how you would respond. What symptoms have you witnessed? What are the apparent causes? What suggestions might you offer?

As an orientation to fieldwork in human service, we hope this revised text provides you with the tools necessary to continue on your journey as a professional helper. Our intent has been threefold: (1) to contribute to a sound knowledge base that you can continue to build on, (2) to expose you to helping skills that you can master in your fieldwork and throughout your career, and (3) to provide you with new and updated information on the many issues in the human services field. Most important, however, is our emphasis on self-awareness. We advocate remaining open and conscious of yourself, as this will remain an essential ingredient in your overall success as a helper.

By being a human service provider, you will undoubtedly experience both sides of life: the beauty of positive growth and change as well as the sadness of human suffering. The former is the reason why many of us entered this field to begin with, while the latter is possibly the most challenging and difficult aspect of our role as helpers. Some of you may also find certain agency systems frustrating and difficult to maneuver. Despite these stressors and potential setbacks, we hope you will maintain an optimistic, positive outlook. Viewing negative situations as challenges can only enhance your growth and understanding.

Learning to blend idealism and realism is another challenge you will face. It is important to learn to cope with disappointments along the way, to accept your limitations and abilities, and to value your role as a helper. Remain persistent, determined, and committed to working toward your ideals, but be open to redefining your dreams and exploring ways to make them a reality.

Where can you go from here?

Each of you will choose your own path to follow. May it be a safe journey—one filled with years of learning, growing, and career satisfaction. As you advance in your career, we encourage you to remain committed to education and self-awareness. We hope that this text has aided your journey in some significant way. We trust that we have helped you look more closely at yourselves, both personally and professionally, and we hope that whatever you pursue is closest to your heart and desires.

Process Recording Form
Outline 1
Matrix Format

Student's Name _____

Client's Name _____

Date of Interview _____ Session # _____

I. INTERVIEW CONTENT

Using the format presented below, briefly provide a verbatim account of a small portion of your interview with the client at the beginning, middle, and ending of your session.

Content of Interview	Client's Feelings	Client's Thoughts	Student's Feelings	Student's Thoughts	Supervisor's Comments
Beginning of Interview					
Middle of Interview					

Content of Interview	Client's Feelings	Client's Thoughts	Student's Feelings	Student's Thoughts	Supervisor's Comments
End of Interview					

II. ASSESSMENT OF INTERVIEW

How would you assess this interview? Did it go well or were there any problems you encountered? Please describe. _____

_____

_____

_____

III. EVALUATION

Evaluate your interactions with the client and the interventions you used. _____

_____

_____

_____

IV. IMPRESSIONS

Consolidate your impressions by providing a summary of the case. _____

_____

_____

_____

V. FUTURE PLANS

What are your plans for further client contact/interactions? Describe a brief treatment plan here. _____

_____

_____

_____

_____

Process Recording Form
Outline 2
Analytical Type

Student's Name _____

Client's Name _____

Date of Interview _____ Session # _____

I. OBJECTIVES

Describe objective of interview. What is the reason for this client contact?

_____
_____
_____
_____
_____

II. NARRATIVE ACCOUNT OF INTERVIEW

Interview Content	Impression of Client's Feelings	Description of Student's Feelings	Comments of Supervisor

III. OVERALL ASSESSMENT AND IMPRESSIONS OF CASE

(Include mental status information here):

_____
_____
_____
_____
_____

IV. CENTRAL ISSUES AND MAJOR THEMES

Discuss what you perceive to be the central issue(s) and/or major theme(s) that emerged during the session.

_____
_____
_____
_____
_____

V. SIGNIFICANT INTERVENTIONS MADE:

Describe what you perceive to be your most significant intervention with the client. How does this intervention fit in with the overall plan for change that you have negotiated with the client?

_____
_____
_____
_____
_____

VI. STUDENT FEELINGS/PROFESSIONAL USE OF SELF

Describe the major feelings that you were aware of while working with this client. How might you use those feelings to help this client to reach the goals that he or she has set for him or herself? (Include body language, use of space, voice, helper's own feelings and values and how those hindered or helped the process, and an account of how the helper is dealing with his or her feelings, etc.)

_____
_____
_____
_____
_____

VII. FURTHER PLANS

Will treatment continue? What are short- and long-term goals of treatment? Describe specific goals for next session as well as for overall case.

_____
_____
_____
_____
_____

VIII. SUPERVISORY DISCUSSION OF PERTINENT QUESTIONS, CONCERNS, OR ISSUES

Describe one significant question or area of concern that you now face with this client. Is there any specific feedback that you want to receive from your supervisor on this issue?

_____

_____

_____

_____

_____

Process Recording Form
Outline 3
Abbreviated Format

Student's Name _____

Client's Name _____

Date of Interview _____ Session # _____

I. PURPOSE OF INTERVIEW

What are objectives and goals of the interview?

_____
_____
_____
_____

II. DESCRIPTION OF ANY NEW DEVELOPMENTS

Provide any new case information that may have a bearing on your assessment and interventions as well as your evaluation.

_____
_____
_____
_____

III. SIGNIFICANT EVENTS OR CLIENT/HELPER INTERVENTIONS

Provide any specific interview questions and/or responses that were significant during the interview. Were any techniques or interventions used that were pivotal? If so, please describe.

_____
_____
_____
_____
_____

IV. IMPRESSIONS OF CURRENT CASE

What are your impressions of the case?

_____

_____

_____

_____

V. OVERALL EVALUATION OF INTERVENTIONS

How would you evaluate the interventions you used? Which were you most satisfied with and why? Which were you least content with and why?

_____

_____

_____

_____

VI. PLANS FOR FURTHER TREATMENT OR FOLLOW-UP

Describe specific plans for your next interview and for overall treatment of this client.

_____

_____

_____

_____

References

Adams, P. L., & McDonald, N. F. (1968). Cooling out of poor people. *American Journal of Orthopsychiatry, 38*(3), 457–463.

Alberti, R., & Emmons, M. (1995). *Your perfect right: A guide to assertive living* (7th ed.). San Luis Obispo, CA: Impact.

Alle-Corliss, L., & Alle-Corliss, R. (1999). *Advanced practice in human service agencies: Issues, trends, and treatment perspectives.* Belmont, CA: Brooks/Cole-Wadsworth.

American Association of Retired Persons (AARP). (2003). *Public Policy Institute FACT SHEET.*

American Association of Retired Persons Public Policy Institute. (1993). *Elder abuse and neglect.* Washington, DC: Author.

American Association of Suicidology. (1997). *Youth suicide fact sheet.* Washington, DC: Author.

American Association of Suicidology. (1999). *Youth suicide fact sheet.* Washington, DC: Author.

American Counseling Association. (1995). *Code of ethics and standards of practice.* Alexandria, VA: Author.

American Counseling Association. (1999). *Ethical standards for Internet online counseling.* Retrieved December 29, 2004, from http://www.counseling.org/resources/internet_standards.htm

American Counseling Association. (2000). *Code of ethics and standards of practice.* Alexandria, VA: Author.

American Counseling Association. (2002). *Code of ethics and standards of practice* (rev. ed.). Retrieved December 29, 2004, from http://www.counseling.org/Content/NavidgationMenu/RESOURCES/ETHICS/ACA_Stan

American Counselor Education and Supervision. (1990). Standards for counseling supervisors. *Journal of Counseling and Development, 69*(1), 30–32.

American Heritage Dictionary. (1995). CD-ROM Reference Tool for Windows. Cambridge, MA: Softkey.

American Psychiatric Association. (2000). *Diagnostic and statistical manual of mental disorders* (4th ed.). Washington, DC: Author.

American Psychological Association. (2002). *Ethical principles of psychologists and code of conduct.* Retrieved December 29, 2004, from http://www.apa.org/ethics/code2002.html

Anderson, M. J., & Ellis, T. (1988). On the reservation. In N. A. Vacem, H. Wittmer, & S. B. DeVaney (Eds.), *Experiencing and counseling multicultural and diverse populations* (2nd ed., pp. 107–126). Muncie, IN: Accelerated Development.

Angelou, M. (1993). A day away. In *Wouldn't take nothing from my journey now* (pp. 136–139). New York: Random House.

Arthur, G. L., & Swanson, C. D. (1993). *Confidentiality and privileged communication.* Alexandria, VA: American Counseling Association.

Asch, A. (1984). The experience of disability: A challenge for psychology. *American Psychologist, 39,* 529–536.

Association for Counselor Education and Supervision. (1995). Ethical guidelines for counseling supervisors. *Counselor Education and Supervision, 34*(3), 270–276.

Atkinson, D., Morten, G., & Sue, D. W. (1993). *Counseling American minorities: A cross-cultural perspective* (4th ed.). Madison, WI: WCB Brown and Benchmark.

Austin, M. J., Skelding A. H., & Smith, P. L. (1977). *Delivering human services: An introductory programmed text.* New York: Harper & Row.

Baird, B. N. (1996). *The internship, practicum, and field placement handbook: A guide for helping professions.* Upper Saddle River, NJ: Prentice Hall.

Baird, B. N. (2000). *The internship, practicum, and field placement handbook: A guide for the helping professions* (2nd ed.). Upper Saddle River, NJ: Prentice Hall.

Baird, B. N. (2002). *The internship practicum and field placement handbook: A guide for the helping professions* (3rd ed.). Upper Saddle River, NJ: Prentice Hall.

Barker, R. L. (1991). *The social work dictionary.* Silver Springs, MD: National Association of Social Workers.

Bass, D. D., & Yep, R. (Eds.). (2002). *Terrorism, trauma, and tragedies: A counselor's guide to preparing and responding.* Alexandria, VA: American Counseling Association.

Beck, A., Kovacs, M., & Weissman, A. (1979). Assessment of suicidal intention. *Journal of Consulting and Clinical Psychology, 47,* 343–352.

Beck, A., Resnik, H., & Lettieri, D. (Eds.). (1974). *The prediction of suicide.* Bowier, MD: Charles Press.

Beneke, S. (1998). *The essentials of California mental health law: A straightforward guide for clinicians of all disciplines.* New York: W. W. Norton & Company.

Berg, I. K., & Miller, S. D. (1992). *Working with the problem drinker: A solution-focused approach.* New York: Norton.

Berger, N., & Graff, L. (1995). Making good use of supervision. In D. G. Martin & A. D. Moore, *First steps in the art of intervention: A guidebook for trainers in the helping professions* (pp. 404–448). Pacific Grove, CA: Brooks/Cole.

Berg-Weger, M., & Birkenmaier, J. (2000). *The practical companion for social work: Integrating class and field work.* Needham Heights, MA: Allyn & Bacon.

Berman, A. L., & Jobes, D. A. (1991). *Adolescent suicide: Assessment and intervention.* Washington, DC: American Psychological Association.

Bersani, C. A., & Chen, H.-T. (1988). Sociological perspective in family violence. In V. B. Van Hasselt, R. L. Morrison, A. Bellack, & M. Hersen (Eds.), *Handbook of family violence* (pp. 57–84). New York: Plenum Press.

Blood, P., Tuttle, A., & Lackey, G. (1992). Understanding and fighting sexism: A call to men. In M. L. Anderson & P. H. Collins (Eds.), *Race, class, and gender: An anthology* (2nd ed., pp. 154–161). Belmont, CA: Wadsworth.

Blumefield, W. J. (1992). *How we all pay the price.* Boston: Beacon Press.

Bolman, L. G., & Deal, T. E. (1991). *Reframing organizations: Artistry, choice and leadership.* San Francisco: Jossey-Bass.

Bolton, R. (1979). *People skills.* New York: Simon & Schuster.

Borders, L. D., & Leddick, G. R. (1987). *Handbook of counseling supervision.* Alexandria, VA: American Association for Counseling and Development.

Bordin, E. S. (1979). The generalizability of the psychoanalytic concept of the working alliance. *Psychotherapy: Theory, Research and Practice, 16,* 252–260.

Brager, G., & Holloway, S. (1978). *Changing human service organizations.* New York: Free Press.

Brammer, L. M. (1988). *The helping relationship: Process and skills* (4th ed.). Englewood Cliffs, NJ: Prentice Hall.

Brammer, L. M., & Shostrom, E. L. (1982). *Therapeutic psychology: Fundamentals of counseling and psychotherapy* (4th ed.). Englewood Cliffs, NJ: Prentice Hall.

Brashears, F. (1995). Supervision as social work practice: A reconceptualization. *Social Work, 40*(5), 577–720.

Brill, N. (1990). *Working with people* (4th ed.). White Plains, NY: Longman.

Brill, N. (1995). *Working with people* (5th ed.). White Plains, NY: Longman.

Brill, N., & Levine, J. (2002). *Working with people: The helping process* (7th ed.). Boston: Allyn & Bacon.

Brownsworth, V. (1993). Not invisible to attack. In V. Cyrus (Ed.), *Experiencing race, class, and gender in the United States* (pp. 323–328). Mountain View, CA: Mayfield.

Buckingham, S. L., & Van Gorp, W. G. (1988). Essential knowledge about AIDS dementia. *Social Work, 33,* 112–115.

Burchum, J. L. R. (2002). Cultural competence: An evolutionary perspective. *Nursing Forum, 37*(4), 5–15.

Butler, R. N. (1975). *Why survive? Being old in America.* New York: Harper & Row.

Butler, R. N. (1998). Ageism. In G. L. Maddox (Ed.), *Encyclopedia of aging* (pp. 22–23). New York: Springer.

Butler, R. N., & Lewis, M. I. (1977). *Aging and mental health* (2nd ed.). St. Louis, MO: Mosby.

California Department of Justice. (1976). Information Pamphlet 8. *Child abuse: The problem of the abused and neglected child.* Los Angeles: Information Services, California Department of Justice.

California Penal Code 13700. (2004). In P. Lenahan (Ed.), *Intimate Partner Violence.* Irvine: Department of Family Medicine, University of California, Irvine.

California Penal Code 11162, 11161.9, 11162.5, & AB 1642. (1993). Family violence resources for healthcare providers: Domestic violence laws in state of California. In P. Lenahan (Ed.), *Intimate partner violence.* Irvine: Department of Family Medicine, University of California, Irvine.

Caplan, G. (1970). *The theory and practice of mental health consultation.* New York: Basic Books.

Cappell, C., & Heiner, R. B. (1990). The intergenerational transmission of family aggression. *Journal of Family Violence, 5,* 135–152.

Capuzzi, D., & Gross, D. R. (2003). *Counseling and psychotherapy: Theory and interventions* (3rd ed.). Upper Saddle River, NJ: Merrill Prentice Hall.

Casey, J. A. (1999). Computer-assisted simulation for counselor training of basic skills. *Journal of Technology in Counseling, 1*(1), 1–7.

Cavanaugh, M. E. (1990). *The counseling experiences.* Prospect Heights, IL: Waveland.

Centers for Disease Control. (2002). *Division of HIV/AIDS prevention.* Retrieved May 23, 2002, from http://www.cdc.gov/nchtp/od./news/At-a-Glance.pdf

Chaidez, L. (Ed.). (1991). *The California child abuse and neglect reporting law: Issues and answers for health practitioners.* Sacramento, CA: State Department of Social Services, Office of Child Abuse Prevention.

Cherniss, C., & Krantz, D. L. (1983). The ideological community as an antidote to burnout in human services. In B. A. Farber (Ed.), *Stress and burnout in the human service professions* (pp. 198–212). New York: Pergamon Press.

Cherry, R. (1993). Institutionalized discrimination. In V. Cyrus (Ed.), *Experiencing race, class, and gender in the United States* (pp. 134–140). Mountain View, CA: Mayfield.

Colburn, D. R., & Pozzeta, G. E. (1979). *America and the new ethnicity.* Port Washington, NY: Kennicat.

Cole, J. B. (1992). Commonalities and differences. In M. L. Anderson & P. H. Collins (Eds.), *Race, class, and gender: An anthology* (2nd ed., pp. 148–154). Belmont, CA: Wadsworth.

Collins, D., Thomlison, B., & Grinnell, R. M., Jr. (1992). *The social work practicum: A student guide.* Itasca, IL: Peacock.

Combs, A. (1982). *A personal approach to teaching: Beliefs that make a difference.* Boston: Allyn & Bacon.

Compton, B. R., & Galaway, B. (1984). *Social work processes* (3rd ed.). Pacific Grove, CA: Brooks/Cole.

Condidaris, M. G., Ely, D. F., & Erikson, J. T. (1990). *California state laws for psychotherapists.* Gardena, CA: Harcourt Brace Jovanovich Legal and Professional Publications.

Congress, E. P. (2000). What social workers should know about ethics: Understanding and resolving ethical dilemmas. *Advances in Social Work, 1*(1), 1–25.

Corey, G. (1996). *Theory and practice of counseling and psychotherapy* (5th ed.). Pacific Grove, CA: Brooks/Cole.

Corey, G. (2004). *Theory and practice of group counseling* (6th ed.). Belmont, CA: Thomson, Brooks/Cole.

Corey, G. (2005). *Theory and practice of counseling and psychotherapy*. (7th ed.). Belmont, CA: Thomson, Brooks/Cole.

Corey, G., & Corey, M. S. (1993). *Becoming a helper* (2nd ed.). Pacific Grove, CA: Brooks/Cole.

Corey, G., & Corey, M. S. (2002). *I never knew I had a choice: Explorations in personal growth* (7th ed.). Pacific Grove, CA: Brooks/Cole.

Corey, G., & Corey, M. S. (2003). *Becoming a helper* (4th ed.). Pacific Grove, CA: Brooks/Cole.

Corey, G., Corey, M. S., & Callanan, P. (1993). *Issues and ethics in the helping professions* (4th ed.). Pacific Grove, CA: Brooks/Cole.

Corey, G., Corey, M. S., & Callanan, P. (2003). *Issues and ethics in the helping professions* (6th ed.). Pacific Grove, CA: Brooks/Cole.

Corey, G., Corey, M, & Haynes, R. (2003). *Ethics in action* [CD-ROM]. Pacific Grove, CA: Brooks/Cole.

Corey, G., & Herlihy, B. (1997). Dual/multiple relationships: Toward a consensus of thinking. In *The Hatherleigh guide to ethics in therapy* (pp. 193–205). New York: Hatherleigh Press.

Cormier, S., & Cormier, B. (1998). *Interviewing strategies for helpers: Fundamental skills and cognitive behavioral interventions* (4th ed.). Pacific Grove, CA: Brooks/Cole.

Cormier, S., & Hackney, H. (2005). *Counseling strategies and interventions* (6th ed.). Boston: Pearson, Allyn & Bacon.

Cormier, S., & Nurius, P. (2003). *Interviewing and change strategies for helpers: Fundamental skills and cognitive behavioral interventions* (5th ed.). Pacific Grove, CA: Brooks/Cole.

Corsini, R. J., & Wedding, D. (2005). *Current psychotherapies* (7th ed.). Belmont, CA: Brooks/Cole.

Costa, L., & Alterkruse, M. (1994). Duty-to-warn guidelines for mental health counselors. *Journal of Counseling and Development, 72,* 346–350.

Cottone, R. R. (2001). A social constructivism model of ethical decision making in counseling. *Journal of Counseling and Development, 79*(1), 39–45.

Council of Social Work Education. (1992a). *Curriculum policy statement for baccalaureate degree programs in social work education*. Alexandria, VA: Author.

Council of Social Work Education. (1992b). *Curriculum policy statement for master's degree programs in social work education*. Alexandria, VA: Author .

Cournoyer, B. (1996). *The social work skills workbook* (2nd ed.). Pacific Grove, CA: Brooks/Cole.

Cournoyer, B. (2005). *The social work skills workbook* (4th ed.). Belmont, CA: Brooks/Cole.

Cowger, C. D. (1992). Assessment of client strengths. In D. Saleebey (Ed.), *The strengths perspective in social work practice* (pp. 139–147). New York: Longman.

Cull, J. G., & Gill, W. S. (1991). *Suicide probability scale (SPS)*. Los Angeles: Western Psychological Services.

Cyrus, V. (Ed.). (1993). *Experiencing race, class, and gender in the United States*. Mountain View, CA: Mayfield.

Daw, J. (1997). Cultural competency: What does it mean? *Family Therapy News, 28,* 8–9, 27.

Day, S. X. (2004). *Theory and design in counseling and psychotherapy*. Boston: Lahaska Press.

DeJong, P., & Miller, S. D. (1995). How to interview for client's strengths. *Social Work, 40*(6), 729–736.

Department of Mental Health, County of San Bernardino, California. (1992). Gravely disabled due to mental illness. In *Understanding the process of requesting involuntary psychiatric evaluation*. San Bernardino, CA: Author.

Deschner, J. P. (1984). *The hitting habit: Anger control for battering couples*. New York: Free Press.

Deutsch, C. J. (1984). Self-reported sources of stress among psychotherapists. *Professional Psychology: Research & Practice, 15,* 833–845.

Devore, W., & Schlesinger, E. G. (1981). *Ethnic-sensitive social work practice*. St. Louis, MO: Mosby.

Devore, W., & Schlesinger, E. G. (1995). *Ethnic-sensitive social work practice*. St. Louis, MO: Mosby.

Dixon, E., & Taylor, L. (1994, March 17). Workshop on Cultural Diversity in the Workplace. The Center for Human Services Training and Development, University Extension, University of California, Davis.

Dixon, S. M. (1987). *Working with people in crisis* (2nd ed.). Columbus, OH: Merrill.

Douglass, R. (1980). *A study of maltreatment of the elderly and other vulnerable adults*. Institute of Gerontology and School of Public Health. Ann Arbor, MI: University of Michigan.

Drakeford, J. (1967). *The awesome power of the listening ear*. Waco, TX: Word.

Driscoll, M. S., Newman, D. L., & Seals, J. M. (1988). The effect of touch on perception of helpers. *Counselor Education and Supervision, 27,* 113–115.

Edelwich, J., & Brodsky, A. (1982). Training guidelines: linking the workshop experience to needs on and off the job. In W. S. Paine (Ed.), *Job stress and burnout* (pp. 133–154). Beverly Hills, CA: Sage.

Egan, G. (1992). *The skilled helper* (2nd ed.). Pacific Grove, CA: Brooks/Cole.

Egan, G. (1998). *The skilled helper: A problem-management approach to helping.* (6th ed.). Pacific Grove, CA: Brooks/Cole.

Egan, G. (2002). *The skilled helper: A problem-management and opportunity-development approach to helping* (7th ed.). Pacific Grove, CA: Brooks/Cole-Wadsworth.

Ellis, A. (2002). *Overcoming resistance: Rational-emotive therapy with difficult clients* (rev. ed.). New York: Springer.

Ellis, A., & MacLaren, C. (1998). *Rational-emotive behavior therapy: A therapist's guide*. Atascadero, CA: Impact.

Encyclopedia of sociology. (1974). Guilford, CT: Duskin.

Equal Employment Opportunity Commission. (1980, April 11). Guidelines on discrimination because of sex, Title VII, Sec. 703. *Federal Register, 45.*

Erickson, E. (1950). *Childhood and society*. New York: Norton.

Erickson, E. H. (1963). *Childhood and society* (2nd ed.). New York: Norton.

Erickson, E. H. (1968). *Identity: Youth and crisis*. New York: Norton.

Erickson, E. H. (1998). *The life cycle completed*. New York: Norton.

eTherapy. (2004). Retrieved December 29, 2004, from http://www.gingerich.net/etherapy.htm

Evans, D. R., Hearn, M. T., Ulhemann, M. R., & Ivey, A. E. (1994). *Essential interviewing: A programmed approach to effective communication* (2nd ed.). Monterey, CA: Brooks/Cole.

Fairchild, H. P. (1964). *Dictionary of sociology*. Paterson, NJ: Littlefield, Adams.

Faiver, C., Eisengart, S., & Colonna, R. (1995). *The counselor intern's handbook*. Pacific Grove, CA: Brooks/Cole.

Faiver, C., Eisengart, S., & Colonna, R. (2004). *The counselor intern's handbook* (3rd ed.). Belmont, CA: Brooks/Cole.

Farber, B. A. (Ed.). (1983). *Stress and burnout in human service professions*. New York: Pergamon Press.

Farley, L. (1980). *Sexual shakedown: The sexual harassment of women on the job*. New York: Warner Books.

Fellin, P. (1995). *The community and the social worker*. Itasca, IL: Peacock.

Filip, J., Schene, P., & McDaniel, N. (Eds.). (1991). *Helping in child protective services: A casework handbook* (rev. ed.). Englewood, CO: American Association for Protecting Children.

Forney, D. S., Wallace-Schutzman, F., & Wiggers, T. T. (1982). Burnout among career development professionals: Preliminary findings and implications. *Personnel and Guidance Journal, 60*, 435–439.

Forrester-Miller, H., & Davis, T. E. (1995). *A practitioner's guide to ethical decision making*. Alexandria, VA: American Counseling Association.

Foster, S. (1996, December). Characterisitcs of an effective counselor. *Counseling Today*, 21.

Frank, J. D., & Frank, J. B. (1991). *Persuasion and healing* (3rd ed.). Upper Saddle River, NJ: Prentice Hall.

Franklin, A. J. (1983). Therapeutic interventions with urban black adolescents. In E. J. Jones & S. J. Korchin (Eds.), *Minority mental health* (pp. 267–295). New York: Praeger.

Franz, M., McCartie, B.; BCATA Ethics Committee Members: Gold, C., & Skyward, B. (1999). *Technology guidelines: BCATA recommendations for ethical practice* (101 & 123). Vancouver, BC: BCATA Technology Guidelines.

Gabbard, G. O. (Ed.). (2001). *Treatment of psychiatric disorders* (3rd ed.). Washington, DC: American Psychiatric Publishing.

Gelso, C. J., & Carter, J. A. (1985). The relationship in counseling and psychotherapy: components, consequences and theoretical antecedents. *The Counseling Psychologist, 13*, 155–243.

Gibson, W. T., & Pope, K. S. (1993). The ethics of counseling: A national survey of certified counselors. *Journal of Counseling and Development, 71*, 330–336.

Gilliland, B., & James, R. (1990). *Crisis intervention strategies*. Pacific Grove, CA: Brooks/Cole.

Gilman, D., & Koverola, C. (1995). Cross-cultural counseling. In D. G. Martin & A. D. Moore (Eds.), *First steps in the art of interventions: A guidebook for trainers in the helping professions* (pp. 378–391). Pacific Grove, CA: Brooks/Cole.

Gladding, S. T. (2004). *Counseling: A comprehensive approach*. Columbus, OH: Pearson, Merrill Prentice Hall.

Gladding, S. T. (2005). *Counseling theories: Essential concepts and applications*. Upper Saddle River, NJ: Pearson, Merrill Prentice Hall.

Glosoff, H. L., Herlihy, S. B., & Spence, E. B. (2000). Privileged communication in the counselor-client relationship. *Journal of Counseling and Development, 78*(4), 454–462.

Goldberg, I. I. (1978). *Oppression and social intervention*. Chicago: Nelson-Hall.

Goldman, D. (1995). *Emotional intelligence*. New York: Bantam Books.

Goldstein, H. (1993). Starting where the client is. *Social Casework, 64*(5), 267–275.

Gondolf, E. W., & Russell, D. (1986). The case against anger control treatment programs for batterers. *Response, 9*(3), 2–5.

Goodman, M., Brown, J., & Dietz, P. (1992). *Managing managed care: A mental health practitioner's survival guide*. Washington, DC: American Psychiatric Press.

Gordon, G. R., McBride, R. B., & Hage, H. H. (2001). *Criminal justice internships: Theory into practice* (4th ed.). Cincinnati, OH: Anderson Publishing.

Gordon, T. (1970). *Parent effectiveness training*. New York: Peter H. Wyden.

Greenberg, H. M. (1980). *Coping with job stress*. Englewood Cliffs, NJ: Spectrum.

Guy, J. D. (1987). *The personal life of the psychotherapist*. New York: Wiley.

Hackney, H., & Cormier, L. S. (1988). *Counseling strategies and interventions* (3rd ed.). Englewood Cliffs, NJ: Prentice Hall.

Hackney, H., & Cormier, L. S. (2005). *The professional counselor: A process guide to helping* (5th ed.). Boston: Pearson, Allyn & Bacon.

Hackney, H., & Wrenn, C. G. (1990). The contemporary counselor in a changed world. In H. Hackney (Ed.), *Changing contexts for counselor preparation in the 1990s* (pp. 1–20). Alexandria, VA: Association for Counselor Education and Supervision.

Haney, H., & Leibsohn, H. (1999). *Basic counseling responses: A multimedia learning system for the helping professional*. Pacific Grove, CA: Brooks/Cole.

Hansen, J., Stevic, R., & Warner, R. (1986). *Counseling: Theory and process*. Boston: Allyn & Bacon.

Hansenfeld, Y. (1983). *Human service organizations*. Englewood Cliffs, NJ: Prentice Hall.

Haynes, R., Corey, G., & Mouton, P. (2003). *Clinical supervision in the helping professions: A practical guide*. Pacific Grove, CA: Brooks/Cole.

Hedges, L. E. (1993, May/June). In praise of dual relationships. *The California Therapist*, 46–50.

Hedges, L. E. (1993, July/August). In praise of dual relationships. Part II: Essential dual relatedness in developmental psychotherapy. *The California Therapist*, 42–46.

Hepworth D. H., & Larsen, J. A. (1990). *Direct social work practice: Theory and skills* (3rd ed.). Belmont, CA: Wadsworth.

Hepworth D. H., Rooney, R. H., & Larsen, J. A. (2002). *Direct social work practice: Theory and skills* (6th ed.). Pacific Grove, CA: Brooks/Cole.

Herlihy, B., & Corey, G. (1993). *Dual relationships in counseling*. Alexandria, VA: American Association for Counseling and Development.

Herlihy, B., & Corey, G. (1996). *ACA ethical standards casebook* (5th ed.). Alexandria, VA: American Counseling Association.

Herlihy, B., & Corey, G. (1997). *Boundary issues in counseling: Multiple roles and responsibilities*. Alexandria, VA: American Counseling Association.

Himle, D. P., Jayaratne, S., & Thyness, P. A. (1989). The buffering effects of four types of supervisory support on work stress. *Administration in Social Work, 13*(1), 19–34.

Ho, D. Y. F. (1991). Cultural values and professional issues in clinical psychology: Implications from the Hong Kong experience. *American Psychologist, 40*(11), 1212–1218.

Hoff, L. A. (1989). *People in crisis: Understanding and helping* (3rd ed.). Redwood City, CA: Addison-Wesley.

Hoffman, K. S., & Salee, A. L. (1994). *Social Work Practice: Bridges to Change.* Boston: Allyn & Bacon.

Hohenshil, T. H. (2000). High-tech counseling. *Journal of Counseling and Development, 78,* 365–368.

Holroyd, J., & Brodsky, A. (1980). Does touching patients lead to sexual intercourse? *Professional Psychology, 11,* 807–811.

Homan, M. S. (1994). *Promoting community change: Making it happen in the real world.* Pacific Grove, CA: Brooks/Cole.

Homan, M. S. (2004). *Promoting community change: Making it happen in the real world.* (3rd ed.). Belmont, CA: Brooks/Cole.

Horejsi, C. R., & Garthwait, C. L. (1999). *The social work practicum: A guide and workbook for students.* Needham Heights, MA: Allyn & Bacon.

Horton, P., Leslie, G. R., & Larson, R. F. (1988). *The sociology of social problems* (9th ed.). Englewood Cliffs, NJ: Prentice Hall.

Houston-Vega, M. K., Nuehring, F. M., & Daguio, E. R. (1997). *Prudent practice: A guide for managing malpractice risk.* Washington, DC: NASW Press.

Hudson, M. F. (1986). Elder mistreatment: Current research. In K. A. Pillemer & R. S. Wolf (Eds.), *Elder abuse: Conflict in the family* (pp. 125–166). Boston: Auburn House.

Hughes, H. M. (1988). Psychological and behavioral correlates of family violence in child witness and victims. *American Journal of Orthopsychiatry, 58,* 77–90.

Hutchins, D. E. (1979). Systematic counseling: The T-F-A model for the counselor intervention. *Personnel and Guidance Journal, 57*(10), 529–531.

Hutchins, D. E., & Cole, C. G. (1992). *Helping relationships and strategies* (2nd ed.). Pacific Grove, CA: Brooks/Cole.

Hutchins, D. E., & Cole, C. G. (1997). *Helping relationships and strategies* (3rd ed.). Pacific Grove, CA: Brooks/Cole.

Ivey, A. E., & Ivey, M. B. (1999). *Intentional interviewing and counseling: Facilitating client development in a multicultural society* (4th ed.). Pacific Grove, CA: Brooks/Cole.

Jacobs, C. (1989, April). The truth about abuse. *Shape,* p. 95.

Jaffe, P., Wolfe, D., Wilson, S., & Zak, L. (1986). Similarities in behavioral and social maladjustment among child victims and witnesses to family violence. *American Journal of Orthopsychiatry, 56,* 142–146.

Janosik. E. H. (1984). *Crisis counseling: A contemporary approach.* Belmont, CA: Wadsworth.

Jenkins, Y. M. (1993). Diversity and social esteem. In J. L. Chin, V. De La Cancela, & Y. M. Jenkins (Eds.), *Diversity in psychotherapy: The politics of race, ethnicity, and gender* (pp. 45–63). Westport, CT: Praeger.

Jensen, J. P., & Bergin, A. E. (1988). Mental health values of professional therapists: A national interdisciplinary survey. *Professional Psychology: Research and Practice, 19*(3), 290–297.

Jobes, D. A., Berman, A. L., & Martin, C. E. (2000). Adolescent suicidality and crisis intervention. In A. L. Roberts (Ed.), *Crisis intervention handbook* (2nd ed., pp. 131–157). New York: Oxford University Press.

Johnson, W. (1986). *The social services: An introduction* (2nd ed.). Itasca, IL: Peacock.

Jordan, A. E., & Maera, N. M. (1990). Ethics and the professional practice of psychologists: The role of virtues and principles. *Professional Psychology: Research and Practice, 21*(2), 107–114.

Jung, M. (1998). *Chinese American family therapy: A new model for clinicians.* San Francisco: Jossey-Bass.

Kalichman, S. (1999). *Mandated reporting of suspected child abuse: Ethics, law, and policy* (2nd ed.). Washington, DC: American Psychological Association.

Kalichman, S. (2001). Reporting suspected child abuse. In E. R. Welfel & R. E. Ingersoll (Eds.), *The mental health desk reference* (pp. 471–477). New York: John Wiley & Sons.

Kaiser Permanente. (2004). *Diversity Newsletter.* Fontana: Southern California Permanente Medical Group.

Kamerman, S. B., & Khan, A. J. (1976). *Social services in the United States: Policies and programs.* Philadelphia: Temple University Press.

Kettner, P., & Martin, L. (1994). Privatization? In M. Austin & I. J. Lowe (Eds.), *Controversial issues in communities and organizations* (pp. 165–172). Boston: Allyn & Bacon.

Kinnier, R. T. (1997). What does it mean to be psychologically healthy? In D. Capuzzi & D. R. Gross (Eds.), *Introduction to the counseling profession* (2nd ed., pp. 48–63). Boston: Allyn & Bacon.

Kiser, P. M. (2000). *Getting the most from your human service internship: Learning from experience.* Belmont, CA: Wadsworth/Thomson Learning.

Kleinke, C. L. (1994). *Common principles of psychotherapy.* Pacific Grove, CA: Brooks/Cole.

Knapp, M. L., & Hall, J. (1997). *Nonverbal communication in human interaction.* (5th ed.). Orlando, FL: Holt, Rinehart and Winston.

Knapp, M. L., & VandeCreek, L. (1982). *Tarasoff:* Five years later. *Professional Psychology, 13*(4), 511–516.

Kock, L., & Kock, J. (1980, Jan. 27). Parent abuse—a new plague. *Parade,* p. 14.

Kopels, S., & Kagle, J. D. (1993, March). Do social workers have a duty to warn? *Social Service Review,* 101–126.

Kottler, J. F. (1986). *On being a therapist.* San Francisco: Jossey-Bass.

Kramer, R. M. (1981). *Voluntary agencies in the welfare state.* Berkeley: University of California Press.

Kurpius, D. J. (1978). Consultation theory and process: An integrated model. *Personnel and Guidance Journal, 56,* 335–338.

Lambert, M. J., & Bergin, A. E. (1983). Therapist characteristics and their contribution to psychotherapy outcome. In C.E. Walker (Ed.), *Handbook of psychotherapy and behavior change* (pp. 205–241). Homewood, IL: Dow Jones-Irwin.

Lange, A. J., & Jakubowski, P. (1976). *Responsible assertive behavior: Cognitive/behavioral procedures for trainers.* Champaign, IL: Research Press.

Lauffer, A. (1984). *Understanding your social agency.* Beverly Hills, CA: Sage.

Lazarus, A. A. (1989). *The practice of multimodal therapy.* Baltimore, MD: Johns Hopkins University Press.

Lazarus, A. A. (1992a). The multimodal approach to treatment of minor depression. *American Journal of Psychotherapy, 46*(1), 50–57.

Lazarus, A. A. (1992b). Multimodal therapy: Technical eclecticism with minimal integration. In J. C. Norcross & M. R. Goldfried (Eds.), *Handbook of psychotherapy integration* (pp. 231–263). New York: Basic Books.

Lazarus, A. A. (1997a). *Brief but comprehensive psychotherapy: The multimodal way.* New York: Springer.

Lazarus, A. A. (1997b). Can psychotherapy be brief, focused, solution-oriented, and yet comprehensive? A personal evolutionary perspective. In J. K. Zeig (Ed.), *The evolution of psychotherapy: The third conference* (pp. 83–94). New York: Brunner/Mazel.

Lazarus, A. A. (2000). Multimodal strategies with adults. In J. Carlson & L. Sperry (Eds.), *Brief therapy with individuals and couples* (pp. 106–124). Phoenix, AZ: Zeig & Tucker.

Lee, C. C., & Waltz, G. R. (Eds.). (1998). *Social action: A mandate for counselors.* Alexandria, VA: American Counseling Association.

Lenahan, P. (2004). *Intimate partner violence.* Irvine: Department of Family Medicine, University of California, Irvine.

Leslie, R. S. (1989). Confidentiality. *The California Therapist.* July/August, p. 5.

Leslie, R. S. (1993). Dual relationships. *The California Therapist,* January/February, p. 6.

Leslie, R. S. (1994). Reporting of domestic violence? *California Therapist,* November/December.

Locke, D. C. (1992). *Increasing multicultural understanding: A comprehensive model.* Newbury Park, CA: Sage.

Long, V. O. (1996). *Facilitating personal growth in self and others.* Pacific Grove, CA: Brooks/Cole.

Longres, J. F. (1990). *Human behavior in the social environment.* Itasca, IL: Peacock.

Los Angeles Child Abuse Hotline. (1985). *Child abuse reporting law summary.* Los Angeles: Author.

Lowenberg, F., & Dolgoff, R. (1992). *Ethical decisions for social work practice* (4th ed.). Itasca, IL: Peacock.

Lum, D. (1996). *Social work practice and people of color* (3rd ed.). Pacific Grove, CA: Brooks/Cole.

Lum, D. (2004). *Social work practice and people of color* (5th ed.). Belmont, CA: Brooks/Cole.

Lurie, B. D. (1985). *A summary of California mental health confidentiality laws.* Los Angeles: Patient's Rights and Advocacy Services Program, Los Angeles County Department of Mental Health.

MacDonald, J. M. (1967). Homicidal threats. *American Journal of Psychiatry 124*(4), 474–482.

MacKinnon, C. A. (1979). *Sexual harassment of working women: A case of sex discrimination.* New Haven, CT: Yale University Press.

Maholick, L. T., & Turner, D. W. (1979). Termination: That difficult farewell. *American Journal of Psychotherapy, 33,* 583–591.

Malka, S. (1989). Managerial behavior, participation and effectiveness in social welfare organizations. *Administration in Social Work, 13*(2), 47–65.

Mandel, B. R., & Shram, B. (1985). *Human services: Introductions and interventions.* New York: Macmillan.

Mansergh, G. (2004). *FilmTX Spousal/partner abuse: Assessment, detection & intervention.* Novato, CA: Toy Box Press.

Margolin, G., Sibner, L. G., & Gleberman, L. (1988). Wife battering. In V. B. Van Hasselt, R. L. Morrison, A. Bellack, & M. Hersen (Eds.), *Handbook of family violence.* New York: Plenum Press.

Martelli, L. J. (1987). *When someone you know has AIDS: A practical guide.* New York: Crown.

Martin, D. G., & Moore, A. D. (Eds.). (1995). *First steps in the art of intervention: A guidebook for trainers in the helping professions.* Pacific Grove, CA: Brooks/Cole.

Maslack, C. (1982). Understanding burnout: Definitional issues in analyzing a complex phenomenon. In W. S. Paing (Ed.), *Job stress and burnout* (pp. 29–40). Beverly Hills, CA: Sage.

Maslow, A. (1968). *Toward a psychology of being* (2nd ed.). New York: Harper & Row.

Maypole, D. E., & Skaine, R. (1983, Sept./Oct.) Sexual harassment in the workplace. *Social Work, 28*(5), 385–390.

McClam, T., & Woodside, M. (1994). *Problem solving in the helping professions.* Pacific Grove, CA: Brooks/Cole.

McGill, D. W. (1992). The cultural story in multiculural family therapy. *Families in Society, 73,* 339–349.

McGoldrick, M. (1982) Ethnicity and family therapy: An overview. In M. McGoldrick, J. K. Pearce, & J. Giorda (Eds.), *Ethnicity and family therapy* (pp. 3–30). New York: Guilford Press.

McLeer, S. (1988). Psychoanalytic perspective on family violence. In V. B. Van Hasselt, R. L. Morrison, A. Bellack, & M. Hersen (Eds.), *Handbook of family violence* (pp. 11–25). New York: Plenum Press.

McMahon, S. (1996). *The portable problem solver: Having healthy relationships.* New York: Dell.

McQuade, W., & Aikman, A. (1974). *Stress.* New York: Bantam Books.

Meara, N. M., Schmidt, L. D., & Day, J. D. (1996). Principles and virtues. A foundation for ethical decisions, policies, and character. *The Counseling Psychologist, 24*(1), 4–77.

Meier, S. T., &. Davis, S. R. (1993). *The elements of counseling* (2nd ed.). Pacific Grove, CA: Brooks/Cole.

Meier, S. T., & Davis, S. R. (2005). *The elements of counseling* (5th ed.). Belmont, CA: Brooks/Cole.

Mehrabian, A. (1971). *Silent messages.* Belmont, CA: Wadsworth.

Menninger, R. W., & Modlin, H. C. (1972). Individual violence prevention in the violence-threatening patient. In J. Fawcett (Ed.), *Dynamics of violence.* Chicago: American Medical Association.

Meyer, M. C., & Oestreich, J. (1981). *The power pinch: Sexual harassment in the workplace.* Northbrook, IL: M. T. I. Teleprograms.

Miles, R. E. (1975). *Theories of management: Implications for organizational behavior and development.* New York: McGraw-Hill.

Miller, J. B., & Stiver, I. P. (1997). *The healing connection: How women form relationships in therapy and in life.* Boston: Beacon Press.

Mills, M. J., Sullivan, G., & Eth, S. (1987). Protecting third parties: A decade after *Tarasoff. American Journal of Psychiatry, 144*(1), 68–74.

Minuchin, S. (1974). *Families and family therapy.* Cambridge, MA: Harvard University Press.

Monahan, J. (1993). Limiting therapist exposure to *Tarasoff* liability. *American Psychologist, 48*(3), 242–250.

Moore, A. D., & Hiebert-Murphy, D. (1995). Paperwork and writing reports. In D. G. Martin & A. D. Moore (Eds.), *First steps in the art of intervention: A guidebook for trainers in the helping professions* (pp. 404–448). Pacific Grove, CA: Brooks/Cole.

Morales, A. T., & Sheafor, B. W. (1995). *Social work: A profession of many faces* (7th ed.). Needham Heights, MA: Simon & Schuster.

Murdock, N. L. (2004). *Theories of counseling and psychotherapy: A case approach.* Upper Saddle River, NJ: Pearson, Merrill, Prentice Hall.

Myers, J. E., Sweeney, T. J., & Witmer, J. M. (2000). The wheel of wellness counseling for wellness: A holistic model for treatment planning. *Journal of Counseling and Development, 78,* 251–266.

NASW National Committee on Racial and Ethnic Diversity. (2001). *NASW standards for cultural competence in social work practice.* Retrieved May 15, 2003, from http://www.socialworkers.org/sections/credentials/cultural_comp.asp

National Association of Social Workers. (1981). *Standards for classification of social work practice.* (Policy Statement 4, p. 9). Silver Springs, MD: Author.

National Association of Social Workers. (1999). *Code of ethics.* Retrieved December 29, 2004, from http://www.socialworkers.org/pubs/coe/code.asp

National Association of Social Workers. (2000). *Code of ethics.* Washington, DC: Author.

National Association of Social Workers. (2001). *Code of ethics.* Washington, DC: Author.

National Association of Social Workers. (2004). *Code of ethics.* Washington, DC: Author.

National Association of Social Workers Insurance Trust. (2004). *The distance counseling practice pointers: Cyberfrontier, part II.* Washington, DC: Author.

National Center for Health Statistics. (1998). *Vital statistics of the United States.* (Vol. 47, No. 9). Washington, DC: U.S. Government Printing Office.

National Organization of Human Service Education. (2000). Ethical standards of human service professionals. *Human Service Education, 20*(1), 61–68.

National Organization for Human Service Education. (2004). *Ethical standards of human service professionals.* Council for Standards in Human Service Education. Retrieved December 29, 2004, from http://www.hohse.com/ethics.html

National Women Abuse Prevention Program. *Effects of domestic violence on children.* FACT SHEET.

Neighbors, I., & McNeair, M. (1996). *Domestic violence: The law is not enough.* Presentation given at Patton State Hospital, Highland, CA.

Nelson, T. (1984). *Handbook for field instructors in social work.* San Bernardino: California State University, Department of Social Work.

Nelson, T. (1993). *Field manual.* San Bernardino: Department of Social Work, California State University, San Bernardino.

Nelson-Jones, R. (1993). *Lifeskills helping.* Pacific Grove, CA: Brooks/Cole.

Netting, E. F., Kettner, P. M., & McMurtry, S. (1993). *Social work macro practice.* New York: Longman.

Neugeboren, B. (1985). *Organization, policy, and practice in the human services.* New York: Longman.

Neukrug, E. (1991). Computer-assisted live supervision in counselors in empathic responding. *New Hampshire Journal for Counseling and Development, 15*(1), 15–19.

Neukrug, E. (1994). *Theory, practice, and trends in human services.* Pacific Grove, CA: Brooks/Cole.

Neukrug, E. (2002). *Skills and techniques for human service professionals: Counseling environment, helping skills, treatment issues.* Pacific Grove, CA: Brooks/Cole.

Neukrug, E. (2003). *The world of the counselor* (2nd ed.). Pacific Grove, CA: Brooks/Cole.

Neukrug, E. (2004). *Theory, practice, and trends in human services* (3rd ed.). Pacific Grove, CA: Brooks/Cole.

Neukrug, E., Milliken, T., & Shoemaker, J. (2001). Counselor-seeking behaviors of NOHSE practitioners, educators, and trainers. *Human Service Education, 21.*

Neukrug, E., & Williams, G. (1993, October). Counseling counselors: A survey of values. *Counseling and Values, 38,* 51–62.

Newman, B., & Newman, P. (1995). *Development through life: A psychosocial approach* (6th ed.). Pacific Grove, CA: Brooks/Cole.

Newman, B., & Newman, P. (2003). *Development through life: A psychosocial approach* (8th ed.). Belmont, CA: Wadsworth, Thomson Learning.

Norcross, J. C., Strausser, D. J., & Faltus, F. J. (1988). The therapist's therapist. *American Journal of Psychotherapy, 42*(1), 53–66.

Nugent, F. A. (1990). *An introduction to the profession of counseling.* Columbus, OH: Merrill.

Office of Attorney General of California. (1988). *Child abuse prevention handbook.* Sacramento, CA: Crime Prevention Center.

Ogloff, J. R. P. (1995). Navigating the quagmire: Legal and ethical guidelines. In D. Martin & A. D. Moore (Eds.), *First steps in the art of intervention* (pp. 349–373). Pacific Grove, CA: Brooks/Cole.

Older, J. (1982). *Touching is healing.* New York: Stein & Day.

O'Leary, K. D. (1988). Physical aggression between spouses: A social learning theory perspective. In V. B. Van Hasselt, R. L. Morrison, A. Bellack, & M. Hersen (Eds.), *Handbook of family violence* (pp. 32–52).New York: Plenum Press.

O'Malley, H. (1979). *Elder abuse: A review of recent literature.* Boston: Legal Research and Services for the Elderly.

Osofsky, J. (1995). The effects of exposure to violence on young children. *American Psychologist, 50*(9), 782–788.

Pachis, B., Rettman, S., & Gotthoffer, D. (2001). *Counseling on the net, 2001.* Boston: Allyn & Bacon.

Packard, T. (1991). Participation in decision-making, performance, and job satisfaction in a social work bureaucracy. *Administration in Social Work, 13*(1), 59–73.

Paine, W. S. (1982). Overview of Burnout Stress Syndromes and the 1980s. In W. S. Paine (Ed.), *Job stress and burnout* (pp. 11–25). Beverly Hills, CA: Sage.

Parad, H. J. (1965). *Crisis intervention: Selected readings.* New York: Family Service Association of American.

Patterson, L. E., & Welfel, E. R. (2000). *Counseling process* (5th ed.). Pacific Grove, CA: Brooks/Cole.

Patterson, W., Dohn, H., Bird, J., & Patterson, G. (1983). Evaluation of suicidal patients: The SAD PERSONS scale. *Psychosomatics, 24,* 343–349.

Peck, M. S. (1998). *People of the lie.* New York: Simon & Schuster.

Pederson, P. (1988). *A handbook for developing multicultural awareness*. Alexandria, VA: American Association for Counseling and Development.

Pederson, P. (2000). *A handbook for developing multicultural awareness* (3rd ed.). Alexandria, VA: American Counseling Association.

Pederson, P. B. (1991). Multiculturalism as a generic approach to counseling. *Journal of Multicultural Counseling and Development, 15,* 16–24.

Penal Code of California 11165.6. (1991). Welfare and Institutions Code, Section 300. In *The California child abuse and neglect reporting law: Issues and answers for health practitioners.* Sacramento, CA: State Department of Social Services, Office of Child Abuse Prevention.

Peters, K. D., & Murphy, S. L. (1998). Deaths: Final data for 1996. *National Vital Statistics Report, 47*(9). Hyattsville, MD: National Center for Health Statistics.

Phinney, J. S. (1996). Understanding ethnic diversity: The role of ethnic identity. *American Behavioral Scientist, 40,* 143–152.

Pines, A., & Aronson, E. (1988). *Career burnout: Causes and cures.* New York: Free Press.

Pines, A., & Aronson, E., with Krafty, D. (1981). *Burnout: From tedium to personal growth.* New York: Free Press.

Pope, K. S. (1985, April). Dual relationships: A violation of ethical, legal, and clinical standards. *California State Psychologist, 20*(3), 3–5.

Pope, K. S., & Tabachnick, B. G. (1994). Therapists as patients: A national survey of psychologists' experiences, problems, and beliefs. *Professional Psychology Research and Practice, 25*(3), 247–258.

Pope, K. S., & Vasquez, M. J. T. (1991). *Ethics in psychotherapy and counseling: A practical guide for psychologists.* San Francisco: Jossey-Bass.

Pope, K. S., & Vasquez, M. J. T. (1998). *Ethics in psychotherapy and counseling: A practical guide for psychologists* (2nd ed.). San Francisco: Jossey-Bass.

Popple, P. R., & Leighninger, L. H. (1990). *Social work, social welfare, and American society.* Boston: Allyn & Bacon.

Potter-Efron, R., & Potter-Efron, P. (1995). *Letting go of anger.* Oakland, CA: New Harbinger.

Prochaska, J. O., & Norcross, J. C. (1983). Contemporary psychotherapists: A national survey of characteristics, practices, orientations, and attitudes. *Psychotherapy: Theory, Research and Practice, 20*(2), 161–173.

Proctor, E. K., & Davis, L. E. (1994). The challenge of racial differences: Skills for clinical practice. *Social Work, 39,* 314–323.

Project on the Status and Education of Women. (1986). Supreme Court rules on sexual harassment. *On Campus with Women 16*(2), 5.

Reamer, F. G. (1991). AIDS, social work, and the "duty to protect." *Social Work, 36*(1), 56–60.

Reamer. F. G. (1994). *Social work malpractice and liability: Strategies for prevention.* New York: Columbia University Press.

Remely, T. P., & Herlihy, B. (2001). *Ethical, legal, and professional issues in counseling.* Upper Saddle River, NJ: Merrill/Prentice Hall.

Ridley, C. R. (1989). Racism and counseling as an adverse behavioral process. In P. Pedersen, J. Draguns, W. Lonner, & U. Trimble (Eds.), *Counseling across cultures* (3rd ed., pp. 55–77). Honolulu: University of Hawaii Press.

Riggar, T. F. (1985). *Stress burnout: An annotated bibliography.* Carbondale: Southern Illinois University Press.

Riley, S. (1994). *Integrative approaches to family art therapy.* Chicago: Magnolia Street Publishers.

Roberts, A. R. (1998). *Battered women and their families* (2nd ed.). New York: Springer.

Roberts, A. R. (Ed) (2000). *Crisis intervention handbook: Assessment, treatment and research* (2nd ed.). New York: Oxford University Press.

Roberts A. R., & Roberts, B. S. (2000). A comprehensive model for crisis intervention with battered women and their children. In A. R. Roberts (Ed.), *Crisis intervention handbook: Assessment, treatment, and research* (2nd ed.). New York: Oxford University Press.

Roberts, J. (1990). *Crisis intervention handbook: Assessment, treatment and research.* Belmont, CA: Wadsworth.

Roberts, J. (2000). *Crisis intervention handbook: Assessment, treatment and research* (2nd ed.). Oxford, England: Oxford University Press.

Robinson , T. (1997). Insurmountable opportunities. *Journal of Counseling and Development, 76,* 6–7.

Robinson, V. P. (1978). *The development of a professional self.* New York: AMS Press.

Rogers, C. (1959). A theory of therapy, personality, and interpersonal relationships as developed in the client-centered framework. In S. Kock (Ed.), *Psychology: A study of a science* (Vol. 3, pp. 210–211). New York: McGraw-Hill.

Rogers, C. R. (1951). *Client-centered therapy.* Boston: Houghton Mifflin.

Rogers, C. R. (1957). The necessary and sufficient conditions of therapeutic personality change. *Journal of Consulting Psychology, 21,* 95–103.

Rogers, C. R. (1961). *On becoming a person.* Boston: Houghton Mifflin.

Rothman, J. (1968). Three models of community organization practice. In F. Cox, J. L. Erlich, J. Rothman, & J. E. Tropman (Eds.), *Strategies of community organization: A book of readings* (pp. 3–26). Itasca, IL: Peacock.

Rowe, W., Bennet, S. K., & Atkinson, D. R. (1994). White racial identity models: A critique and alternative proposal. *The Counseling Psychologist, 22,* 129–146.

Royse, D., Dhooper, S. S., & Rompf, E. L. (1999). *Instruction: A guide for social work students* (3rd ed.). New York: Longman.

Rush, A. J., & Beck, A. T. (1978). Cognitive therapy of depression and suicide. *American Journal of Psychotherapy, 32,* 201–219.

Russell-Chapin, L. A., & Ivey, A. E. (2004). *Your supervised practicum and internship: Field resources for turning theory into action.* Belmont, CA: Thomson, Brooks/Cole.

Safran, J. D., & Maran, J. C. (2000). *Negotiating the therapeutic alliance: A relational treatment guide.* New York: Guilford.

St. Germaine, J. (1993). Dual relationships: What's wrong with them? *American Counselor, 2*(3), 25–30.

Saleeby, D. (1992). *The strengths perspective in social work practice.* New York: Longman.

Saleeby, D. (Ed.). (1997). The *strengths perspective in social work practice* (2nd ed.). New York: Longman.

Saltsman, A. (1988). Hands off at the office. *U.S. News and World Report.* January/February, p. 10–11.

Sampson, J. P., Jr. (2000). Using the Internet to enhance testing in counseling. *Journal of Counseling and Development, 78,* 348–356.

Sampson, J. P., Kolodinsky, R. W., & Greeno, B. P. (1997). Counseling on the information highway: Future possibilities and potential problems. *Journal of Counseling and Development, 75,* 203–212.

Sandler, B. R. (1993). Sexual harassment: A hidden issue. In V. Cyrus (Ed.), *Experiencing race, class, and gender in the United States* (pp. 225–229). Mountain View, CA: Mayfield.

Satir, V. (1972). *Peoplemaking.* Palo Alto, CA: Science and Behavior Books.

Satir, V. (1983). *Conjoint family therapy* (3rd ed.). Palo Alto, CA: Science and Behavior Books.

Schechter, S. (1982). *Women and male violence: The visions and struggles of the battered women's movement.* Boston: South End.

Schmolling, P., Jr., Youkeles, M., & Burger, W. R. (1993). *Human services in contemporary America* (3rd ed.). Pacific Grove, CA: Brooks/Cole.

Scott, G. G. (1990). *Resolving conflict with others and within yourself.* Oakland, CA: New Harbinger.

Seligman, L. (2004). *Technical and conceptual skills for mental health professionals.* Upper Saddle River, NJ: Pearson, Merrill Prentice Hall.

Sexton, T., & Whiston, S. C. (1991). A review of the empirical basis for counseling: Implications for practice and training. *Counselor Education and Supervision, 30,* 333–354.

Shannon, J. W., & Woods, W. J. (1995). Affirmative psychotherapy for gay men. In D. R. Atkinson & H. Hackett (Eds.), *Counseling diverse populations* (pp. 307–324). Madison, WI: Brown & Benchmark.

Shart, R. S. (2004). *Theories of psychotherapy and counseling: Concepts and cases* (3rd ed.). Pacific Grove, CA: Thomson, Brooks/Cole.

Shea, S. (1998). *Psychiatric interviewing: The art of understanding* (2nd ed.). Philadelphia: Saunders.

Shea, S. (1999). *The practical art of suicide assessment: A guide for mental health practitioners and substance abuse counselors.* New York: John, Wiley & Sons.

Sherry, P. (1991). Ethical issues in the conduct of supervision. *The Counseling Psychologist, 19*(4), 566–584.

Shertzer, B., & Stone, S. (1980). *Fundamentals of counseling* (3rd ed.). Boston: Houghton and Mifflin.

Shneidman, E. S., Farberow, N. L., & Litman, R. E. (1976). *The psychology of suicide.* New York: Aronson.

Slater, J. R., & Spetalnick, D. (2001). Compassion fatigue: A personal perspective. In T. McClam & M. Woodside (Eds.), *Human service challenges in the 21st century* (pp. 215–224). Birmingham, AL: Ebsco Media.

Smith, D. S. (1982). Trends in counseling and psychotherapy. *American Psychologist, 37,* 802–809.

Soloman, B. (1976). *Black empowerment: Social work in oppressed communities.* New York: Columbia University Press.

Sommers, T., & Shields, L. (1987). *Women take care.* Gainesville, FL: Triad.

Sperry, L., Carlson, J., & Kjos, D. (2003). *Becoming an effective therapist.* Boston: Allyn & Bacon.

Steinmetz, S. (1978). Battered parents. *Society, 15,* 54–55.

Straus, M. A., Gelles, R. J., & Steinmetz, S. K. (1980). *Behind closed doors: Violence in the American family.* New York: Anchor Press.

Sue, D. W. (2001). Multidimensional facets of cultural competence. *The Counseling Psychologist, 29,* 790–821.

Sue, D. W., Arredondo, P., & McDavis, R. J. (1992). Multicultural counseling competencies and standards: A call to the profession. *Journal of Counseling and Development, 70*(4), 477–486.

Sue, D. W., Bernier, J. E., Durran, A., Feinberg, L., Pedersen, P., Smith, E. J., et al. (1982). Position paper: Cross-cultural counseling competencies. *The Counseling Psychologist, 10*(2), 45–52.

Sue, D. W., & Sue, D. (1990). *Counseling the culturally different: Theory and practice* (2nd ed.). New York: Wiley.

Sue, D. W., & Sue, D. (2003). *Counseling the culturally diverse: Theory and practice* (4th ed.). New York: Wiley.

Sweitzer, H. F., & King, M. A. (1995). The internship seminar: A developmental approach. Annual conference, National Organization for Human Services Education, Seattle, WA.

Sweitzer, H. F., & King, M. A. (2004). *The successful internship: Transformation and empowerment in experiential learning* (2nd ed.). Belmont, CA: Brooks/Cole.

Szasz, T. (1986). The case against suicide prevention. *American Psychologist, 41*(7), 806–812.

Tafoya, T. (1996, June). *New heights in human services: Multiculturalism.* Keynote address at National Organization of Human Services Annual Conference, St. Louis, MO.

Tarasoff v. *Board of Regents of the University of California.* (1976). 131 *California Reporter* 14, 551 P.2d 334.

Taylor, R. L. (1979). Black ethnicity and the persistence of ethnogenesis. *American Journal of Sociology, 84,* 1401–1423.

Taylor, R. L. (Ed.). (1994). *Minority families in the United States: A multicultural perspective.* Englewood Cliffs, NJ: Prentice Hall.

Teyber, E. (2000). *Interpersonal process in psychotherapy: A relational approach* (4th ed.). Belmont, CA: Brooks/Cole.

Thompson, A. (1983). *Ethical concerns in psychotherapy and their legal ramifications.* Lanham, MD: University Press of America.

Thornton, S., & Garrett, K. (1995). Ethnography as a bridge to multicultural practice. *Journal of Social Work Education, 31*(1), 67–74.

Tjaden, P., & Thoennes, N. (1998). Battering in America: Findings from the National Violence against Women Survey. *Research in Brief* (60–66). Washington, DC: National Institute of Justice, U.S. Department of Justice.

Tomm, K. (1993, Jan./Feb.). The ethics of dual relationships. *The California Therapist,* pp. 7–19.

Torrance, E. P. (1974). *Torrance tests of creative thinking.* Bensonville, IL: Scholastic Testing Service.

Toughy, W. (1974). World Health Agency zeros in on suicide. *Los Angeles Times,* October 25, 1979, pp. 1–3.

Travers, A. W. (1993). *Supervision: Techniques and new dimensions* (2nd ed.). Englewood Cliffs, NJ: Regents/Prentice Hall.

Turner, J. H., Singleton, R., Jr., & Museik, D. (1994). *Oppression: A socio-history of black-white relations in America.* Chicago: Nelson-Hall.

U.S. Census Bureau. (2001a). *Table 1. Prevalence of disability by age, sex, race, and Hispanic origin: 1997.* Retrieved June 1, 2001, from http://www.census.gov/hhes/www/disable/sipp/disab97/ds97tl.html

U. S. Census Bureau. (2001b). *Resident population estimates of the United States by sex, race, and Hispanic origin: April 1, 1990 to July 1, 1999, with short-term projection in November 1, 2000.* Retrieved June 1, 2001, from http://www.census/gov/population/estimates/nation/intfile3-txt

U. S. Census Bureau. (2001c). *Projections of the resident population by race, Hispanic origin, and nativity: middle series, 2050 to 2070.* Retrieved June 1, 2001, from http://www.ins.usdoj.gov/graphics/aboutins/statistics/imm98.pdf

U.S. Senate Subcommittee on Aging. (1985). *Elder abuse: A national disgrace.* Washington, DC: U.S. Government Printing Office.

Valentine, P. V., Roberts, A. R., & Burgess, A. W. (1998). The stress-crisis continuum: Its application to domestic violence. In A. R. Roberts (Ed.), *Battered women and their families* (2nd ed., pp. 29–57). New York: Springer.

VandeCreek, L., Knapp, S., & Herzog, C. (1988). Privileged communication for social workers. *Social Casework, 69,* 28–34.

Vergeer, G. E. (1995). Therapeutic applications of humor. *Directions in Mental Health Counseling, 5*(3), 4–11.

von Bertalanffy, L. (1934). *Modern theories of development. An orientation to theoretical biology.* London: Oxford University Press.

von Bertalanffy, L. (1968). *General systems theory.* New York: Braziller.

Walker, L. (1979). *The battered woman.* New York: Harper & Row.

Walker, L. E. (1984). *The battered woman syndrome.* New York: Springer.

Ward, D. E. (1983). The trend toward eclecticism and the development of comprehensive models to guide counseling and psychotherapy. *Personnel and Guidance Journal, 67,* 154–157.

Watkins., E. C. Jr. (1985, Feb.).Countertransference: Its impact on the counseling situation. *Journal of Counseling and Development, 63,* 356–359.

Weber, M. (1946). In H. Gerth & C. W. Mills (Eds.), *From Max Weber: Essays in sociology.* New York: Oxford University Press.

Weeks, D. (1992). *The eight essential steps to conflict resolution.* New York: Putnam.

Weick, A., Rapp, C., Sullivan, P., & Kisthardt, W. (1989). A strengths perspective for social work practice. *Social Work, 34,* 350–354.

Weiner, M. E. (1990). *Human services management: Analysis and applications* (2nd ed.). Belmont, CA: Wadsworth.

Weiten, W., & Lloyd, M. A. (2000). *Psychology applied to modern life: Adjustment at the turn of the century* (6th ed.). Belmont, CA: Wadsworth.

Welfel, E. R. (2002). *Ethics in counseling and psychotherapy: Standards, research, and emerging issues* (2nd ed.). Pacific Grove, CA: Brooks/Cole.

Wenar, C. (1994). *Developmental psychopathology from infancy through adolescence* (3rd ed.). New York: McGraw-Hill.

Whiston, S. C., & Coker, J. K. (2000). Reconstructing clinical training: Implications for research. *Counselor Education and Supervision, 39,* 228–253.

Willison, B. G., & Masson, R. L. (1986). The role of touch in therapy: An adjunct to communication. *Journal of Counseling and Development, 64,* 497–500.

Wilson, S. J. (1978). *Confidentiality in social work: Issues and principles.* New York: Free Press.

Wilson, S. J. (1980). *Recordings: Guidelines for social workers.* New York: Free Press.

Wintrob, R. M., & Harvey, Y. K. (1991). The self-awareness factor in intercultural psychotherapy: Some reflections. In P. P. Pedersen, J. G. Draguns, W. J. Lonner, & J. E. Trimble (Eds.), *Counseling across cultures: Revised and expanded edition* (p. 11). Honolulu: University Press of Hawaii.

Woodside, M., & McClam, T. (1994). *An introduction to human services* (2nd ed.). Pacific Grove, CA: Brooks/Cole.

Wrenn, C. G. (1962). The culturally encapsulated counselor. *Harvard Educational Review, 32,* 444–449.

Wrenn, C. G. (1985). Afterward: The culturally encapsulated counselor revisited. In Pederson (Ed.), *Handbook of cross-cultural counseling and therapy* (pp. 323–329). Westport, CT: Greenwood.

Yarhouse, M. A., & VanOrman, B. T. (1999). When psychologists work with religious clients: Applications of the general principles of ethical conduct. *Professional Psychology: Research and Practice, 30*(6), 557–562.

Young, M. E. (2001). *Learning the art of helping: Building blocks and techniques* (2nd ed.). Upper Saddle River, NJ: Pearson, Merrill Prentice Hall.

Zastrow, C. (1990). *Introduction to social welfare* (4th ed.). Belmont, CA: Wadsworth.

Zastrow, C. (1995). *The practice of social work* (3rd ed.). Pacific Grove, CA: Brooks/Cole.

Zastrow, C., & Kirst-Ashman, K. (1994). *Understanding human behavior and the social environment* (3rd ed.). Chicago: Nelson-Hall.

Index